SALTERS ADVANCED CHEMISTRY

Chemical
Ideas

Central Team

George Burton

John Holman

Gwen Pilling

David Waddington

Heinemann

Heinemann Educational Publishers,
Halley Court, Jordan Hill, Oxford OX2 8EJ
a division of Reed Educational & Professional Publishing Ltd

MELBOURNE AUCKLAND FLORENCE
PRAGUE MADRID ATHENS SINGAPORE
TOKYO SÃO PAULO CHICAGO PORTSMOUTH (NH)
MEXICO IBADAN GABORONE JOHANNESBURG
KAMPALA NAIROBI

© University of York Science Education Group 1994

First published 1994

98 97
10 9 8 7 6

ISBN 0 435 63105 5

Designed, illustrated and typeset by Gecko Limited,
Bicester, Oxon.

Printed in the UK by The Bath Press, Bath

ACKNOWLEDGEMENTS

The authors and publishers are grateful to the following for permission to reproduce text extracts:

Abstracted with permission from American Chemical Society *What's happening in Chemistry* 1992, (Question 7, p.18); Aldrich Chemical Co Ltd, (Fig. 40, p. 121); Schering Agrochemicals Limited, (Fig. 51, p.126).

The authors and publishers are grateful to Dr R E Hubbard/Department of Chemistry, University of York for permission to reproduce the cover photographs.

The publishers have made every effort to trace the copyright holders but if they have inadvertently overlooked any, they will be pleased to make the necessary arrangements at the first opportunity.

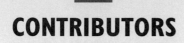

CONTRIBUTORS

Many people have contributed to the Salters Advanced Chemistry course, and a full list of contributors is given in the Teacher's Guide. They include the following.

Central Team

George Burton	Cranleigh School and University of York
John Holman (Project Director)	Watford Grammar School and University of York
Margaret Ferguson (1990-1991)	King Edward VI School, Louth
Gwen Pilling	University of York
David Waddington (Chairman of Steering Committee)	University of York

Advisory Committee

Dr Peter Doyle	Zeneca
Dr Tony Kirby, FRS	University of Cambridge
Professor The Lord Lewis, FRS (Chairman)	University of Cambridge
Sir Richard Norman, FRS	University of Oxford
Mr John Raffan	University of Cambridge

Sponsors

Many industrial companies have contributed time and expertise to the development of the Salters Advanced Chemistry Course. The work has been made possible by generous donations from the following:

The Salters' Institute for Industrial Chemistry
The Association of the British Pharmaceutical Industry
BP Chemicals
British Steel
Esso UK
ICI Agrochemicals
The Royal Society of Chemistry
Shell UK

CONTENTS

INTRODUCTION FOR STUDENTS

The Salters Advanced Chemistry course is made up of 13 self-contained units, together with a structured industrial visit and an Individual Investigation. The book of **Chemical Storylines** provides the backbone of each unit, and this book of **Chemical Ideas** needs to be used alongside the Storylines. You will also be given sheets to guide you through the **Activities**.

Each unit is driven by the Storyline. You work through the Storyline, making 'excursions' to Activities and Chemical Ideas at appropriate points.

Excursions to Chemical Ideas

As you work through the Storylines, you will find references to sections in this book of Chemical Ideas. These sections cover the chemical principles that are needed to understand each particular part of the Storyline, and you will need to study that section of the Chemical Ideas book before you can go much further. The Chemical Ideas are written to give you a clear, concise explanation of the principles.

As you study the Chemical Ideas you will find problems to tackle. These are designed to check and consolidate your understanding of the chemical principles involved.

Building up the Chemical Ideas

Salters Advanced Chemistry has been planned so that you build up your understanding of chemical ideas gradually. For example, the idea of chemical equilibrium is introduced in a simple, qualitative way in *The Atmosphere* unit. A more detailed, quantitative treatment is given in *Engineering Proteins* and *Aspects of Agriculture*, and applied to acids and precipitation in *The Oceans*. By the time you have finished the course, you should have a good understanding of what equilibrium is all about – and if necessary you can go back to the earlier sections to remind yourself.

It is important to bear in mind that the Chemical Ideas book is not the only place where chemistry is covered! The Chemical Ideas cover chemical principles that are needed in more than one unit of the course. Chemistry that is specific to a particular unit is dealt with in the Storyline itself and in related Activities.

How much do you need to remember?

The syllabus for Salters Advanced Chemistry defines what you have to remember. Each unit concludes with a 'Check your Notes' activity, which you can use to check that you have mastered all the required knowledge, understanding and skills for that unit. 'Check your Notes' tells you whether a topic is to be found in the Chemical Ideas, Chemical Storylines or Activities.

We hope that you will learn as much about the fascinating world of chemistry from the Salters Advanced Chemistry books as we have learned writing them.

George Burton   John Holman   Gwen Pilling   David Waddington

MEASURING AMOUNTS OF SUBSTANCE

1.1 *Amount of substance*

Relative atomic mass

Imagine you have a bag containing 10 golf balls and 20 table tennis balls. The golf balls would make up most of the mass, but the table tennis balls would make up most of the contents.

Chemists encounter a comparable situation when they try to work out the composition of water, or any other substance. Oxygen makes up nearly 90% of the mass of water, but two thirds of the atoms are hydrogen atoms. The difference between the two situations is that, whereas golf balls and table tennis balls can be picked up and counted, the chemist is unable to pick up and count the atoms in water molecules – they are far too small.

You know the formula of water is H_2O, and this means there are *two* hydrogen atoms combined with *one* oxygen atom in *each* water molecule. The link between the mass of an element and the number of atoms it contains is the **relative atomic mass** (A_r) of the element. It is this link which allows chemists to work out chemical formulae.

The relative atomic mass scale is used to compare the masses of different atoms. The hydrogen atom, the lightest atom, was originally assigned an A_r value of 1, and A_r values of other atoms were compared with this. The reference that is used now is the carbon-12 isotope (^{12}C) which is assigned a relative atomic mass of exactly 12. (You will learn more about isotopes later.) In chemistry, approximate relative atomic masses are used most of the time, and you can simply think of carbon atoms, set at $A_r = 12$, as being the reference point for the relative atomic mass scale. The approximate relative atomic masses of several elements are listed in Table 1.

Table 1 Some approximate relative atomic masses

Element	Symbol	Relative atomic mass, A_r
hydrogen	H	1
carbon	C	**12**
oxygen	O	16
magnesium	Mg	24
sulphur	S	32
calcium	Ca	40
iron	Fe	56
copper	Cu	64
iodine	I	127
mercury	Hg	200

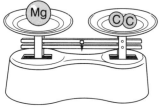

Figure 1 Two ^{12}C atoms have the same mass as one Mg atom. So the relative atomic mass of Mg is $2 \times 12 = 24$

Notice that A_r values have no units. Copper atoms *do not* have a mass of 64 g or 64 'anythings'. They are just 64 times heavier than hydrogen atoms, or 4 times heavier than oxygen atoms, or twice as heavy as sulphur atoms, and so on.

Chemical quantities

Suppose you have two bottles containing equal masses of copper ($A_r = 64$) and sulphur ($A_r = 32$). You would know that you had twice as many sulphur atoms as copper atoms because sulphur atoms have only half the mass of copper atoms. If you had a bottle of mercury ($A_r = 200$) that was five times heavier than a bottle of calcium ($A_r = 40$), you would know that both bottles contained equal numbers of atoms because each mercury atom has five times the mass of each calcium atom.

12 g of carbon, 1 g of hydrogen and 16 g of oxygen all contain equal numbers of atoms because these masses are in the same ratio as the relative atomic masses. This amount of each of these three elements has a special significance in chemistry. It is called a **mole** (or **mol** for short). The mole is the unit that measures **amount of substance** in such a way that equal

Some definitions

The **relative atomic mass, A_r,** of an element is the mass of its atom relative to $^{12}C = 12$.

The **relative molecular mass, M_r,** of a compound is the mass of its molecule relative to $^{12}C = 12$.

A **mole** (1 **mol**) of a substance is the **amount of substance** which contains as many elementary entities (atoms, molecules etc) as there are atoms in 12 g of ^{12}C.

The **molar mass** of a substance is the mass of substance which contains 1 mol.

amounts of elements consist of equal numbers of atoms. The mole is the unit that measures amount in the same way as the kilogram is the unit that measures mass.

Chemical amounts are defined so that the mass of one mole (the **molar mass**) is equal to the relative atomic mass in grams. Thus, the molar mass of carbon is $12\,\mathrm{g\,mol^{-1}}$.

If you had 6 g of carbon you would have 0.5 mol of carbon atoms; 4 g of oxygen would contain 0.25 mol of oxygen atoms and 3 g of hydrogen would contain 3 mol of hydrogen atoms. You can work out these amounts like this:

6 g of carbon contains $6\,\mathrm{g} \div 12\,\mathrm{g\,mol^{-1}} = 0.5\,\mathrm{mol}$ of carbon atoms

4 g of oxygen contains $4\,\mathrm{g} \div 16\,\mathrm{g\,mol^{-1}} = 0.25\,\mathrm{mol}$ of oxygen atoms

3 g of hydrogen contains $3\,\mathrm{g} \div 1\,\mathrm{g\,mol^{-1}} = 3\,\mathrm{mol}$ of hydrogen atoms

Each of these calculations uses the relationship:
mass in grams ÷ molar mass = amount in moles of atoms

The Avogadro constant

The number of atoms in a mole of an element is a constant. It is called the **Avogadro constant** after the Italian scientist, Amedeo Avogadro (1776–1856) who did some work which was fundamental to this method of counting atoms. The value of the Avogadro constant (symbol L) is 6.02×10^{23} atoms $\mathrm{mol^{-1}}$. (If you are counting molecules it is 6.02×10^{23} molecules $\mathrm{mol^{-1}}$; for electrons it is 6.02×10^{23} electrons $\mathrm{mol^{-1}}$, and so on.) The number is so huge that it is difficult to comprehend. Figure 2 shows one way to illustrate how big it is.

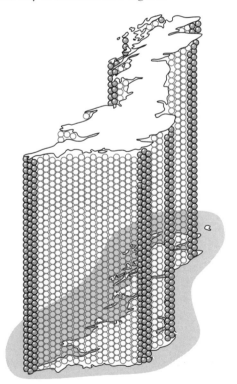

Figure 2 If the 6.02×10^{23} atoms in 12 g of carbon were turned into marbles, the marbles could cover Great Britain to a depth of 1500 km!

If you want to make sure you have the same number of atoms of different elements, you don't have to count out incredibly small atoms, neither do you have to work with numbers as big as the Avogadro constant. You just have to measure out a mole. This is easily done by weighing out the molar mass, and is the reason why chemists find the mole so useful.

Relative molecular mass

We can use the mole to deal with compounds as well as elements. A molecule of methane, CH_4, is formed when one carbon atom combines with four hydrogen atoms: this also means that one *mole* of methane molecules is formed when one *mole* of carbon atoms combines with four *moles* of hydrogen atoms.

Just as there are relative atomic masses, A_r, for atoms, so chemists use relative molecular masses, M_r, for molecules. M_r is simply the sum of all the A_r values for the atoms in the molecule. M_r gives the relative mass of the molecule on the same scale as for atoms; the scale which has the relative atomic mass of ^{12}C set at 12.

For example,

$$
\begin{aligned}
M_r \text{ for } CH_4 &= A_r(C) + 4A_r(H) \\
&= 12 + (4 \times 1) \\
&= 16
\end{aligned}
$$

$$
\begin{aligned}
M_r \text{ for } C_6H_6 &= 6A_r(C) + 6A_r(H) \\
&= (6 \times 12) + (6 \times 1) \\
&= 72 + 6 \\
&= 78
\end{aligned}
$$

Chemical formulae

We use moles when we work out chemical formulae. If you did a laboratory experiment to determine the formula of magnesium oxide, you would burn a known mass of magnesium and find out what mass of oxygen combined with it. Here are some specimen results and, from them, a calculation of the formula:

mass of magnesium = 0.84 g   mass of oxygen = 0.56 g

amount of Mg atoms   amount of O atoms
$= 0.84\,g \div 24\,g\,mol^{-1} = 0.035\,mol$ $= 0.56\,g \div 16\,g\,mol^{-1} = 0.035\,mol$

ratio of moles of atoms of Mg : O in magnesium oxide = 1 : 1

formula of magnesium oxide is MgO

Another way of analysing a compound is to find the *percentage mass* of each element it contains. Here are some specimen results for methane:

% mass of carbon = 75%   % mass of hydrogen = 25%

mass of carbon in   mass of hydrogen in
100 g of methane = 75 g   100 g of methane = 25 g

amount of C atoms   amount of H atoms
$= 75\,g \div 12\,g\,mol^{-1} = 6.25\,mol$ $= 25\,g \div 1\,g\,mol^{-1} = 25\,mol$

ratio of moles of atoms of C : H in methane = 1 : 4

formula of methane = CH_4

In a methane molecule there is a central carbon atom surrounded by four hydrogen atoms. The simple ratio of atoms from which it is formed is the same as in the formula of the molecules.

```
        H
        |
   H — C — H
        |
        H     a methane molecule
```

But this isn't the case for all substances. Ethane molecules contain eight atoms: two carbon atoms and six hydrogen atoms.

$$H-C-C-H$$

an *ethane* molecule

The **molecular formula** of ethane is C_2H_6, but the simplest ratio for the moles of atoms of $C:H$ is $1:3$. So a calculation from percentage masses would lead you to a formula CH_3. Chemists call this type of formula an **empirical formula**. Table 2 shows you some more examples of the two types of formulae.

> The **empirical formula** of a substance tells you the *ratio* of the numbers of different types of atom in the substance.
>
> The **molecular formula** tells you the *actual numbers* of different types of atom.

Substance	Molecular formula	Empirical formula
ethene	C_2H_4	CH_2
benzene	C_6H_6	CH
butane	C_4H_{10}	C_2H_5
phosphorus(V) oxide	P_4O_{10}	P_2O_5
oxygen	O_2	O
bromine	Br_2	Br

Table 2 Some molecular formulae and empirical formulae

PROBLEMS FOR 1.1

1 How many times heavier are
  a Mg atoms than H atoms?
  b Mg atoms than C atoms?
  c Mg atoms than O atoms?
  d O atoms than H atoms?
  e O atoms than C atoms?
  f Hg atoms than H atoms?
  g Hg atoms than Ca atoms?

2 How many moles of atoms are contained in the following masses?
  a 32 g of sulphur   **f** 32 g of copper
  b 20 g of calcium   **g** 3 g of carbon
  c 8 g of sulphur   **h** 0.1 g of hydrogen
  d 4 g of calcium   **i** 112 g of iron
  e 6 g of magnesium   **j** 1 kg of mercury

3 Fill in the blanks in the partially completed table which follows.

4 a Work out the empirical formulae of the compounds formed from the following masses and elements:
  i 32 g of copper and 8 g of oxygen
  ii 10 g of calcium and 4 g of oxygen
  iii 8 g of sulphur and 8 g of oxygen
  iv 1.2 g of magnesium and 0.1 g of hydrogen
  v 12.7 g of iodine and 4.0 g of oxygen

b Work out the empirical formulae of the compounds with the following percentage composition by mass:
  i 92.3% carbon and 7.7% hydrogen
  ii 52.2% carbon, 13.0% hydrogen and 34.8% oxygen

Elements	Molar mass/g mol^{-1}	Mass of sample/g	Amount of sample/mol	Number of atoms
hydrogen	1	1.00	1.00	6.02×10^{23}
carbon	12	12.00	1.00	
carbon	12	24.00		12.04×10^{23}
iron	56		1.00	
calcium	40		2.00	
iodine	127			3.01×10^{23}

5 What are the empirical formulae of the following substances?

a C_3H_6 d Al_2Cl_6
b $C_{10}H_8$ e S_8
c P_4O_6 f P_4

6 You are given the empirical formulae and the relative molecular masses of three compounds, I, II, III. Write down their molecular formula.

Compound	Empirical formula	Relative molecular mass
I	HO	34
II	CO	28
III	CH	26

7 Calculate the molar mass in $g\,mol^{-1}$ of

a benzene, C_6H_6
b copper(II) carbonate, $CuCO_3$
c potassium manganate(VII), $KMnO_4$
d sodium thiosulphate pentahydrate, $Na_2S_2O_3.5H_2O$.

8 How many moles of molecules are contained in

a 560 g of but-1-ene, $CH_3CH_2CH{=}CH_2$?
b 2.92 g of sulphur hexafluoride, SF_6?
c 1.00 tonne of water?

1.2 *Balanced equations*

A balanced chemical equation tells you the reactants and products in a reaction, and the relative amounts involved. The equation is balanced so that there are equal numbers of each type of atom on both sides. For example, the equation

$$CH_4(g) + 2O_2(g) \rightarrow CO_2(g) + 2H_2O(l)$$

tells you that one molecule of methane reacts with two molecules of oxygen to form one molecule of carbon dioxide and two molecules of water. These are also the *amounts in moles* of substances involved in the reaction.

The number written in front of each formula in a balanced equation tells you the number of formula units involved in the reaction. A formula unit may be a molecule, or another species such as an atom or an ion. The small subscript numbers are part of the formulae, and cannot be changed.

Writing balanced equations

The only way to be sure of the balanced equation for a reaction is to do experiments to find out what is formed in the reaction and what quantities are involved. But chemists use equations a lot, and it isn't possible to do experiments every time. Fortunately, if we know the reactants and products, we can usually find their formulae and *predict* the balanced equation.

The steps for predicting balanced equations
We will illustrate the steps by using the reaction that occurs when propane, C_3H_8, burns in air.

Step 1 Decide what the reactants and products are.
 ▶ propane + oxygen → carbon dioxide + water
Step 2 Write formulae for the substances involved. State symbols should be included.
 ▶ $C_3H_8(g) + O_2(g) \rightarrow CO_2(g) + H_2O(l)$
This equation is *unbalanced* because there are different numbers of each type of atom on each side.
Step 3 Balance the equation so that there are the same numbers of each type of atom on each side.
 ▶ $C_3H_8(g) + 5O_2(g) \rightarrow 3CO_2(g) + 4H_2O(l)$

Equations can only be balanced by putting numbers in front of the formulae. You cannot balance them by altering the formulae themselves, because that would create different substances.

> **State symbols**
>
> State symbols are included in chemical equations to show the *physical state* of the reactants and products:
>
> (g)   gas
> (l)   liquid
> (s)   solid
> (aq) aqueous solution

> Atoms are not created or destroyed in chemical reactions. They are simply rearranged …
>
> … so equations *must* balance.

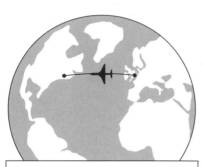

A flight from London to New York takes 7 hours. A jumbo jet carrying 375 people uses 10 tonnes of fuel per hour.

Figure 3 What mass of CO_2 is produced during one transatlantic flight? (See question 6)

1.3 *Using equations to work out reacting masses*

A balanced equation tells you the *amount in moles* of each substance involved in the reaction. For example, the equation

$$CH_4(g) + 2O_2(g) \rightarrow CO_2(g) + 2H_2O(l)$$

tells you that one mole of CH_4 reacts with two moles of O_2 to give one mole of CO_2 and two moles of H_2O.

The mass of one mole of CH_4	is	$12g + (4 \times 1)g$	$= 16g$
The mass of one mole of O_2	is	$(2 \times 16)g$	$= 32g$
The mass of one mole of CO_2	is	$12g + (2 \times 16)g$	$= 44g$
The mass of one mole of H_2O	is	$(2 \times 1)g + 16g$	$= 18g$

So, $16g$ CH_4 react with $64g$ O_2 to give $44g$ CO_2 and $36g$ H_2O.

(Note that the *total* mass on each side of the equation must always be the same.)

This means that chemists can use equations to work out the masses of reactants and products involved in a reaction, without having to do an experiment.

The steps for working out reacting masses

Step 1 Write a balanced equation.
Step 2 In words, state what the equation tells you about the amount in moles of the substances you are interested in.
Step 3 Change amounts in moles to masses in grams.
Step 4 Scale the masses to the ones in the question.

Example

What mass of carbon dioxide is produced when $64g$ of methane is burned in a plentiful supply of air?

Step 1	$CH_4(g)$	$+ 2O_2(g)$	$\rightarrow$	$CO_2(g)$	$+ 2H_2O(l)$
Step 2	1 mole			1 mole	
Step 3	$16g$			$44g$	
Step 4	$64g$			$(44 \times 4)g$	
				$= 176g$	

So, $176g$ CO_2 are produced when $64g$ CH_4 is burned.

PROBLEMS FOR 1.2 AND 1.3

You will need to consult the table of relative atomic masses in the Data Sheets when you do these questions.

1 Write balanced equations for the following reactions:
   a ethane, C_2H_6, burning in oxygen to form carbon dioxide and water
   b ethane burning in oxygen to form carbon monoxide and water
   c hexane, C_6H_{14}, burning in oxygen to form carbon dioxide and water
   d ethanol, C_2H_5OH, burning in oxygen to form carbon dioxide and water
   e nitrogen reacting with oxygen to form nitrogen monoxide, NO
   f nitrogen reacting with oxygen to form nitrogen dioxide, NO_2.

2 1 kg of charcoal is burned in a plentiful supply of air. Assuming charcoal is pure carbon, what mass of CO_2 will be formed?

3 a Write an equation for the complete combustion of octane.
   b What mass of oxygen would be needed for the complete combustion of 50 kg of octane?
   c What mass of carbon dioxide would be produced by the complete combustion of 50 kg of octane?

4 Calcium carbonate is decomposed by heating to give calcium oxide and carbon dioxide,

$$CaCO_3(s) \rightarrow CaO(s) + CO_2(g)$$

What mass of calcium carbonate would be needed to produce 1.4 g of calcium oxide?

5 A coal-burning power station burns coal which contains 1% sulphur by mass. if the power station burns 5 600 tonnes of coal a day, what mass of sulphur dioxide gas is released into the air each day?

6 Use the information in Figure 3 to work out, for one transatlantic flight,

   a the total mass of CO_2 produced

   b the mass of CO_2 produced per person.

(Jet fuel is a mixture of hydrocarbons, but you can assume it is typically $C_{12}H_{26}$).

7 When iron ore is reduced by carbon monoxide in a blast furnace, the equation for the reaction is

$$Fe_2O_3(s) + 3CO(g) \rightarrow 2Fe(s) + 3CO_2(g)$$

   a What is the M_r of Fe_2O_3?

   b Calculate the mass of iron produced from 1 mole of Fe_2O_3.

   c Calculate the mass of iron produced from 1000 tonnes of Fe_2O_3.

   d How many tonnes of Fe_2O_3 would be needed to produce 1 tonne of iron?

   e If the iron ore contains 12% of iron(III) oxide, how many tonnes of ore are needed to produce 1 tonne of iron?

1.4 *Calculations involving gases*

Chemists use chemical equations to work out the masses of reactants and products involved in a reaction. If one or more of these is a gas, it is sometimes more useful to know its volume rather than it mass.

Molar volume

The number of molecules in one mole of any gas is always 6.02×10^{23}. This quantity is known as **Avogadro's constant**, *L*. The molecules in a gas are very far apart so that the actual size of each molecule has negligible effect on the total volume the gas occupies. So, one mole of any gas always occupies the same volume, no matter which gas it is. Avogadro realised this as long ago as 1811, when he put forward his famous law (sometimes called Avogadro's hypothesis).

    The volume occupied by one mole of any gas at stp is $22.4\,dm^3$. This volume is called the **molar volume**. The letters stp mean the measurement was made at *standard temperature and pressure*. This is $0\,°C$ (273 K) and 1 atmosphere pressure (101.3 kPa)

    At room temperature, around $25\,°C$ (298 K), and 1 atmosphere pressure, the volume of a mole of any gas is about $24\,dm^3$ (Figure 4).

Reacting volumes of gases

We can use molar volumes to work out the volumes of gases involved in a reaction.

    For example, consider the manufacture of ammonia from nitrogen and hydrogen:

$$N_2(g) + 3H_2(g) \rightarrow 2NH_3(g)$$

From the equation:

   1 mole N_2 +   3 moles H_2 →   2 moles NH_3

Using the idea that one mole of each gas occupies $22.4\,dm^3$ at stp, we can write:

   $22.4\,dm^3\,N_2$ + $(3 \times 22.4)\,dm^3\,H_2$ → $(2 \times 22.4)\,dm^3\,NH_3$

or $1\,dm^3\,N_2$ + $3\,dm^3\,H_2$ → $2\,dm^3\,NH_3$

or 1 volume N_2 + 3 volumes H_2 → 2 volumes NH_3

Avogadro's law (1811)

Equal volumes of all gases at the same temperature and pressure contain an equal number of molecules.

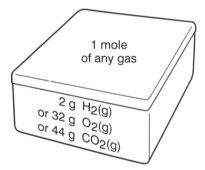

1 mole
of any gas

2 g $H_2(g)$
or 32 g $O_2(g)$
or 44 g $CO_2(g)$

Figure 4 A mole of any gas at room temperature and pressure occupies $24\,dm^3$ (about the volume of a large biscuit tin)

Measuring volumes in chemistry

A decimetre is a tenth of a metre, ie 10 cm. A cubic decimetre is therefore $10\,cm \times 10\,cm \times 10\,cm$, or $1000\,cm^3$, which is 1 litre.

In chemistry we normally use the cubic decimetre (dm^3) and the cubic centimetre (cm^3) as units of volume, rather than the litre (l) and the millilitre (ml).

So if we had $10 \, cm^3$ nitrogen it would react with $30 \, cm^3$ hydrogen to form $20 \, cm^3$ ammonia, *provided all the measurements were taken at the same temperature and pressure.*

For a reaction *involving only gases*, we can convert a statement about the numbers of *moles* of each substance involved to the same statement about *volumes*.

$$N_2(g) \quad + \quad 3H_2(g) \quad \rightarrow \quad 2NH_3(g)$$

1 mole	3 moles	→	2 moles
1 volume	3 volumes	→	2 volumes

PROBLEMS FOR 1.4

1 Methane burns in oxygen to produce carbon dioxide and water:

$$CH_4(g) + 2O_2(g) \rightarrow CO_2(g) + 2H_2O(g)$$

   a What are the relative volumes of the reactants and products involved (assuming all measurements are made at 500 °C and 1 atmosphere pressure)?

   b What volume of oxygen would be needed for the complete combustion of $10 \, cm^3$ of methane?

   c What would be the *total* volume of the products formed?

   d What would be the *change* in volume as the reaction is carried out?

   e How would your answer to **d** differ if the initial and final volume measurements were made at room temperature?

2 **a** Write an equation for the complete combustion of pentane, C_5H_{12}, in air.

   b Suppose 18 g of pentane are burned completely in a car engine, forming carbon dioxide and water. Calculate

    i the amount in moles of pentane that is burned.

    ii the amount in moles of oxygen that is needed to burn the pentane.

    iii the *volume* of oxygen that is needed.

    iv the volume of *air* that is needed (assume air is 20% oxygen by volume).

    v the volume of carbon dioxide that is produced.

3 What volume of oxygen is required to react completely with a mixture of $10 \, cm^3$ hydrogen and $20 \, cm^3$ carbon monoxide? (Assume all volumes are measured at the same temperature and pressure.)

4 $10 \, cm^3$ of a gaseous hydrocarbon react completely with exactly $40 \, cm^3$ of oxygen to produce $30 \, cm^3$ of carbon dioxide. What is the empirical formula of the hydrocarbon? (All volumes are measured at the same temperature and pressure.)

5 Unlike gases, the molar volumes of liquids and solids are far from constant. Why do you think this is so?

1.5 *The Ideal Gas law*

Compressing gases

We picture a gas as a collection of randomly moving particles which are continuously colliding with one another and with the walls of their container. The particles make such frequent collisions with the container walls that they exert a constant pressure on the walls. Pressure is defined as force per unit area.

Figure 5 When you reduce the volume of a gas, the pressure increases because the particles collide with the walls more frequently

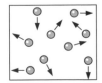
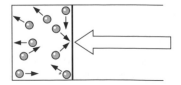

Gases are easy to compress because there is so much empty space between their particles. When the volume of a sample of gas is reduced, then the number of collisions made per second against the vessel walls increases, and so the pressure exerted by the gas increases (Figure 5).

Pressure is measured in newtons per square metre. One **pascal** is defined as a pressure of 1 newton per square metre.

$$1\,Pa = 1\,Nm^{-2}$$

Standard atmospheric pressure is 101.325 kPa.

$$1\,atm = 101.325\,kPa.$$

Robert Boyle measured the volumes of gases at different pressures. He found that, at a fixed temperature, the volume of a given amount of gas is inversely proportional to the applied pressure.

Boyle's law (stated in 1662):

$$pV = constant$$
(for a fixed amount of gas at a constant temperature).

We know that the law is obeyed only approximately, but it is still widely used. It gives a simple way of predicting the volumes of gases as the pressure on them is changed, and it is a good approximation for most gases, except when the temperature is low and the pressure high.

Example
One mole of a gas occupies a volume of 24 dm^3 at room temperature and a pressure of 1 atm. What does this volume become if the pressure is increased to 15 atm?

Answer   initial pressure × initial volume = final pressure × final volume
$$1\,atm \times 24\,dm^3 = 15\,atm \times V_{final}$$
$$V_{final} = 1.6\,dm^3$$

The effect of temperature
The volume of a gas is also affected by the temperature: gases expand when you heat them. There is another law relating the volume of a gas to the temperature: it is called **Charles' law**, and it says that the volume of a given amount of gas is proportional to its absolute temperature, provided the pressure remains constant. In symbols:

$$\frac{V}{T} = constant$$

Note that T is the *absolute* temperature, measured in kelvins.

If we combine Boyle's law and Charles' law, we get a single law that relates the volume, pressure and temperature of a gas to one another.

$$\frac{PV}{T} = constant, \text{ for a fixed amount of gas.}$$

For one mole of gas, we give the constant the symbol R. If we have n moles of gas, the relationship becomes

$$\frac{PV}{T} = nR$$

So

$PV = nRT$

This expression is called the **Ideal Gas law**. If we measure P in Pa (Nm^{-2}), V in m^3 and T in kelvins, the value of R is 8.31 J K^{-1}mol^{-1}.

The Ideal Gas law is only obeyed approximately, by the real gases that we meet in everyday life. Even so, it is useful for calculating the volume and pressure of gases under different conditions, and it is widely used by chemists.

For one mole of gas at standard temperature and pressure, $n = 1.00\,mol$, $T = 273\,K$ and $P = 101\,325\,Pa$. So

$$PV = nRT$$
$$\Rightarrow 101\,325\,Pa \times V = 1.00\,mol \times 8.31\,J\,K^{-1}\,mol^{-1} \times 273\,K$$
$$\Rightarrow \qquad\qquad V = 0.02239\,m^3$$
$$= 22.4\,dm^3$$

This is the volume of one mole of a gas at standard temperature and pressure.

PROBLEMS FOR 1.5

1 a   Use the Ideal Gas law to calculate the volume occupied by one mole of a gas at room temperature and standard pressure. Assume room temperature is 20 °C.

b   When one mole of TNT is detonated in an evacuated 1000 cm³ container, 23 moles of gaseous products are formed at a temperature of 727 °C (assuming that the only products are carbon, carbon dioxide, water and nitrogen). Calculate the pressure generated in the container (assuming that it does not burst).

2   Hydrogen has been proposed as a fuel to replace petrol. One problem with hydrogen is that it occupies a large volume. It would need to be compressed to make it fit the volume of a typical fuel tank. Many fuel tanks in cars have volumes of about 45 dm³. If pressure is applied to 45 dm³ of hydrogen the volume occupied by the gas decreases. Assuming that the temperature of the hydrogen remains constant

a   what volume would be occupied by the hydrogen if the pressure on it increased from 1 atm to
   i   15 atm?
   ii 70 atm?

b   what *mass* of hydrogen can be contained in a volume of 45 dm³ at room temperature and at pressures of
   i   1 atm
   ii 30 atm?
   iii 200 atm?
   (Assume one mole of hydrogen occupies 24 dm³ at room temperature and pressure.)

c   what problems arise in practice if high pressures are used to increase the quantity of hydrogen which can be supplied in cylinders or tanks?

3   In the Ideal Gas law, n, the amount in moles of gas, may be replaced with m/M where m is the mass of gas with molar mass M.

a   Rewrite the Ideal Gas law to give an expression relating M to m, P, V, T and the gas constant R.

b   A sample of a volatile alkane found in unleaded petrol has a mass of 0.100 g. At 373 K and standard pressure this sample produces 52.9 cm³ of vapour. Use your expression in part a to calculate the relative molecular mass of the organic liquid.

c   This alkane contains 82.8% carbon and 17.2% hydrogen by mass. Calculate the empirical formula of the compound.

d   What is the molecular formula of this alkane?

e   Draw two structural formulae corresponding to this alkane.

f   Suggest which of the two structural isomers that you have drawn might be found in unleaded petrol and give a reason for your choice of structure.

1.6 *Concentrations of solutions*

Chemists often carry out reactions in solution. When you are using a solution of a substance, it is important to know how much of the substance is dissolved in a particular volume of solution.

Concentrations are sometimes measured in grams per litre or grams per cubic decimetre (**Section 1.4**). A solution containing 80 g of sodium hydroxide in 1 dm³ of solution has a concentration of 80 g dm⁻³.

However, chemists usually prefer to measure out quantities in moles rather than in grams, because working in moles tells you about the number of particles present. So the preferred units for measuring concentrations in chemistry are moles per cubic decimetre, or mol dm⁻³.

To convert grams per cubic decimetre to moles per cubic decimetre, you need to know the molar mass of the substance involved. For example, the molar mass of sodium hydroxide, NaOH, is 40 g mol⁻¹. So a solution containing 80 g dm⁻³ has a concentration of

$$\frac{80\,\text{g dm}^{-3}}{40\,\text{g mol}^{-1}} = 2\,\text{mol dm}^{-3}$$

In general,

$$\text{concentration/mol dm}^{-3} = \frac{\text{concentration/g dm}^{-3}}{\text{molar mass/g mol}^{-1}}$$

In some books you may see 'mol dm⁻³' abbreviated to 'M'. This is quick to write, and was once widely used, but nowadays mol dm⁻³ is preferred.

When we make a solution, its concentration will depend on:
- the amount of solute
- the volume of solution it is dissolved in (Figure 6).

If you know the concentration of a solution, you can work out the amount of solute in a particular volume.

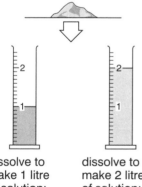

1 mol copper sulphate, CuSO₄

dissolve to make 1 litre of solution: concentration =1 mol dm⁻³

dissolve to make 2 litres of solution: concentration =0.5 mol dm⁻³

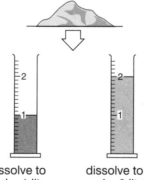

2 mol copper sulphate, CuSO₄

dissolve to make 1 litre of solution: concentration =2 mol dm⁻³

dissolve to make 2 litres of solution: concentration =1 mol dm⁻³

Figure 6 The concentration of a solution depends on the amount *of solute and the* volume *of solvent*

Example
Suppose we have 250 cm³ of a solution of sodium hydroxide whose concentration is 2 mol dm⁻³. How many moles of NaOH do we have?

Answer   250 cm³ is 0.25 dm³, and there are 2 mol in each dm³. So in 0.25 dm³ we must have (2 mol dm⁻³ × 0.25 dm³)
= 0.5 mol of NaOH.

In general,

amount/mol = (concentration of solution/mol dm^{-3}) × (volume of solution/dm^3)

Using concentrations in calculations

If you are carrying out a chemical reaction in solution, and you know the equation for the reaction, you can use the concentrations of the reacting solutions to predict the volumes you will need.

Example

Consider the reaction of sodium hydroxide with hydrochloric acid. Suppose the concentrations of both solutions are 2 mol dm^{-3}. If you have 0.25 dm^3 of sodium hydroxide solution, what volume of hydrochloric acid would be needed to neutralise it?

Answer   The equation for the reaction is

$$NaOH(aq) + HCl(aq) \rightarrow NaCl(aq) + H_2O(l)$$

The equation shows that one mole of NaOH reacts with one mole of HCl.

Amount of sodium hydroxide = 2 mol dm^{-3} × 0.25 dm^3 = 0.5 mol
$\Rightarrow$ amount of HCl needed = 0.5 mol
$\Rightarrow$ volume of HCl needed = $\dfrac{0.5\,mol}{2\,mol\,dm^{-3}}$ = 0.25 dm^3

You could probably have worked out this simple example without going through these stages, but it isn't so easy when the solutions are of differing concentrations, and when they don't react in a 1 : 1 ratio. In general, the steps for working out the reacting volumes of solutions are as follows (compare these steps with those used for calculating reacting masses in **Section 1.3**).

Step 1   Write a balanced equation
Step 2   Say what the equation tells you about the amounts in moles of the substances you are interested in
Step 3   Use the known concentrations of the solutions to change amounts in moles to volumes of solutions
Step 4   Scale the volumes of solutions to the ones in the question.

PROBLEMS FOR 1.6

1   How many moles of solute are dissolved in each of the following solutions?
   a   1 dm^3 of 0.5 mol dm^{-3} KCl(aq)
   b   2 dm^3 of 2 mol dm^{-3} NaOH(aq)
   c   250 cm^3 of 4 mol dm^{-3} HCl(aq)
   d   100 cm^3 of 2 mol dm^{-3} Na$_2$CO$_3$(aq)
   e   250 cm^3 of 0.2 mol dm^{-3} H$_2$SO$_4$(aq)
   f   50 cm^3 of 0.04 mol dm^{-3} NaI(aq)

2   What is the concentration (in mol dm^{-3}) of the solutions with the following compositions?
   a   2 moles of KOH in 1 dm^3
   b   1 mol of NaCl in 500 cm^3
   c   0.5 mol of HCl in 100 cm^3
   d   2 mol of HCl in 10 dm^3
   e   0.1 mol of HNO$_3$ in 25 cm^3
   f   0.05 mol of CaCl$_2$ in 250 cm^3

   g   10 g of NaOH in 1 dm^3
   h   4 g of NaOH in 500 cm^3
   i   100 g of CaBr$_2$ in 100 cm^3
   j   2 g of CaBr$_2$ in 25 cm^3
   k   8 g of CuSO$_4$ in 100 cm^3
   l   40 g of CuSO$_4$ in 2 dm^3

3   Complete the following table for the concentrations of ions in Dead Sea water.

Ion	Concentration/ g dm^{-3}	Concentration/ mol dm^{-3}
Cl$^-$	183.0	
Mg^{2+}	36.2	
Na$^+$		1.37
Ca^{2+}		0.335
K$^+$	6.8	
Br$^-$	5.2	
SO$_4{}^{2-}$		0.00625

4 A 25.0 cm^3 sample of 0.0500 mol dm^{-3} phosphoric acid, H_3PO_4, was titrated against 0.125 mol dm^{-3} sodium hydroxide. Using phenolphthalein as an indicator, 20.0 cm^3 of the sodium hydroxide were required. When screened methyl orange was used as an indicator, only 10.0 cm^3 of the sodium hydroxide were required.
Complete the following table and use the information produced to write equations for the different reactions whose end points are detected by the two indicators.

Volume 0.0500 mol dm^{-3} H_3PO_4/cm^3	Volume 0.125 mol dm^{-3} NaOH/cm^3	Amount H_3PO_4/mol	Amount NaOH/mol	Reacting mole ratio H_3PO_4 : NaOH
25.0	20.0			
25.0	10.0			

5 Calculate the volume of 0.100 mol dm^{-3} $AgNO_3$ required to precipitate all the chloride ions in 5 cm^3 of 0.5 mol dm^{-3} NaCl. (The reaction involved is $Ag^+(aq) + Cl^-(aq) \rightarrow AgCl(s)$)

6 10.0 cm^3 of a sample of domestic bleach was added to an excess of acidified aqueous potassium iodide. The bleach contains sodium chlorate(I) and, in the presence of acid, the chlorate(I) ion, $ClO^-(aq)$, oxidises the iodide ion to iodine, the reaction mixture turning yellow orange.
$$ClO^-(aq) + 2I^-(aq) + 2H^+(aq) \rightarrow Cl^-(aq) + I_2(aq) + H_2O(l)$$
The iodine formed was titrated against 0.500 mol dm^{-3} sodium thiosulphate. 20.0 cm^3 of the sodium thiosulphate was required. The thiosulphate ion, $S_2O_3^{2-}(aq)$, reduces the iodine back to iodide ions.
$$2S_2O_3^{2-}(aq) + I_2(aq) \rightarrow S_4O_6^{2-}(aq) + 2I^-(aq)$$
a Calculate the amount in moles of thiosulphate ion required to reduce the iodine and hence, using the second equation, the amount in moles of iodine present.
b What is the amount in moles of chlorate(I) needed to produce this amount of iodine?
c This amount of chlorate(I) was present in the 10.0 cm^3 sample of bleach. What is the concentration of chlorate(I) in the domestic bleach?

2 ATOMIC STRUCTURE

2.1 A simple model of the atom

What is inside atoms?

No one has yet been able to look directly inside atoms to see what they are really like, but a considerable amount of experimental evidence has given us a good working model of atomic structure. Scientific models should not be considered as the 'truth' or the 'right answer'. They are useful because they help to explain our observations of nature and they guide our thinking in productive directions.

Sometimes we can explain things by using a simplified version of a model. For example, regarding atoms as tiny snooker balls is sufficient to explain the states of matter, but not detailed enough to describe how chemical bonds are formed.

Many chemical and nuclear processes can be explained by a simple model of atomic structure in which atoms are thought to be made from three types of *sub-atomic particle*: **protons**, **neutrons** and **electrons**. This is not the whole story, but it's enough for present purposes. Protons and neutrons form the nucleus (or centre) of atoms. Electrons move around the nucleus in a way which is described in **Sections 2.3** and **2.4**. Figure 1 shows a very simple model of the atom. It is not to scale. The nucleus is tiny compared with the volume occupied by the electrons. If you imagined the atom to be the size of Wembley Stadium, the nucleus would be the size of a pea on the centre spot! Since most atoms have a radius of 0.1 nm–0.2 nm (1×10^{-10} m–2×10^{-10} m), the nucleus must be very, very small – about 10^{-15} m radius, in fact.

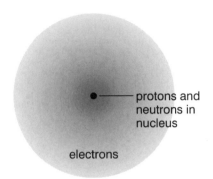

Figure 1 A simple model of the atom – not to scale

Sub-atomic particles

Some properties of protons, neutrons and electrons are summarised in Table 1.

Particle	Mass on relative atomic mass scale	Charge (relative to proton)	Location in atom
proton	1	+1	in nucleus
neutron	1	0	in nucleus
electron	0.00055	−1	around nucleus

Table 1 Some properties of sub-atomic particles

Protons and electrons have equal but opposite electrical charges. Neutrons have no charge. Protons and neutrons have almost equal masses, and are much more massive than electrons. The nucleus accounts for almost all the mass of the atom and hardly any of its volume. Most of the atom is empty space.

It is the outer parts of atoms which interact together in chemical reactions, so for chemists, electrons are the most important particles.

Nuclear symbols

The nucleus can be described by just two numbers – the **atomic number** (symbol Z) and the **mass number** (symbol A).

The atomic number is the number of protons in the nucleus. It is also numerically equal to the charge on the nucleus. The atomic number is the same for every atom of an element: for example, $Z = 6$ for all carbon atoms.

The mass number is the number of protons plus neutrons in the nucleus. If the number of neutrons is given the symbol N, then

$$A = Z + N$$

Nuclear symbols identify the mass number and the atomic number as well as the symbol for the element. Figure 2 gives an example.

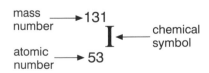

Figure 2 The nuclear symbol for iodine–131

Sometimes the atomic number is omitted since the chemical symbol tells us which element it is anyway. Thus $^{12}_{6}$C is simplified to ^{12}C.

What are isotopes?

Atoms of the same element which have different mass numbers are called **isotopes**. Since the number of protons is the same for all atoms of an element, the differences in mass must arise from different numbers of neutrons.

Most elements exist naturally as a mixture of isotopes. The relative atomic mass is an average of the masses of the isotopes which also takes account of their abundances. Isotopes of some elements are given in Table 2, together with their percentage abundances in a naturally occurring sample. From the figures for chlorine, we can work out that the relative atomic mass of chlorine is $(35 \times 75) + (37 \times 25)/100 = 35.5$.

Element	Isotope	Abundance
chlorine	^{35}Cl	75%
	^{37}Cl	25%
iron	^{54}Fe	5.8%
	^{56}Fe	91.7%
	^{57}Fe	2.2%
	^{58}Fe	0.3%
bromine	^{79}Br	50%
	^{81}Br	50%
calcium	^{40}Ca	96.9%
	^{42}Ca	0.7%
	^{43}Ca	0.1%
	^{44}Ca	2.1%
	^{48}Ca	0.2%

Table 2 Isotopes of some elements

PROBLEMS FOR 2.1

1 Copy and complete the following table.

Isotope	Symbol	Atomic number	Mass number	Number of neutrons
carbon-12	$^{12}_{6}$C	6	12	6
carbon-13			13	7
oxygen-16	$^{16}_{8}$O	8	16	
strontium-90	$^{90}_{38}$Sr	38		52
iodine-131	$^{131}_{53}$I			

2 How many protons, neutrons, and electrons are present in the following atoms?

  a $^{79}_{35}$Br

  b $^{81}_{35}$Br

  c $^{35}_{17}$Cl

  d $^{37}_{17}$Cl

3 Calculate the relative atomic masses of Br and Fe from the data in Table 2.

4 The A_r of iridium is 192.2. Iridium occurs naturally as a mixture of iridium-191 and iridium-193. What is the percentage of each isotope in naturally-occurring iridium?

2.2 Radioactive decay

Emissions from radioactive substances

Some isotopes of some elements are unstable. Their nuclei break down *spontaneously*, and are described as being **radioactive**. As these nuclei break down, they emit rays and particles called **emissions**. This breakdown (or **radioactive decay**) occurs *of its own accord*: it isn't triggered off by something we do. Some isotopes decay very quickly, for others the process takes thousands of years.

Not all unstable atoms decay in the same way. Three different kinds of emissions have been identified. They are called α, β and γ emissions. All three are capable of knocking electrons off the atoms they collide with, which ionises the atoms. Because of this, these emissions are sometimes referred to as **ionising radiation**. Some of the properties of α, β and γ emissions are summarised in Table 3.

Property	Type of emission		
	α (alpha)	β (beta)	γ (gamma)
relative charge	+2	−1	0
relative mass	4	0.00055	0
nature	2 protons + 2 neutrons (He nucleus)	electron (produced by nuclear changes)	very high frequency electromagnetic radiation
range in air	few centimetres	few metres	very long
stopped by	paper	aluminium foil	lead sheet
deflection by electrical field	low	high	nil

Table 3 Some properties of α, β and γ emissions

Nuclear equations

Nuclear equations summarise the processes which produce α and β radiation. They include the **mass number** (number of protons plus neutrons), **nuclear charge** (usually indicated by the **atomic number**, or number of protons) and **chemical symbol** for each particle involved. Both mass and charge must balance in a nuclear equation.

α-**decay** involves the emission of α-particles, which are helium nuclei, $^4_2$He. α-decay is common among heavier elements with atomic numbers greater than 83. α-decay reduces the mass of these heavy nuclei. The isotope produced from α-decay will have a mass number 4 units lower and a nuclear charge 2 units lower than the original atom. An example is

$$^{238}_{92}\text{U} \rightarrow ^{234}_{90}\text{Th} + ^4_2\text{He}$$

β-**decay** involves the emission of electrons, written as $^{\;\;0}_{-1}$e. β-decay is common among lighter elements where the isotopes contain a relatively large numbers of neutrons. For example,

$$^{14}_6\text{C} \rightarrow ^{14}_7\text{N} + ^{\;\;0}_{-1}\text{e}$$

During β-decay, the mass number (protons and neutrons) remains constant but the nuclear charge (number of protons) increases by one unit. This means that a neutron, $^1_0$n, is converted into a proton, $^1_1$p, plus an electron which is ejected from the nucleus:

$$^1_0\text{n} \rightarrow ^1_1\text{p} + ^{\;\;0}_{-1}\text{e}$$

Notice that α- and β- decay result in the production of a different element. For example, when uranium-238 undergoes α-decay, it turns into thorium-234; when carbon-14 undergoes β-decay, it turns into nitrogen. γ-**decay** is different: it is the emission of energy from a nucleus which is changing from a high energy level to a lower one. γ radiation often accompanies the emission of α and β particles. γ-rays have much higher energy and frequency than ultra violet or visible light, but you could think of their origin from the nucleus as similar to the way atomic emission spectra arise from changes in electron energy levels.

Half-lives

Radioactive decay is a random process. Each nucleus in a sample of an isotope decays at random, regardless of what the other nuclei are doing and regardless of outside conditions such as temperature and pressure. Gradually, as more and more of the nuclei in the sample decay, the sample becomes less and less radioactive.

The time for half of the nuclei to decay is called the **half-life**. For any given isotope, the half-life is fixed – it doesn't matter how much of the isotope is present, or what is the temperature or pressure. In the course of one half-life, half of the radioactivity of the sample always disappears. But the sample never completely disappears: that's why we can still detect radioactive isotopes even though the Earth was formed over four billion years ago.

For example, strontium-90 has a half-life of 28 years. If you started now with 8 g of strontium-90, in 28 years' time half the nuclei would have decayed, and only 4 g of strontium-90 would be left. In 56 years' time, only 2 g would be left. By that time the isotope would have gone through two half-lives: the original quantity would have been halved and halved again (Figure 3).

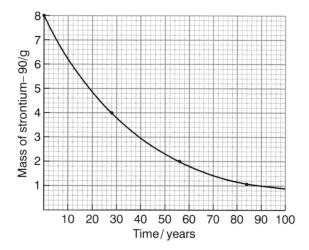

Figure 3 The half-life of strontium-90 is 28 years. This means that the amount of strontium-90 halves every 28 years

Isotope	Half-life
uranium-238	4.5×10^9 years
carbon-14	5.7×10^3 years
strontium-90	28 years
iodine-131	8.1 days
bismuth-214	19.7 minutes
polonium-214	1.5×10^{-4} seconds

Table 4 shows you some examples of isotopes and their half-lives.

Table 4 Some isotopes and their half-lives

_____ **PROBLEMS FOR 2.2** _____

1 Write nuclear equations for the α-decay of

   a $^{238}_{94}\text{Pu}$ **b** $^{221}_{87}\text{Fr}$ **c** $^{230}_{90}\text{Th}$

   (You will need to refer to the Periodic Table to identify the symbols for the elements produced.)

2 Write nuclear equations for the β-decay of

   a $^{90}_{38}\text{Sr}$ **b** $^{131}_{53}\text{I}$ **c** $^{231}_{90}\text{Th}$

3 Write a nuclear equation for each of the following:

   a a $^{7}_{3}\text{Li}$ nucleus absorbs a colliding proton and then disintegrates into two identical fragments.

   b the production of carbon-14 by collision of a neutron with an atom of nitrogen-14

   c the collision of an α-particle with an atom of nitrogen-14 releasing a proton and forming another element (Rutherford observed this

reaction in 1919. It was the first nuclear reaction ever observed).

   d the collision of an α-particle with an atom of aluminium-27 to form phosphorus-30.

4 Part of the decay series involving radon-222 is shown below. Copy it out and fill in the missing symbols for atomic number, mass number and type of decay.

$$_{88}\text{Ra} \longrightarrow \ ^{222}_{86}\text{Rn} \ \underset{\alpha}{\longrightarrow} \ \text{Po} \longrightarrow \ ^{214}\text{Pb}$$

5 $^{232}_{90}\text{Th}$ emits a total of six α and four β particles in its natural decay series. What is the atomic number and the mass number of the final product?

6 This question refers to the isotopes in Table 4 and assumes you begin with a 10 g sample of each isotope.

a How much uranium-238 would be left after 4.5×10^9 years?

b How much ^{214}Bi would be left after 78.8 minutes?

c How long would it take for you to be left with 1.25 g of ^{214}Po?

d How much ^{14}C would be left after 4.56×10^4 years?

7 A rock sample from the rugged coast of northern Labrador in Canada has recently been dated at 4.0×10^9 years. As the Earth is believed to have been formed 4.6×10^9 years ago, this sample provides a glimpse of the Earth's crust as it was soon after its formation.

Geochemist Kenneth Collerson of the University of California dated the sample by measuring the ratio of two rare earth element isotopes, samarium-147 and neodymium-143. $^{147}_{62}$Sm decays to $^{143}_{60}$Nd by alpha decay with a half life of 1.06×10^{11} years.

a Write a nuclear equation for the decay of samarium-147.

b Use the half-life to plot a graph of the percentage of samarium-147 remaining against time. Use your graph to determine the percentage of samarium in the Labrador rock sample. What was the ratio of samarium-147 to neodymium-143 found by Collerson?

2.3 *Electronic structure: shells*

Section 6.1 describes how Neils Bohr's theory of the hydrogen atom explained the appearance of the emission spectrum of hydrogen. However, Bohr's theory only worked for the simple hydrogen atom, with its one electron. It needed extending to make sense of the structure of atoms which contained several electrons.

We now know that, for these atoms, energy levels 2, 3, 4 … have a more complex structure than the single levels which exist in hydrogen. You will find out more about this in **Section 2.4**. It is more appropriate to talk about the first, second, third … **electron shell** rather than energy level 1, 2, 3 … . The shells are labelled by giving each one a **principal quantum number**, *n*. For the first shell, $n = 1$; for the second shell $n = 2$, and so on. The higher the value of *n*, the higher the energy associated with the shell. Electron shells can hold more than one electron. The maximum numbers of electrons which can be held in the first three shells are

first shell	($n = 1$)	2 electrons
second shell	($n = 2$)	8 electrons
third shell	($n = 3$)	18 electrons

A shell which contains its maximum number of electrons is called a **filled shell**. Electrons are arranged so that the lowest energy shells are filled first. Table 5 shows the **electron shell configuration**, for elements 1 to 20. The electron shell configuration of sodium, for example, is written as 2.8.1.

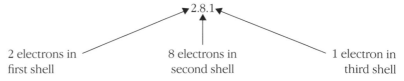

2 electrons in first shell / 8 electrons in second shell / 1 electron in third shell

Notice that the filling of shell 3 is not straightforward. This is because of the pattern of energy levels within shells 3 and 4. Electrons 19 and 20 have lower energy if they are placed in shell 4 rather than in shell 3. (This is explained in **Section 2.4**).

Chemists can explain many of the properties of atoms without needing to use a detailed theory of atomic structure. Much chemistry is decided only

Element	Atomic number	1st shell	2nd shell	3rd shell	4th shell
hydrogen	1	1			
helium	2	2			
lithium	3	2	1		
beryllium	4	2	2		
boron	5	2	3		
carbon	6	2	4		
nitrogen	7	2	5		
oxygen	8	2	6		
fluorine	9	2	7		
neon	10	2	8		
sodium	11	2	8	1	
magnesium	12	2	8	2	
aluminium	13	2	8	3	
silicon	14	2	8	4	
phosphorus	15	2	8	5	
sulphur	16	2	8	6	
chlorine	17	2	8	7	
argon	18	2	8	8	
potassium	19	2	8	8	1
calcium	20	2	8	8	2

Table 5 Electron shell configurations for the first 20 elements

by the outer shell electrons, and one very useful model treats the atom as being composed of a **core** of nucleus plus inner shells, surrounded by an **outer shell**. Figure 4 shows how a sodium atom would be thought of on this model.

Electronic structure and the Periodic Table

The arrangement of elements by rows and columns in the Periodic Table is a direct result of the electronic structure of atoms.

The *number of outer shell* electrons determines the **Group** number. So, for example, Li, Na, K all have one outer shell electrons and are in Group 1; C and Si have four outer shell electrons and are in Group 4. The noble gases all have filled outer shells, and are in Group 0 (the zero refers to the shell *beyond* the filled outer one, which does not have any electrons in it yet.)

The number of the *shell* which is being filled determines the **Period** to which an element belongs. Thus from Li to Ne, the outer electrons are being placed into shell 2, so these elements belong to Period 2. The more detailed description of electronic structure in **Section 2.4** also explains how the elements are assigned to *blocks* in the Periodic Table.

Figure 5 shows how first ionisation enthalpy varies with atomic number for elements 1–56. (Look at **Section 6.1** for an explanation of ionisation enthalpies). Notice how the elements at the peaks are all in Group 0 (the noble gases). These elements, with filled outer shells, have high ionisation enthalpies; they are difficult to ionise and very unreactive. The elements at the troughs are all in Group 1 (the alkali metals). These elements, with only one outer shell electron, have low ionisation enthalpies; they are easy to ionise and very reactive.

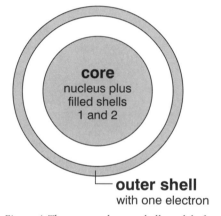

core
nucleus plus filled shells 1 and 2

outer shell
with one electron

Figure 4 The core and outer shell model of a sodium atom (2.8.1)

Figure 5 Variation of first ionisation enthalpy with atomic number for elements with atomic numbers 1 to 56

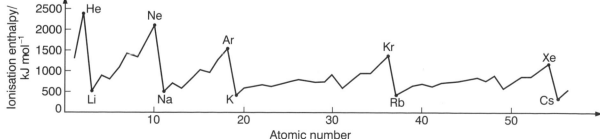

_____ **PROBLEMS FOR 2.3** _____

1   For each of the following electron configurations
    give
    a   the Group
    b   the Period,
    to which you would expect the element to belong.
        i 2.8.7
        ii 2.2
        iii 2.5
        iv 2.8.3
        v 2.8.6

2   Look carefully at Figure 5.
    a   What is the pattern of the change in the first
            ionisation enthalpy as you go down Group 1
            from Li to Cs? How is this linked to the pattern
            of changing reactivity in this Group?
    b   After Ca (element number 20) there are 10
            elements all of which have similar first
            ionisation enthalpies.
            i What are these elements known as?
            ii What might be the significance of the fact
                   that they have similar first ionisation
                   enthalpies?

2.4 *Electronic structure: sub-shells and orbitals*

Sub-shells of electrons

Section 2.3 describes how the electrons in atoms are arranged in shells.
Much of our knowledge of electron shells has come from studying the
emission spectrum of hydrogen (**Section 6.1**). The hydrogen atom has
only one electron and its spectrum is relatively simple to interpret. When we
come to look at elements other than hydrogen, we find their spectra are
much more complex. electron shells are not the whole story. The shells are
themselves split up into **sub-shells**.

The sub-shells are labelled **s**, **p**, **d** and **f**. The $n = 1$ shell has only an s
sub-shell. The $n = 2$ shell has two sub-shells, s and p, then $n = 3$ shell has
three sub-shells, s, p and d.

The different types of sub-shells can hold different numbers of electrons.
These are given in Table 6.
Thus,

Sub-shell	Maximum number of electrons
s	2
p	6
d	10
f	14

*Table 6 Number of electrons in the sub-shells
s, p, d and f*

the $n = 1$ shell can hold   2 electrons in the s sub-shell

the $n = 2$ shell can hold   2 electrons in its s sub-shell
                             6 electrons in its p sub-shell
                                          a total of 8 electrons

the $n = 3$ shell can hold   2 electrons in its s sub-shell
                             6 electrons in its p sub-shell
                             10 electrons in its d sub-shell
                                          a total of 18 electrons

the $n = 4$ shell can hold   2 electrons in its s sub-shell
                             6 electrons in its p sub-shell
                             10 electrons in its d sub-shell
                             14 electrons in its f sub-shell
                                          a total of 32 electrons

In atoms other than hydrogen, the sub-shells within a shell have different
energies. Figure 6 shows the relative energies of the sub-shell for each of
the shells $n = 1$ to $n = 4$ in a typical many-electron atom. Note the overlap
of energy between the $n = 3$ and $n = 4$ shells in Figure 6. This has
important consequences which you will meet later.

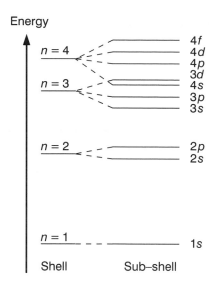

Energy

n = 4
n = 3
n = 2
n = 1

Shell   Sub–shell

4f
4d
4p
3d
4s
3p
3s

2p
2s

1s

Figure 6 Energies of electron sub-shells from n = 1 to n = 4 in a typical many-electron atom. The energy of a sub-shell is not fixed, but falls as the charge on the nucleus increases as you go from one element to the next in the Periodic Table. The order shown in the diagram is correct for the elements in Period 3 and up to nickel in Period 4. After nickel the 3d sub-shell has lower energy than 4s

Atomic orbitals

The *s*, *p*, *d*, and *f* sub-shells are themselves divided further into **atomic orbitals**. An electron in a given orbital can be found in a particular region of space around the nucleus.

An *s* sub-shell always contains *one*   *s* atomic orbital
A *p* sub-shell always contains *three* *p* atomic orbitals
A *d* sub-shell always contains *five*   *d* atomic orbitals
An *f* sub-shell always contains *seven* *f* atomic orbitals

You can see the energy levels associated with each of these atomic orbitals in Figure 7. In an isolated atom, orbitals in the same sub-shell have the same energy.

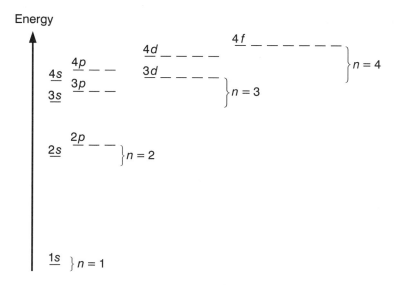

Energy

4s
3s

4p
3p

2s

2p

1s } n = 1

4d
3d

4f

} n = 4

} n = 3

} n = 2

Figure 7 Energy levels of atomic orbitals in the n = 1 to n = 4 shells

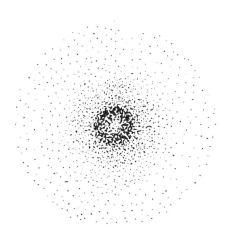

Do not think of an atomic orbital as a fixed electron orbit. The position of an electron cannot be mapped exactly. For an electron in a given atomic orbital, we only know the *probability* of finding the electron at any point – Figure 8 shows one way of representing this for a 1s orbital.

Figure 8 One way of representing a 1s electronic orbital. The dots represent the probability of finding the electron in that position. The denser the dots, the higher the probability of finding the electron

Each atomic orbital can hold a maximum of two electrons. Electrons in atoms have a **spin**, which you can picture as a spinning motion in one of two directions. Every electron spins at the same rate in either a clockwise (↑) or anticlockwise (↓) direction. Electrons can only occupy the same orbital if they have opposite or **paired** spins. We can write this as

where the box represents the atomic orbital and the arrows the electrons.

To describe accurately the state of an electron in an atom you need to supply four pieces of information about it – a bit like an address:

- the electron shell it is in
- its sub-shell
- its orbital within the sub-shell
- its spin

Filling up atomic orbitals

The arrangement of electrons in shells and orbitals is called the **electronic configuration** of an atom. The orbitals are filled in a definite order to produce the lowest energy arrangement possible.

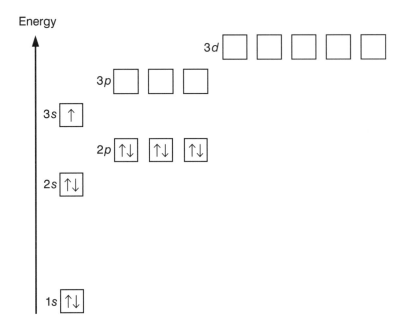

Figure 9 Arrangement of electrons in atomic orbitals in a ground state sodium atom

The orbitals are filled in order of increasing energy. Where there is more than one orbital with the same energy, these orbitals are first occupied singly by electrons. This keeps the electrons in an atom as far apart as possible. Only when every orbital is singly occupied do the electrons pair up in orbitals. For the lowest energy arrangement, electrons in singly occupied orbitals have parallel spins.

Using these rules, you can now assign the 11 electrons in a sodium atom to atomic orbitals. This is done in Figure 9.

The electron configuration of a sodium atom can be represented in shorthand notation as

$$1s^2\,2s^2\,2p^6\,3s^1$$

The large numbers show the principal quantum number of each shell, the letters show the sub-shells and the small superscripts indicate the numbers of electrons in each sub-shell.

Building up the Periodic Table

Hydrogen is the simplest element, with atomic number $Z = 1$. It has one electron which will occupy the s orbital of the $n = 1$ shell.

| ↑ | $1s$ H $1s^1$

The next element, helium ($Z = 2$) has 2 electrons which both occupy the $1s$ orbital with paired spins.

$1s$ | ↑↓ | He $1s^2$

Lithium ($Z = 3$) has 3 electrons. The third electron cannot fit in the $n = 1$ shell and so occupies the next lowest orbital, the $2s$ orbital,

$1s$ | ↑↓ | $2s$ | ↑ | Li $1s^2 2s^1$

and so on across the first short period. Nitrogen ($Z = 7$) has 7 electrons

$1s$ | ↑↓ | $2s$ | ↑↓ | $2p$ | ↑ | ↑ | ↑ | N $1s^2 2s^2 2p^3$

The three electrons occupying the $2p$ sub-shell must occupy the three separate p orbitals singly and their spins must be parallel.

Oxygen ($Z = 8$) has 8 electrons

$1s$ | ↑↓ | $2s$ | ↑↓ | $2p$ | ↑↓ | ↑ | ↑ | O $1s^2 2s^2 2p^4$

The electronic configurations of elements with atomic numbers 1 to 20 are shown in Table 7.

Atomic number (Z)	Element	Electronic configuration
1	hydrogen	$1s^1$
2	helium	$1s^2$
3	lithium	$1s^2 2s^1$
4	beryllium	$1s^2 2s^2$
5	boron	$1s^2 2s^2 2p^1$
6	carbon	$1s^2 2s^2 2p^2$
7	nitrogen	$1s^2 2s^2 2p^3$
8	oxygen	$1s^2 2s^2 2p^4$
9	fluorine	$1s^2 2s^2 2p^5$
10	neon	$1s^2 2s^2 2p^6$
11	sodium	$1s^2 2s^2 2p^6 3s^1$
12	magnesium	$1s^2 2s^2 2p^6 3s^2$
13	aluminium	$1s^2 2s^2 2p^6 3s^2 3p^1$
14	silicon	$1s^2 2s^2 2p^6 3s^2 3p^2$
15	phosphorus	$1s^2 2s^2 2p^6 3s^2 3p^3$
16	sulphur	$1s^2 2s^2 2p^6 3s^2 3p^4$
17	chlorine	$1s^2 2s^2 2p^6 3s^2 3p^5$
18	argon	$1s^2 2s^2 2p^6 3s^2 3p^6$
19	potassium	$1s^2 2s^2 2p^6 3s^2 3p^6 4s^1$
20	calcium	$1s^2 2s^2 2p^6 3s^2 3p^6 4s^2$

Table 7 *Ground state electronic configurations of elements with atomic numbers (Z) 1–20*

To write the electronic configuration of the next element, scandium ($Z = 21$), you need to look back to Figure 6. The energy level of the $3d$ sub-shell lies just above that of the $4s$ sub-shell but just below the $4p$ sub-shell. This means that, once the $4s$ level is filled in calcium ($Z = 20$), the next element, scandium, has the electronic structure $1s^2 2s^2 2p^6 3s^2 3p^6 3d^1 4s^2$. The $3d$ sub-shell continues to be filled across the Period in the elements Sc to Zn. Zinc has the electronic configuration $1s^2 2s^2 2p^6 3s^2 3p^6 3d^{10} 4s^2$.

Note that the $3d$ sub-shell is written along with the other $n = 3$ sub-shells even though it is filled after the $4s$ sub-shell. The filling of the $3d$ sub-shell has an important effect on the chemistry of the elements Sc to Zn.

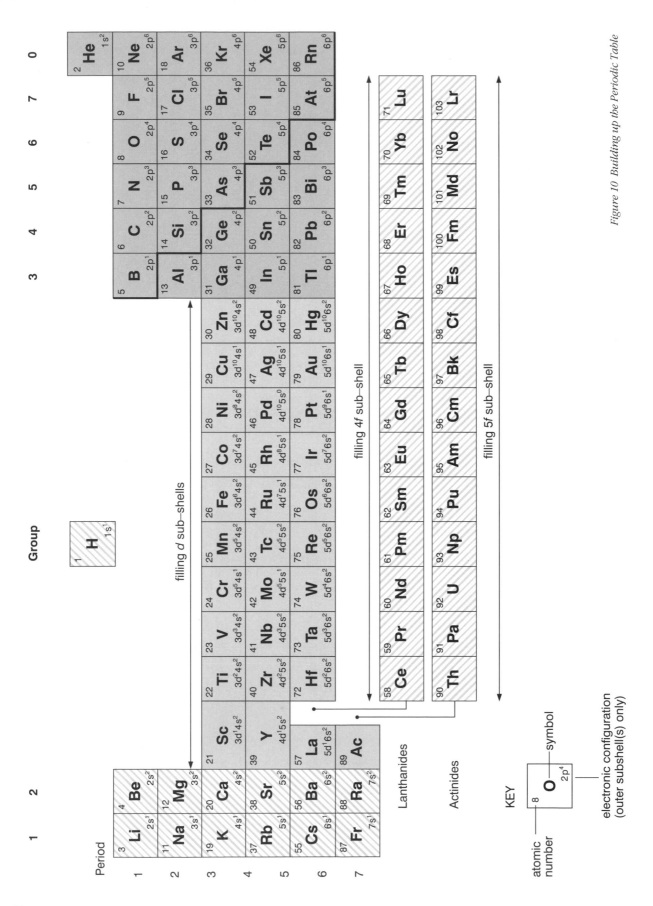

Figure 10 Building up the Periodic Table

Once the $3d$ orbitals are filled, subsequent electrons go into the $4p$ sub-shell.

In this way we can understand the building up of the Periodic Table (see Figure 10). Period 2 (Li to Ne) corresponds to the filling of the $2s$ and $2p$ orbitals, Period 3 (Na to Ar) to the filling of the $3s$ and $3p$ orbitals and so on.

The d block elements (Sc to Zn, Y to Cd and La to Hg) correspond to the filling of the d orbitals ($3d$, $4d$ and $5d$ respectively) and the lanthanides and actinides to the filling of the $4f$ and $5f$ orbitals, respectively.

Chemical properties and electronic structure

The chemical properties of an element are decided by the electrons in the incomplete outer shells. These are the electrons which are involved in chemical reaction.

Compare the electronic arrangement of the noble gases

He $1s^2$
Ne $1s^2 2s^2 2p^6$
Ar $1s^2 2s^2 2p^6 3s^2 3p^6$
Kr $1s^2 2s^2 2p^6 3s^2 3p^6 3d^{10} 4s^2 4p^6$

All have sub-shells fully occupied by electrons. Such arrangements are called **closed shell** arrangements. These are particularly stable arrangements, but note that in Ar and Kr the outer shell is only 'temporarily full' and can expand further by filling d and f sub-shells.

Now compare the electronic configurations of the elements in Group 1. They all have one electron in the outermost s sub-shell and as a result show similar chemical properties. Group 2 elements all have 2 electrons in the outermost s sub-shell. Groups 1 and 2 elements are known as **s block elements.**

In Groups 3, 4, 5, 6, 7 and 0 the outermost p sub-shell is being filled. These elements are known as **p block elements**.

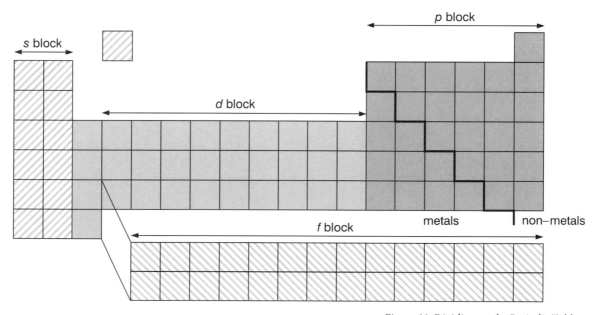

Figure 11 Dividing up the Periodic Table

The elements where a d sub-shell is being filled are **d block elements** and those where an f sub-shell is being filled are **f block elements.** Figure 11 summarises the blocks in the Periodic Table. Dividing it up in this way is very useful to chemists since it groups together elements with similar electronic configurations and similar chemical properties.

PROBLEMS FOR 2.4

1 Look at Figure 11 and classify the s, p, d and f blocks as containing metals, non-metals or a mixture of both.

2 Write out the electronic configuration of the following atom and ions:

   a boron ($Z = 5$)   **d** zinc ($Z = 30$)

   b phosphorus ($Z = 15$)   **e** Cl^- ion ($Z = 17$)

   c potassium ($Z = 19$)   **f** Ca^{2+} ion ($Z = 20$)

3 The electronic configuration of the outermost shell of an atom of element X is $3s^2 3p^4$. What is the atomic number and name of the element?

4 Electronic configurations are sometimes abbreviated by labelling the core of filled inner shells as the electron configuration of the appropriate noble gas. For example, the electronic configuration of neon is $1s^2 2s^2 2p^6$ and that of sodium $1s^2 2s^2 2p^6 3s^1$.

Thus, we can write the electronic configuration of sodium as [Ne] $3s^1$. Name the elements whose electronic configuration may be written as

   a [Ne] $3s^2 3p^5$ **c** [Ar] $3d^2 4s^2$

   b [Ar] $4s^1$ **d** [Kr] $4d^{10} 5s^2 5p^2$

5 Classify the following elements as s, p, d or f block elements.

   a $1s^2 2s^2 2p^6 3s^2 3p^6 3d^{10} 4s^2 4p^6 5s^1$

   b $1s^2 2s^2 2p^6 3s^2 3p^4$

   c $1s^2 2s^2 2p^6 3s^2 3p^6 3d^6 4s^2$

   d $1s^2 2s^2 2p^6 3s^2 3p^6 3d^{10} 4s^2 4p^6 4d^{10} 4f^4 5s^2 5p^6 6s^2$

   e chromium

   f aluminium

   g uranium

   h strontium

3

BONDING, SHAPES AND SIZES

3.1 *Chemical bonding*

Noble gas electron configurations

In 1916, W Kossel and G N Lewis realised that, with the exception of helium, all the noble gases had eight electrons in their outer shells. They linked the chemical stability of the noble gases to this outer shell electron configuration. They suggested that other elements try to achieve eight outer shell electrons by losing or gaining electrons when they react to form compounds. Some light elements are able to achieve stability by reaching the helium configuration with two outer shell electrons.

Ionic bonding

s-Block metal atoms have only one or two outer shell electrons. What makes them metals is that a noble gas configuration is reached if these electrons are *lost* to form positively charged ions called **cations**.

Most non-metal atoms have more than three outer shell electrons. One way they can reach a noble gas configuration is by *gaining* electrons to form negatively charged ions or **anions**.

There are limits to how many electrons an atom can pick up. One electron leads to an anion with a single negative charge. A second electron is repelled by the anion because their charges are the same, so making a doubly charged anion is much harder. Getting a third electron to stick, by the process: $A^{2-} + e^- \rightarrow A^{3-}$, is very difficult and does not often happen. Anions with a charge of 3– are unusual.

It is also hard to *remove* three or more electrons from atoms. Cations with a 4+ charge are almost unknown.

When metals react with non-metals, electrons are *transferred* from the metal atoms to the non-metal atoms. This usually gives both metal and non-metal a stable electronic structure like a noble gas. The cations and anions which are formed are held together by **electrostatic attraction** because of their opposite charges.

We use **electron dot-cross diagrams** to represent the way atoms bond together. In these, the outer shell electrons of one atom are represented by dots, with crosses for the other. Figure 1 shows dot-cross diagrams for the formation of sodium chloride and magnesium fluoride. The dots and crosses represent *only the outer shell electrons*. The numbers underneath the chemical symbols show the arrangement of electrons in the shells.

Each sodium atom loses one electron and each chlorine atom gains one electron, so the compound formed has the formula NaCl. Each magnesium atom loses two electrons but each fluorine atom gains only one electron, so the formula for magnesium fluoride is MgF_2.

Figure 1 Dot-cross diagrams for two ionic compounds

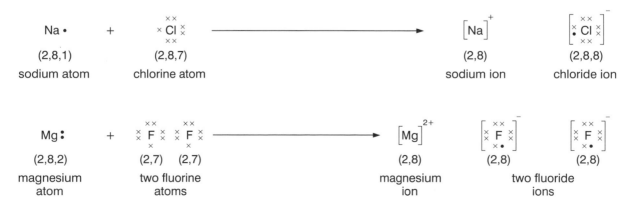

The oppositely-charged ions attract each other strongly: this attraction is an **ionic bond**. In the solid compound, each ion attracts many others of opposite charge, and the ions build up into a giant lattice like the one shown in Figure 2. There is more about this in **Section 5.1**.

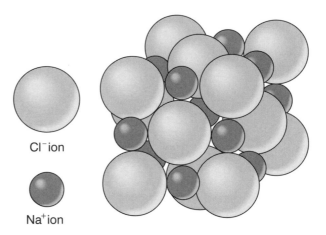

Figure 2 The sodium chloride lattice, built up from oppositely charged sodium ions and chloride ions

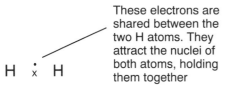

Figure 3 Electron sharing in the hydrogen molecule

Covalent bonding

Ionic compounds are usually formed when metals react with non-metals. But there are many compounds containing only non-metallic elements. These cannot bond ionically because some of the atoms would have to lose too many electrons. Instead, noble gas configurations are achieved by *sharing* electrons. Shared electrons count as part of the outer shell of *both* atoms in the bond. The dot-cross diagram for the H_2 molecule is shown in Figure 3.

Bonds formed by shared electrons are called **covalent bonds**. If a *pair* of electrons is involved, the bond is called a *single* covalent bond, or more simply a **single bond**. The two atoms are bonded together because their nuclei are simultaneously attracted to the shared electrons.

Examples of dot-cross diagrams for two more covalent compounds are shown in Figure 4. Notice that, by sharing electrons in this way, the atoms all achieve stable electron structures like noble gases.

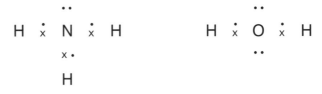

Figure 4 Dot-cross diagrams for NH₃ and H₂O

     ammonia, NH₃   *water,* H₂O

Electron pairs which form bonds are called **bonding pairs**. Pairs of electrons not involved in bonding are called **lone pairs**. Both water (H_2O) and ammonia (NH_3) have lone pair electrons.

When two pairs of electrons form a covalent bond, it is called a **double bond**. The bonds in molecular oxygen and carbon dioxide are double covalent bonds (Figure 5).

Figure 5 Dot-cross diagrams for O₂ and CO₂

    oxygen, O₂   *carbon dioxide,* CO₂

When three pairs of electrons form a bond it is called a **triple bond**. The bonds in nitrogen (N_2) and between carbon and nitrogen in hydrogen cyanide (HCN) are examples (Figure 6).

nitrogen, N_2 　　　　 hydrogen cyanide, HCN

Figure 6 Dot-cross diagrams for N_2 and HCN

Electron dot-cross diagrams are useful for representing individual electrons in chemical bonds. Chemists often draw structures in a simpler way by using lines to represent a pair of electrons shared between two atoms. A single line represents a single covalent bond. Double and triple lines represent double and triple covalent bonds.

Some of the molecules used as examples earlier are shown in this way in Figure 7.

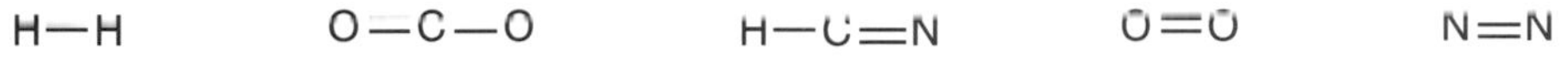

H—H 　　　 O—C—O 　　　 H—C≡N 　　　 O=O 　　　 N≡N

hydrogen 　　 carbon dioxide 　 hydrogen cyanide 　 oxygen 　　 nitrogen

Figure 7 Covalently bonded molecules

Dative covalent bonds

Figure 8 shows the bonding in carbon monoxide, CO. Look carefully at the three pairs of electrons that make the triple bond between the C and O atoms. Two of these are formed by the C and O atoms contributing an electron each to the shared pair – these are ordinary covalent bonds. But in the third pair of electrons, both electrons come from the oxygen atom. This is called a **dative bond**. In a dative bond, *both the bonding electrons come from the same atom*, unlike an ordinary covalent bond where the atoms contribute an electron each to the pair. A dative covalent bond can be shown by an arrow, with the arrow pointing away from the atom that donates the pair of electrons (Figure 8(c)).

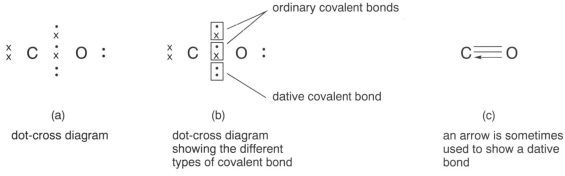

(a) 　　　　　　　　　　 (b) 　　　　　　　　　　 (c)

dot-cross diagram 　　　 dot-cross diagram 　　　 an arrow is sometimes
　　　　　　　　　　 showing the different 　　　 used to show a dative
　　　　　　　　　　 types of covalent bond 　　　 bond

Figure 8 Dative covalent bonding in CO

Why are bonds like bears?

Figure 9 shows a more detailed diagram of the way the hydrogen molecule is bonded.

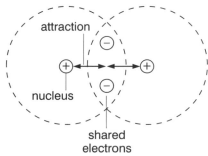

Figure 9 In a hydrogen molecule, the atoms are held together because their nuclei are both attracted to the shared electrons

The atoms are held together because their nuclei are both attracted to the electron pair which is shared between them. Both atoms are identical so the electrons are shared equally.

With atoms more complicated than hydrogen, it is the atomic cores which are attracted to the shared electron pairs. The atomic core is made up of everything except the outer shell electrons. The core of the fluorine atom, for example, is made up from the nucleus with a charge of 9+ and the two electrons in the inner most shell, giving a core with a charge of 7+. The positive charge on the core is the same as the group number of the element. The F_2 molecule is held together because the 7+ core charges are attracted to the negative charges on the shared electrons.

Very often the two atoms bonded together have different sizes. In this case, the core of the smaller atom will be closer to the shared electrons and will exert a stronger pull on them (Figure 10a). A similar situation arises when the atoms are from different groups and have different core charges (Figure 10b).

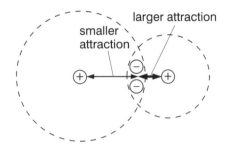

Figure 10a Shared electrons are attracted more strongly by the core of the smaller atom, which is closer

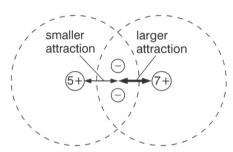

Figure 10b Shared electrons are attracted more strongly by the atom with the greater core charge

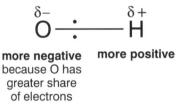

more negative **more positive**
because O has
greater share
of electrons

Figure 11 The O–H bond is polar

In general, different atoms attract bonding electrons unequally. One atom gets a slight negative charge because it has a greater share of the bonding electrons. The other atom becomes slightly positively charged because it has lost some of its share in the bonding electrons. Bonds like this are called **polar bonds**. An example is shown in Figure 11. The small amounts of electrical charge are shown by $\delta-$ and $\delta+$ where δ means 'a small amount off'. Some more examples of **bond polarity** are shown in Table 1.

Bonded atoms	Sizes of atoms	Core charges	Bond polarity
F and Br	F smaller	same	$\overset{\delta-}{F} - \overset{\delta+}{Br}$
O and N	similar	O greater	$\overset{\delta-}{O} - \overset{\delta+}{N}$
F and P	F smaller	F greater	$\overset{\delta-}{F} - \overset{\delta+}{P}$

Table 1 Some examples of polar bonds

So bonds are like bears – some are polar and some are not. The O—H bond is a particularly important example of a polar covalent bond. It has many consequences for the chemistry of water molecules (Figure 12)

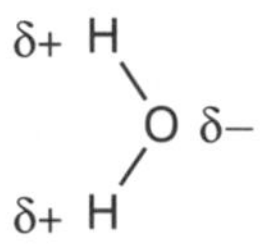

Electronegativity

To decide the polarity of a covalent bond, we need a measure of each atom's 'electron pulling power'. This is called **electronegativity**. Atoms with strong electron pulling power in covalent bonds are said to be highly electronegative. Table 2 lists some atoms in order of their electronegativity

Figure 12 Polar bonding in the water molecule

values: notice that the most electronegative ones are highly reactive non-metal elements. The least electronegative elements are the reactive metals.

We can use electronegativity to decide the polarity of a particular bond. For example, in the C—F bond, F has a higher electronegativity than C so it attracts the shared electrons more strongly, and the polarity is

δ+ δ–
C—F

There is more about electronegativity and bond polarity in **Section 5.3**. Polar bonds are like covalent bonds with a bit of *ionic character* in them. The ionic and covalent models are extreme forms of bonding: polar bonds are somewhere between the two. The bigger the difference in electronegativity between the atoms, the more polar the bond and the greater the ionic character (Figure 13).

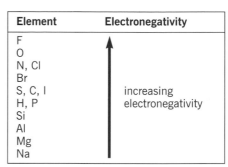

Element	Electronegativity
F	
O	
N, Cl	
Br	
S, C, I	increasing
H, P	electronegativity
Si	
Al	
Mg	
Na	

Table 2 Atoms listed in order of electronegativity

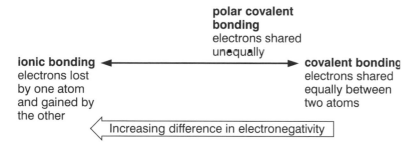

Figure 13 Bonds can vary from ionic to covalent, with polar covalent bonds in between

A cautionary note

Theories of bonding are fundamental to an understanding of chemistry. The dot-cross model is a simple approach to bonding. Like the Bohr theory it is an over-simplified model, but it is nevertheless useful. It needs to be extended before it can do more than explain simple situations. So don't be surprised if dot-cross diagrams sometimes don't work, or if some molecules seem to break the rules.

PROBLEMS FOR 3.1

1. Draw electron dot-cross diagrams for the following ionic compounds.

   a. potassium fluoride, KF
   b. magnesium oxide, MgO
   c. calcium chloride, $CaCl_2$
   d. sodium sulphide, Na_2S

2. Draw electron dot-cross diagrams for the covalent compounds shown in Figure 14.

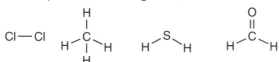

chlorine   methane   hydrogen sulphide   methanal

Figure 14

3. a. Ammonia reacts with hydrogen chloride to give the salt ammonium chloride. This contains ammonium ions, NH_4^+, and chloride ions, Cl⁻. In the ammonium ion, a lone pair of electrons from ammonia forms a dative covalent bond to the hydrogen ion, H⁺, from the hydrogen chloride.

   Draw a dot-cross diagram for
   i ammonia
   ii the ammonium ion.

   b. When hydrogen chloride dissolves in water, it produces hydrochloric acid. An equation representing this reaction is

   $HCl(g) + H_2O(l) \rightarrow H_3O^+(aq) + Cl^-(aq)$

   $H_3O^+(aq)$ is called the oxonium ion. Like the ammonium ion it contains a dative covalent bond. Draw dot-cross diagrams for
   i water
   ii the oxonium ion.

4. Predict the polarity of the bonds in the following molecules:

   a. NO
   b. HF
   c. N_2
   d. HCl
   e. ClF
   f. CO

3.2 *The size of ions*

Section 4.5 looks at the way ions behave in solution and how this affects the solubility of ionic substances. Two things are particularly important in deciding the way ions behave in solution: the *charge* on the ion and its *size*. These two factors are also important in deciding how ions behave during ion-exchange.

If an ion is *highly charged*, like Al^{3+}, it will be strongly attracted to other ions and to polar molecules such as water.

If the ion is *small*, it will be strongly attracted to other ions and to polar molecules, because it can get close.

If an ion is both highly charged and small, we say it has a high **charge density**. Al^{3+} (charge 3+, radius 0.053 nm) has a much higher charge density than K^+ (charge 1+, radius 0.138 nm).

Ions with a high charge density tend to

- attract water molecules strongly and become very hydrated;
- attract other ions strongly, forming ionic lattices with large lattice enthalpies and therefore high melting points.

What decides the sizes of atoms and ions?

In atoms and ions, the electron shells take up nearly all the space. So you might think that more electrons mean a bigger atom or ion. This is true if you look at a group of elements, like the Group 1 elements shown in Table 3, because a new electron shell is added at each new element. Each new electron shell is further from the nucleus than the previous outer shell.

Element	Number of electrons	Atomic radius/nm
Li	3	0.157
Na	11	0.191
K	19	0.235
Rb	37	0.250
Cs	55	0.272

Table 3 Atomic radii of the Group 1 elements

However, size also depends on the number of protons in the nucleus. Table 4 shows the atomic radii of the elements in Period 3.

Element	Number of electrons	Atomic radius/nm
Na	11	0.191
Mg	12	0.160
Al	13	0.130
Si	14	0.118
P	15	0.110
S	16	0.102
Cl	17	0.099
Ar	18	0.095

Table 4 Atomic radii of the Period 3 elements

You can see that going across Period 3 from Na to Ar, the atomic radii actually get smaller, even though electrons are being added. This is because all the electrons are going *into the same shell:* shell 3. Across the Period, the number of protons in the nucleus is steadily increasing, and the increasing positive charge attracts the electrons increasingly strongly. So the atoms get smaller, even though they have more electrons.

At potassium, a new electron shell is started – the fourth shell. This shell is further from the nucleus than the third shell, so K has a much larger radius than Ar.

From atoms to ions

When atoms become ions, electrons are added or taken away, and this has a major effect on the size. Adding an electron to an atom to produce a negative ion like Cl⁻ makes the ion *bigger* than the atom. Adding two electrons (to make, say, O^{2-}) has an even greater effect. Removing an electron to make a positive ion like Na^+ makes the ion *smaller* than the atom. Removing two or three electrons (to make, say, Mg^{2+} or Al^{3+}) has an even greater effect.

The general rule is

$$r_{2+} \quad < \quad r_+ \quad < \quad r_{atom} \quad < \quad r_- \quad < \quad r_{2-}$$

Figure 15 illustrates this.

Hydration changes sizes

The rule just described works well for ions in the solid, liquid and gaseous phases. But as soon as you dissolve an ionic substance in water, the situation is different. The ions attract water molecules and become hydrated (see **Section 4.5**). This makes the ion bigger, because it has layers of water molecules around it. The higher the charge density of the ion, the more water molecules it attracts, and the bigger it becomes (see Figure 16). So you have the situation in which an ion that is small in the absence of water becomes large when it is in the aqueous phase. Figure 17 illustrates this.

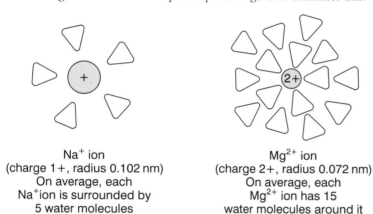

Na⁺ ion
(charge 1+, radius 0.102 nm)
On average, each
Na⁺ion is surrounded by
5 water molecules

Mg²⁺ ion
(charge 2+, radius 0.072 nm)
On average, each
Mg²⁺ ion has 15
water molecules around it

Figure 16 Ions with higher charge densities attract more water molecules (ions are only shown in two dimensions)

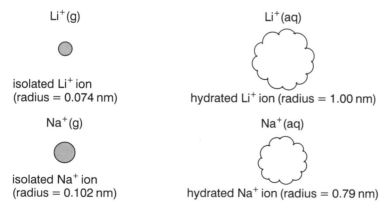

Li⁺(g)

Li⁺(aq)

isolated Li⁺ ion
(radius = 0.074 nm)

hydrated Li⁺ ion (radius = 1.00 nm)

Na⁺(g)

Na⁺(aq)

isolated Na⁺ ion
(radius = 0.102 nm)

hydrated Na⁺ ion (radius = 0.79 nm)

Figure 17 Hydrated ions are much bigger than isolated ions

You need to remember the effects of hydration on size when you are thinking about the way ions behave in solution. Hydration makes apparently small, strongly attracting ions behave as if they are larger and less attracting. You can see this effect when you look at how strongly different ions are attracted to ion exchange resins (**Section 7.5**).

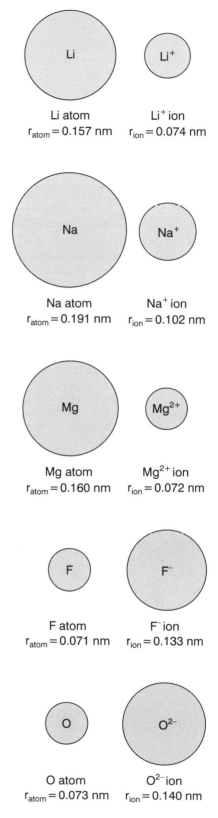

Li atom
$r_{atom} = 0.157$ nm

Li⁺ ion
$r_{ion} = 0.074$ nm

Na atom
$r_{atom} = 0.191$ nm

Na⁺ ion
$r_{ion} = 0.102$ nm

Mg atom
$r_{atom} = 0.160$ nm

Mg²⁺ ion
$r_{ion} = 0.072$ nm

F atom
$r_{atom} = 0.071$ nm

F⁻ ion
$r_{ion} = 0.133$ nm

O atom
$r_{atom} = 0.073$ nm

O²⁻ ion
$r_{ion} = 0.140$ nm

Figure 15 The sizes of atoms and ions

PROBLEMS FOR 3.2

1 For each of the following pairs, say which you would expect to have the larger radius. Explain your reason in each case.
a lithium ion $Li^+(g)$ and lithium atom $Li(g)$
b lithium ion $Li^+(g)$ and hydrated lithium ion $Li^+(aq)$
c hydrogen ion $H^+(g)$ and hydrogen atom $H(g)$
d hydride ion $H^-(g)$ and hydrogen atom $H(g)$

2 Look at the following values of ionic radii. It will help to refer to a copy of the Periodic Table when you answer **a** and **b**.

Ion	Ionic radius/nm
Be^{2+}	0.027
Mg^{2+}	0.072
Ca^{2+}	0.100
Fe^{2+}	0.061
Fe^{3+}	0.055

a Explain why
  i Ca^{2+} is larger than Mg^{2+}
  ii Fe^{2+} is larger than Fe^{3+}
b **i** Calculate the ratio $\dfrac{\text{radius of } Mg^{2+}}{\text{radius of } Be^{2+}}$
  ii Calculate the ratio $\dfrac{\text{radius of } Ca^{2+}}{\text{radius of } Mg^{2+}}$
  iii Compare the ratios and suggest a reason for any difference.

3 Use your knowledge of ionic radii to suggest explanations for the following:
a The lattice enthalpy of NaCl has a more negative value than that of KCl.
b The melting point of MgO is larger than that of CaO.
c In aqueous solution, $Na^+(aq)$ ions move more slowly than $K^+(aq)$ ions.
d In ion exchange, K^+ ions are more strongly held on the resin than Na^+ ions.

4 The following table gives the ionic radii of some ions formed by elements in Period 3 of the Periodic Table.

Ion	Na^+	Mg^{2+}	Al^{3+}	P^{3-}	S^{2-}	Cl^-
Ionic radius/nm	0.095	0.065	0.050	0.212	0.184	0.181

a For each ion, give
  i the number of protons in its nucleus
  ii its electronic structure.
b Plot the values in the table on a bar chart.
c Comment on, and suggest reasons for, the shape of the chart.

3.3 *The shapes of molecules*

Electron pair repulsions

The arrangement of electrons in methane can be represented by a dot-cross diagram (Figure 18). There are *four groups* of electrons around the central carbon atom – the four covalent bonding pairs. Because similar charges repel, these groups of electrons arrange themselves so that they are as far apart as possible.

The farthest apart they can get is when the H—C—H bond angles are 109°. Clearly, this does not correspond to a flat methane molecule shaped like a 'plus sign': in that arrangement the bond angles are only 90°. The 109° angle arises if methane has a **tetrahedral** shape with H atoms at the corners and the C atom in the centre (Figure 19). (If you have trouble visualising this, you should build a model of methane and see how its structure resembles a tetrahedron – or make a 'balloon model' in **Activity EP 2.3**.)

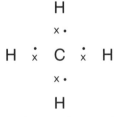

Figure 18 A dot-cross diagram showing the bonding in methane

Figure 19 The methane molecule has a tetrahedral shape

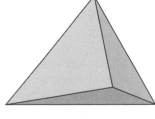

a *tetrahedron*

a *methane* molecule

Chemists have developed a system for drawing three-dimensional structures on two-dimensional paper and this has been used in Figure 19. Bonds which lie in the plane of the paper are drawn in the normal way as solid lines. Bonds which come forwards, towards you, are represented as wedge-shaped lines. Bonds which go backwards, away from you, are shown as dashed lines. (Build a model of methane and arrange it so that one C—H bond comes towards you, two go away and the other goes up and down – to correspond to Figure 19.)

All carbon atoms which are surrounded by four single bonds have a tetrahedral distribution of bonds. Figure 20 shows ethane, in which both C atoms have a tetrahedral arrangement.

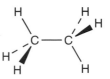

Figure 20 The tetrahedral bonding around the carbon atoms in ethane

Lone pairs count too

A molecule of methane has only bonding pairs in the outer shell of carbon – there are no lone pairs. Molecules like ammonia (NH_3) and water (H_2O) do have lone pairs in their outer shells (Figure 21). These lone pairs repel the bonding pairs of electrons, as in methane. So, ammonia and water adopt the same tetrahedral shape as methane, but in these molecules one or more of the corners of the tetrahedron is occupied by a lone pair of electrons (Figure 22).

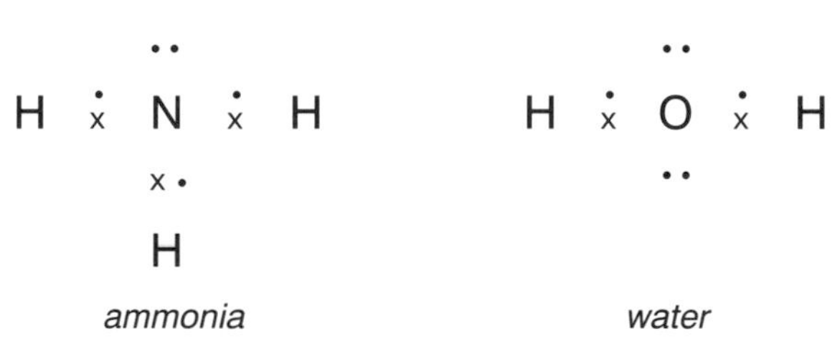

H $\overset{\cdot\cdot}{\underset{\times}{N}}$ H   H $\overset{\cdot\cdot}{\underset{\cdot\cdot}{O}}$ H

$\overset{\times\,\cdot}{H}$

ammonia   *water*

Figure 21 Dot-cross diagrams showing the bonding in ammonia and water

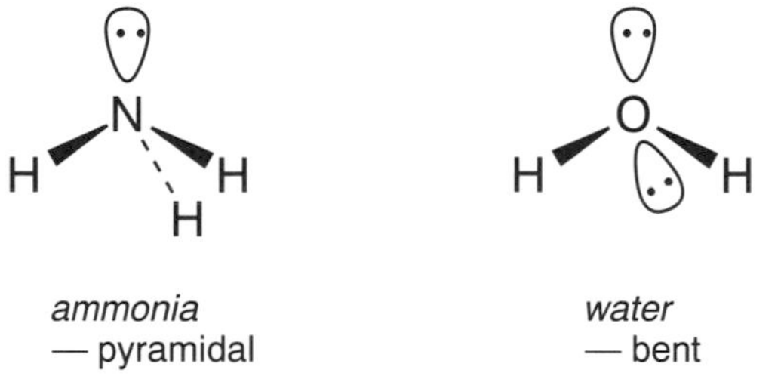

ammonia
— pyramidal

water
— bent

Figure 22 The shapes of molecules of ammonia and water

The H—N—H and H—O—H bond angles are both close to 109°. NH_3 is said to have a *pyramidal* shape; H_2O is described as *bent*. (Again, it is best to build models of these molecules to confirm the structures that are drawn.)

The shapes of covalent molecules are decided by a simple rule: *groups of electrons in the outer shell repel one another and move as far apart as possible*. It doesn't matter if they are groups of bonding electrons or lone pairs – they all repel one another.

Other shapes

Linear molecules

In $BeCl_2$, there are two groups of electrons around the central atom (Figure 23).

Cl $\underset{\times}{\cdot}$ Be $\underset{\times}{\cdot}$ Cl   shape   Cl—Be—Cl

Figure 23 The shape of the BeCl₂ molecule. (The lone pairs around the Cl atom have been omitted for clarity)

Because there are fewer groups of electrons than in methane, they can get further apart. The furthest apart they can get is at an angle of 180°, so $BeCl_2$ is **linear** with a Cl—Be—Cl angle of 180°.

Carbon dioxide and ethyne are other examples of linear molecules: there are only two groups of electrons around the central atoms in these molecules (Figure 24).

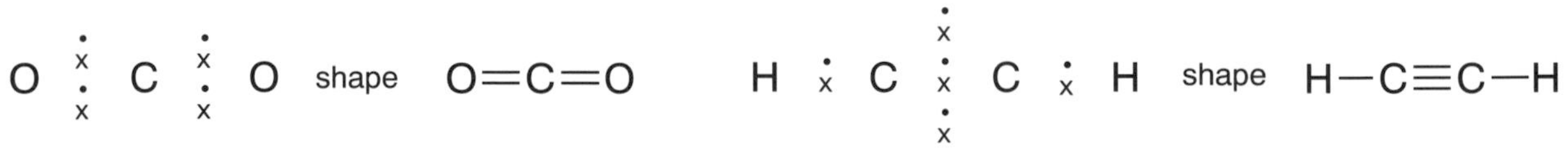

O $\overset{\displaystyle \cdot}{\underset{\displaystyle \times}{\overset{\displaystyle \times}{:}}}$ C $\overset{\displaystyle \times}{\underset{\displaystyle \times}{:}}$ O   shape   O=C=O   H $\overset{}{\underset{\displaystyle \times}{:}}$ C $\overset{\displaystyle \times}{\underset{\displaystyle \times}{:}}$ C $\overset{}{\underset{}{\times}}$ H   shape   H—C≡C—H

carbon dioxide   *ethyne*

Figure 24 Carbon dioxide and ethyne both have linear molecules

Planar molecules

In BF_3 there are *three* not four, groups of electrons around the central atom.

F $\overset{\displaystyle \times \cdot}{\underset{\displaystyle \cdot \times}{\times}}$ B   shape   F—B with F atoms

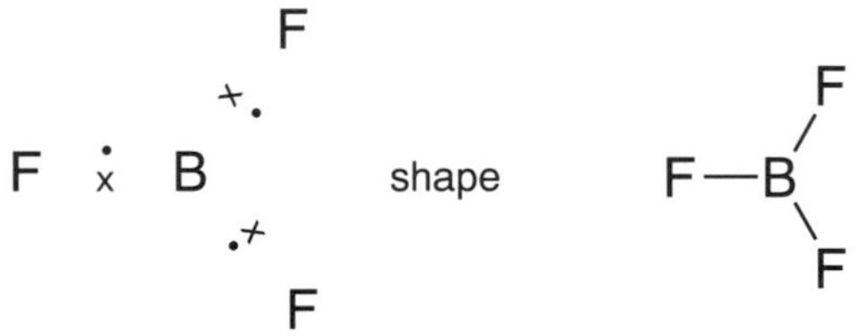

Figure 25 The shape of the BF_3 molecule. (The lone pairs around the F atom have been omitted for clarity)

The furthest the three groups of electrons can get apart is at an angle of 120°, so the F—B—F bond angle is 120°. BF_3 is flat and shaped like a triangle with B at the centre and F atoms at the corners (Figure 25). Its shape is described as **planar triangular**.

There are other molecules with planar structures (Figure 26). In methanal, for example, there are three groups of electrons around the carbon atom, and in ethene *both* carbons have three groups of electrons. Remember it is the number of *groups* of electrons which determine the shape – there are four pairs of electrons around the carbon atoms in these molecules, but the bonds are *not* tetrahedrally directed because there are only three *groups* of electrons.

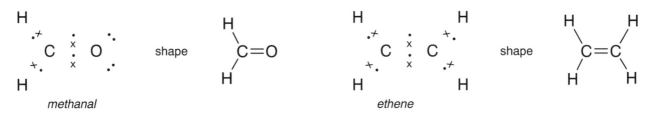

H C $\overset{\displaystyle \times}{\underset{\displaystyle \times}{:}}$ O   shape   C=O

methanal   *ethene*

Figure 26 Methanal and ethene have planar molecules

Whenever you draw diagrams of molecules, try to represent their shapes as accurately as possible. For example, avoid drawing right-angle bonds in alkenes.

Octahedral molecules

Six groups of electrons around a central atom gives rise to an **octahedral** shape, where the electrons are directed to the six corners of an octahedron. Some molecules, such as SF_6, adopt this structure, but it is more commonly found in the octahedral shapes of complexes with six ligands (Figure 27).

Figure 27 Six groups of electrons around a central atom or ion give an octahedral shape

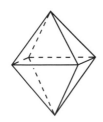

an octahedron   *SF_6 molecule*   *$[Ni(H_2O)_6]^{2+}$ complex ion*

The shapes of ions

We can use the same ideas to work out the shapes of *ions* as we use for *molecules*. We need to include any extra electrons that have been added or taken away when the ion was formed. Figure 28 illustrates the idea for two ions based on ammonia, NH_4^+ and NH_2^-.

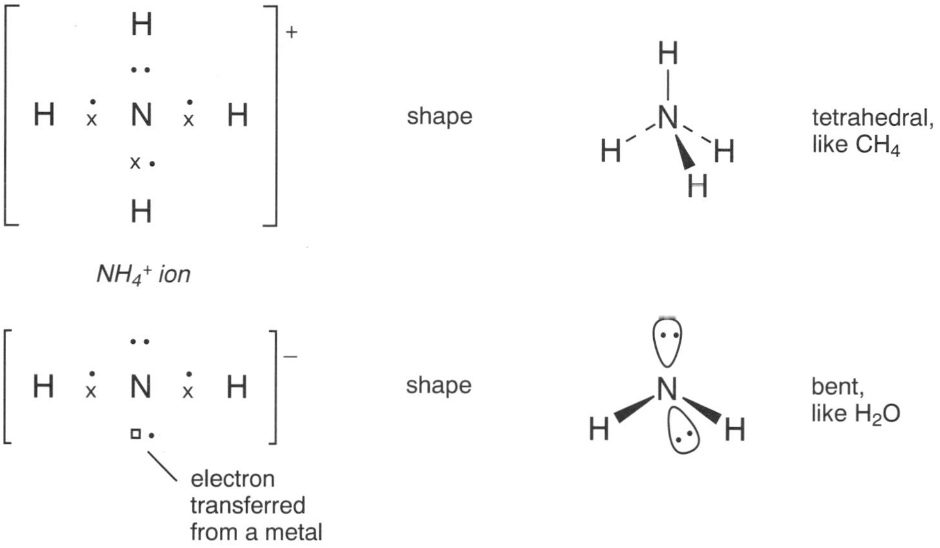

NH₄⁺ ion — shape — tetrahedral, like CH₄

electron transferred from a metal

NH₂⁻ ion — shape — bent, like H₂O

Figure 28 The shapes of NH_4^+ and NH_2^-

PROBLEMS FOR 3.3

1 a Draw dot-cross diagrams for the following molecules
  i SiH_4 **ii** SF_2 **iii** $COCl_2$ **iv** HCN
 b For each molecules draw a diagram to illustrate the shape you would expect it to adopt and indicate the approximate bond angles.

2 Draw diagrams to show the shapes of the following molecules. In each case show bonds and lone pairs, and clearly indicate the approximate bond angles.
  a CH_3NH_2 (methylamine)
  b CH_3OH (methanol)
  c CH_3COCl (ethanoyl chloride)
  d CH_3CN (ethanenitrile)
  e $CH_2{=}CHCl$ (chloroethene)
  f H_2NOH (hydroxylamine)

3 The compound $CH_3N{=}NCH_3$ can exist as two geometric isomers. Draw diagrams to represent the two structures and explain why they correspond to geometric isomers.

4 Explain why the following pairs of structures do *not* represent different compounds.

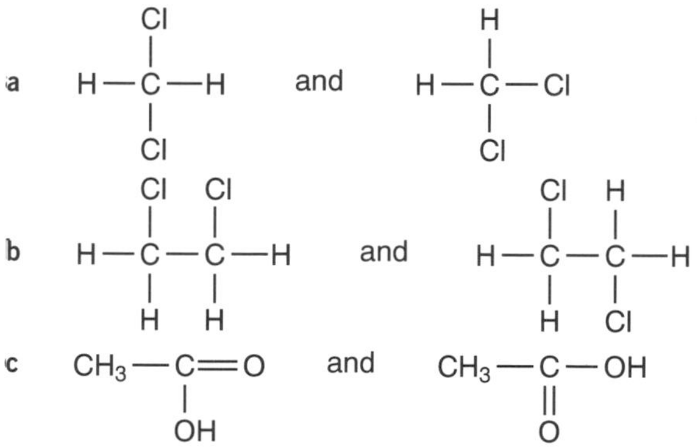

a

b

c $CH_3{-}C{=}O$ with OH and $CH_3{-}C{-}OH$ with O (double bond)

5 a Consider the carbocation CH_3^+.
  i Draw a dot-cross diagram to show the electron structure of this ion (note: it only has 6 electrons in the outer shell of the C atom).
  ii Predict the shape of the ion.
 b Consider the carboanion CH_3^-.
  i Draw a dot-cross diagram for the ion (note: it has 8 electrons in the outer shell of the C atom).
  ii Predict its shape.

3.4 *Structural isomerism*

Two molecules which have the same molecular formula but differ in the way their atoms are arranged are called **isomers**. Isomers are distinct compounds with different physical properties, and often different chemical properties too.

The occurrence of isomers (**isomerism**) is commonest in carbon compounds because of the great variety of ways in which carbon can form chains and rings, but you will meet examples in inorganic chemistry too.

There are two ways in which atoms can be arranged differently in isomers:

either   the atoms are bonded together in a *different order* in each isomer. These are called **structural isomers**

or   the order of bonding in the isomers is the same, but *the arrangement of the atoms in space is different* in each isomer. These are called **stereoisomers**.

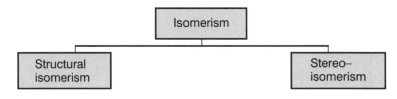

Figure 29 The two main types of isomerism

You can find out about stereoisomers in **Sections 3.5** and **3.6**. For the moment, we shall concentrate on structural isomerism.

Structural isomerism

Structural isomers have the same molecular formula but have atoms bonded together in a different order. They have different structural formulae.

There are various ways in which structural isomerism can arise (Figure 30).

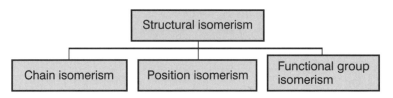

Figure 30 Types of structural isomerism

Chain isomerism

There is only one alkane corresponding to each of the molecular formulae, CH_4, C_2H_6 and C_3H_8. With four or more C atoms in a chain, different arrangements are possible. For example,

butane
boiling point 273 K

methylpropane
boiling point 261 K

Both these compounds have the same molecular formula, C_4H_{10}. Their different structures lead to different properties. For example, the boiling point of methylpropane is 12 K lower than that of butane.

As the number of carbon atoms in an alkane increases, the number of possible isomers increases. There are over four thousand million isomers with the molecular formula $C_{30}H_{62}$!

Position isomerism

This occurs where there is an atom or group of atoms substituted in a carbon chain or ring. These are called **functional groups**.

Isomerism occurs when the functional group is situated in different positions in the molecule. For example,

a Isomers of C_3H_7OH

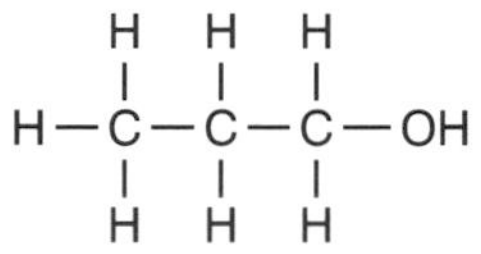

          propan-1-ol 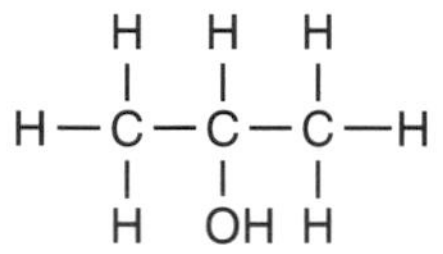   *propan-2-ol*

Here the —OH functional group is situated at two different places on the hydrocarbon chain.

b Isomers of $C_6H_4Cl_2$

1,2-dichlorobenzene   *1,3-dichlorobenzene*   *1,4-dichlorobenzene*

In this example, the —Cl functional groups are situated at different positions on the benzene ring.

Functional group isomerism

It is sometimes possible for compounds with the same molecular formula to have quite different functional groups. For example,

a Molecular formula C_3H_8O

    propan-1-ol   *methoxyethane*
    (an alcohol)   (an ether)
  functional group   functional group
      —OH   —O—

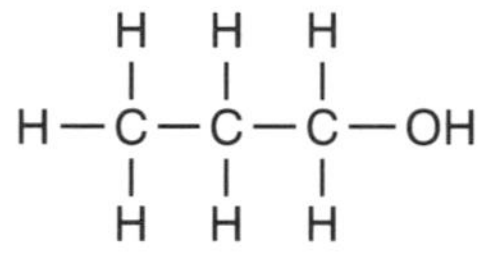

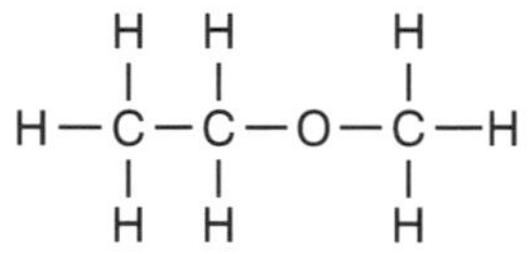

b Molecular formula C_3H_6O

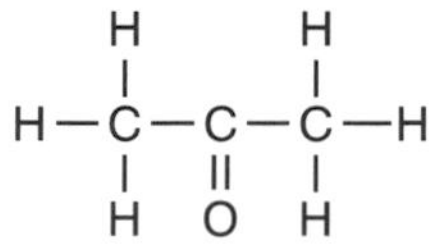

    propanone   *propanal*
    (a ketone)   (an aldehyde)
  functional group   functional group

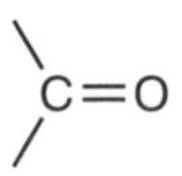

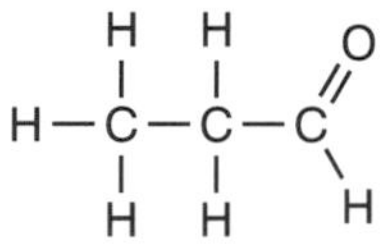

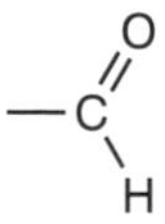

PROBLEMS FOR 3.4

1 Write out structural formulae for isomers of C_5H_{12}. Name each of your isomers.

2 How many isomers are there with molecular formula C_4H_9Br? Draw out their structures.

3 Draw out possible isomers of the molecular formula C_8H_{10} which contain a benzene ring.

4 Which of these pairs of compounds are isomers?

a

$CH_3CH_2CH_2CH_2CH_2CH_3$ and

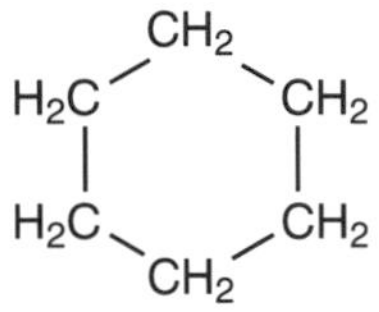

b

$CH_3CH_2CH_2CH_2Cl$ and

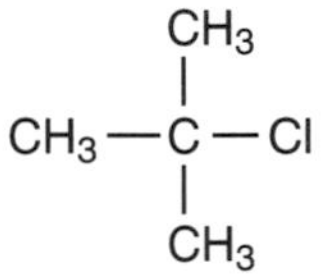

c

$CH_3CH_2CH_2OH$ and CH_3COCH_3

d

and

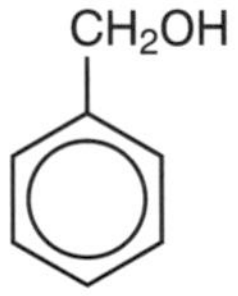

e

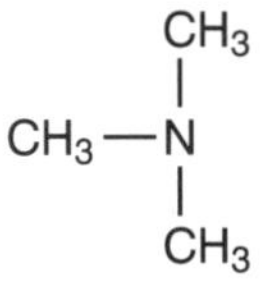

and

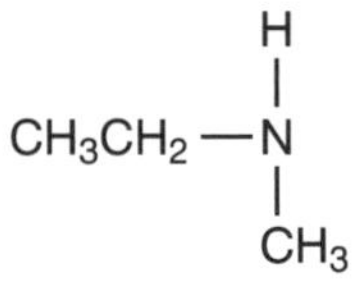

5 An organic compound was subjected to combustion analysis. 0.100 g of the compound formed 0.191 g carbon dioxide and 0.117 g water and no other products.

a Calculate the percentage by mass of carbon and of hydrogen in the compound.

b What other element must be present?

c Calculate the empirical formula of the compound.

d A mass spectrum of the compound showed that its relative molecular mass is 46. Write down its molecular formula.

e Chemical tests showed the molecule is an alcohol, containing an —OH functional group. Draw the full structural formula of the compound.

f Draw the structural formula of a compound that is isomeric with this one.

3.5 *Geometric isomerism*

Geometric isomerism, also known as *cis-trans* isomerism is one type of **stereoisomerism** (Figure 31). **Stereoisomers** have identical molecular formulae, and the atoms are bonded together in the same order, but the arrangement of atoms in space is different in each isomer.

Figure 31 Geometric isomerism is one type of stereoisomerism. (Optical isomerism is covered in **Section 3.6***)*

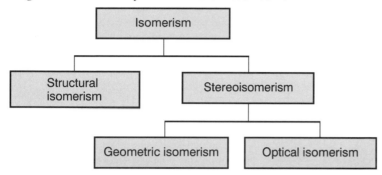

Modelling geometric isomers

Build a model of but-2-ene: $CH_3CH=CHCH_3$. If a model kit is not available you can draw the structure with correctly represented bond angles.

There is another isomer of but-2-ene with a different shape to the first one you built. Build or draw this other form.

The two forms of but-2-ene look like this:

To turn the second form into the first form you would have to spin one end of the molecule round in relation to the other end. This can only be done by first breaking one of the bonds in the double bond. If you have the models available you can easily prove this to yourself. Figure 32 shows what space-filling models of the two isomers look like.

The average bond energy for the bond which has to be broken here is about $+270\,kJ\,mol^{-1}$ and this much energy is not available at room temperature. A covalent bond has to be broken and another reformed in order to interconvert the forms of but-2-ene. In other words the process is an example of a chemical reaction, and the two forms are different chemicals.

Cis *and* trans *isomers*

The two different but-2-enes need different names. The form with the substituent groups (in this case both methyl groups) on the same side of the double bond – is called ***cis***-but-2-ene;

cis-*but-2-ene*

Where the substituents are on opposite sides of the double bond the molecule is said to be in the ***trans*** form (*trans* meaning across, as in trans-Atlantic);

trans-*but-2-ene*

Cis and *trans* forms are isomers with different geometries, which is why this form of isomerism is sometimes called **geometric isomerism**.

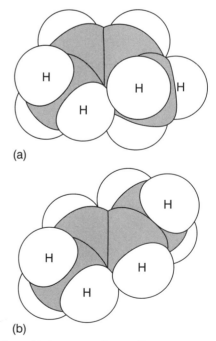

(a)

(b)

Figure 32 Diagrams of space-filling models of (a) cis-*but-2-ene and* (b) trans-*but-2-ene*

Cis and trans isomers are different compounds, so they have different properties. Table 4 gives you some information on isomers which are clearly different substances.

	Melting point/K	Density/g cm^{-3}		Melting point/K	Density/g cm^{-3}
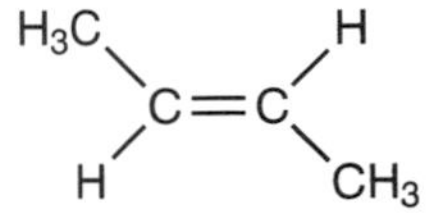	168	0.604	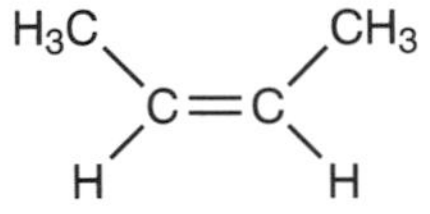	134	0.621
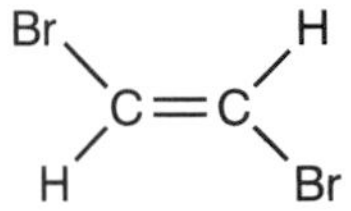	267	2.23	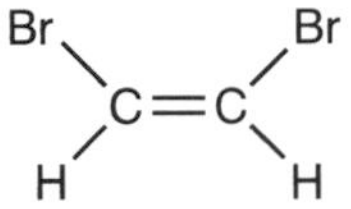	220	2.25
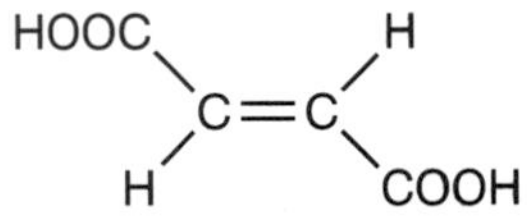	573	1.64	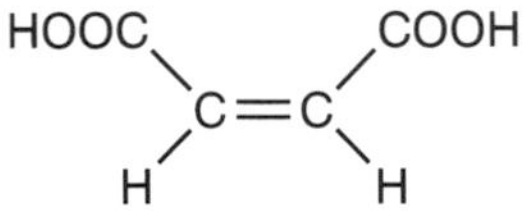	412	1.59

Table 4 Physical properties of some geometric isomers

Geometric isomerism in transition metal complexes

Geometric isomerism can also occur in *inorganic* chemicals. **Section 11.7** describes how transition metal complexes can show this form of isomerism.

PROBLEMS FOR 3.5

1 a Which of the following have *cis* and *trans* forms?
   i fluoroethene
   ii 1,2-difluoroethene
   iii 1,1,2-trifluoroethene
   iv 1,1,2,2-tetrafluoroethene
   b Draw the structures of any geometric isomers which exist for the compounds listed in **a**.

2 Nerol, which occurs in bergamot oil; geraniol – found, for example, in roses; and linalool, which is a consitiuent of the scent of lavender, are three compounds from the *terpene* family. Their structures are shown below:

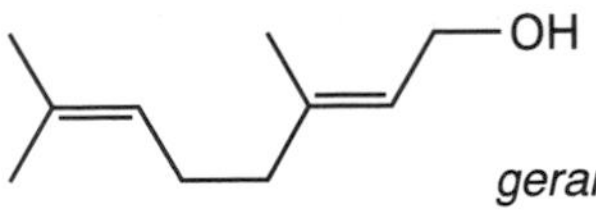

a How are the structures of nerol and geraniol related?
b How many moles of H_2 molecules would be required to saturate one mole of geraniol?
c How are nerol and geraniol related to citronellol which is shown in problem 4 in **Section 12.2**?
d How are the structures of nerol and geraniol related to linalool?

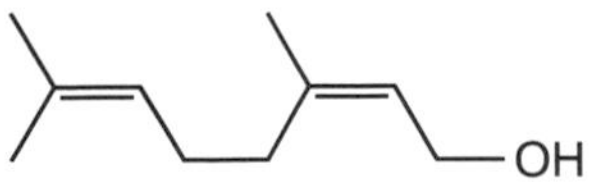

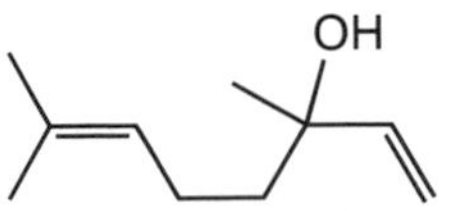

geraniol

nerol

linalool

3.6 *Optical isomerism*

Optical isomerism arises because of the different ways in which you can arrange four groups around a carbon atom. It is a type of stereoisomerism.

It's all done with mirrors

Four single bonds around a carbon atom are arranged tetrahedrally. When four different atoms or groups are attached to these four bonds, the molecule can exist in *two isomeric forms*. Figure 33 illustrates these two forms for the amino acid alanine (2-aminopropanoic acid).

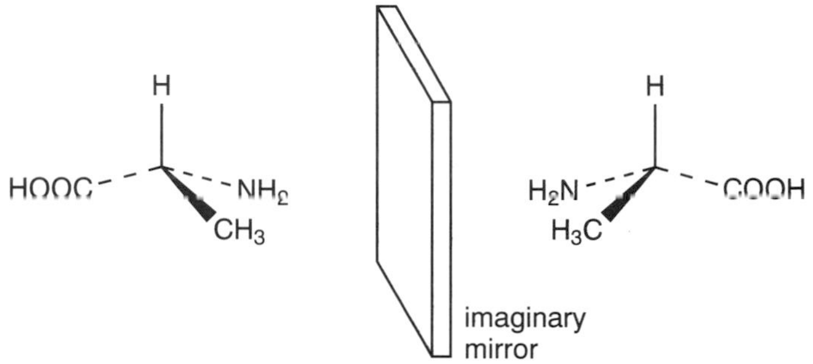

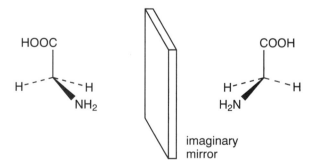

Figure 33 Two isomers of alanine

You may need to build models of these two structures to convince yourself they are different. The way they are different is that the right hand structure in Figure 32 is the **mirror image** of the left hand structure. If you have built models, you can use a mirror to prove this to yourself. Otherwise, imagine a mirror placed between the two forms of alanine: the NH_2 group is near to the mirror so it will be at the front of the reflection; the COOH group is furthest from the mirror so it will be at the back of the reflection, and so on.

All molecules have mirror images, but they don't all exist as two isomers. Glycine (aminoethanoic acid), for example, has only one form (Figure 34).

Figure 34 Glycine, drawn to show that the mirror image is identical with the original structure, if you turn it round

What makes alanine exist in two forms is that the mirror image and the original molecule are **non-superimposable**. If you move the mirror image across to the left, the H atom, C atom and CH_3 group will coincide, but NH_2 and COOH will be in the wrong places. No amount of twisting and turning will put things right! The only way you can make them superimpose is to break the $C—NH_2$ and C—COOH bonds and swap the groups round. Breaking and reforming bonds corresponds to a chemical reaction; and in a reaction, a compound is turned into a new compound. In this example, the new compound is a different isomer of the original one.

The pictures of glycine, on the other hand, will coincide if you move them together. They are superimposable, and there is only one form.

D/L-enantiomers

Molecules like the two forms of alanine are called **optical isomers** or **enantiomers**. You don't always have to build models to find them – whenever a molecule contains a carbon atom surrounded by four different atoms or groups there will be optical isomerism. Molecules that are not superimposable on their mirror images are called **chiral** molecules. A carbon atom surrounded by four different groups is called a **chiral centre**. There are lots of other chiral things in the world – most keys, your shoes, your hands, for example. In fact chirality comes from a Greek word meaning 'handedness'. If you look at the reflection of your *right* hand in a mirror, the mirror image is superimposable on your *left* hand, but not on your right hand.

The proteins in our bodies are built up from only one enantiomer of each amino acid. These are the **L-enantiomers**, and they have the same arrangement of the four groups around the central carbon atom. You can spot them by using the '*CORN rule*' (Figure 35). Stand your model or arrange your diagram with the H atom pointing upwards, and look down on the H atom towards the carbon. The optical isomer which has the sequence

C̲O̲OH, R̲, N̲H₂ (i.e. 'CORN')

in a *clockwise* direction is the **L**-amino acid. The other isomer is called the **D**-enantiomer. In Figure 33, the isomer of alanine on the right is the D-isomer; the other form on the left obeys the 'CORN rule' and is the L-isomer.

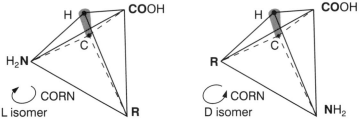

Figure 35 The CORN rule. Look down the H—C bond from hydrogen towards the central carbon atom

Enantiomers behave identically in ordinary test tube reactions. Most of their physical properties, such as melting point, density and solubility, are also the same. But enantiomers behave differently in the presence of other chiral molecules. Our proteins are built up from chiral amino acids, and so our enzymes and protein hormones must be chiral too. D- and L- enantiomers will therefore react differently with enzymes. It's like your left shoe being made for your left foot; your right foot doesn't really fit into it at all.

Here are some examples of how enantiomers differ:

- Enantiomers can interact differently with the chiral 'taste-buds' on your tongue. D-amino acids all taste sweet; L-amino acids are often tasteless or bitter.
- Enantiomers can smell different. For example, the different smells of oranges and lemons are caused by enantiomers (see problem 3 at the end of this section).
- Many beneficial medicines have enantiomers which have little or no pharmacological effect. In the case of one medicine, thalidomide, the apparently non-active enantiomer was found to damage unborn children when the drug was taken during pregnancy. Medicines have been more thoroughly tested since the thalidomide incident, but it is often expensive to separate enantiomers if no dangers are found, and about 80% of synthetic medicines are marketed as a mixture of isomers.

D-amino acids do exist in nature. For example, penicillin works by breaking peptide links which involve D-alanine. These occur in the cell walls of bacteria but not in humans. When its cell wall is broken, the bacterium is killed. So penicillin is very effective at killing bacteria but cannot have the same effect on us – because we don't use D-amino acids.

PROBLEMS FOR 3.6

1 For each of the following molecules state whether optical isomers (enantiomers) can exist or not. Where there are optical isomers, draw diagrams to illustrate their structures.
   a CH_2Cl_2 b CH_2ClBr c $CHClBrI$

2 But-1-ene undergoes an addition reaction with hydrogen bromide to produce a mixture of three isomeric bromobutanes. One is a structural isomer of the other two, which are themselves a pair of optical isomers. Draw structures to represent these three isomers.

3 *Limonene* and *carvone* are examples of compounds for which both enantiomers occur naturally. One enantiomer of limonene smells of oranges, the other of lemons; carvone can smell like caraway seeds or like spearmint.

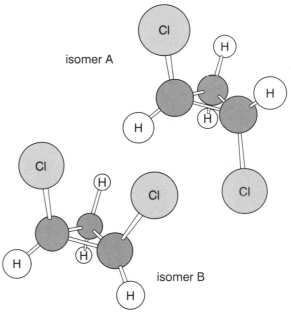

limonene   *carvone*

   a Draw out the full structural formulae of limonene and carvone (formulae which show all the bonds and atoms). Use an asterisk(*) to show the chiral carbon atom in each molecule. Remember, this means a carbon atom with four different groups attached to it.
   b Write down the structure of the compound which is formed when one mole of limonene undergoes addition with two moles of hydrogen molecules.
   c Does the product of the reaction in b exhibit optical isomerism? Explain your answer.

4 Draw diagrams to show the two optical isomers of the α-amino acid proline. Use the 'CORN-rule' to label your isomers D and L.

5 In a recent catalogue of chemicals, the price of 100 g of L-cysteine was quoted as £15. The same company sells 1 g of D-cysteine for £32. Suggest why there is such a price difference between the two enantiomers of cysteine.

6 The presence of *rings* of carbon atoms can lead to both geometric and optical isomerism.
   1,2-dichlorocyclopropane has the structure

$$H_2C \diagup \begin{matrix} CHCl \\ | \\ CHCl \end{matrix}$$

   a Figure 36 shows the geometric isomers of 1,2-dichlorocyclopropane. Which is the *cis* isomer and which is the *trans*?
   b One of the two geometric isomers is chiral, so it also shows optical isomerism. Which isomer is this? Draw structural formulae for the two optical isomers.

Figure 36 The geometric isomers of 1,2-dichlorocyclopropane. (These are 3-dimensional diagrams of molecular models)

4

ENERGY CHANGES AND CHEMICAL REACTIONS

4.1 *Energy out, energy in*

Energy changes are a characteristic feature of chemical reactions. Many chemical reactions give out energy and a few take energy in. A reaction that gives out energy and heats the surroundings is described as **exothermic**. A reaction that takes in energy and cools the surroundings is **endothermic**.

During an exothermic reaction the chemical reactants are losing energy. This energy is used to heat the surroundings – the air, the test tube, the laboratory, the car engine. The products end up with less energy than the reactants had – but the surroundings end up with more, and get hotter. We measure the energy transferred to and from the surroundings as **enthalpy change**. Enthalpy changes can be shown on an **enthalpy level diagram** also called an energy level diagram (Figure 1).

Enthalpy (*H*)

We cannot measure the enthalpy, *H*, of a substance. What we can do is measure the *change* in enthalpy when a reaction occurs

$\Delta H = H_{products} - H_{reactants}$

The enthalpy change in a chemical reaction gives the quantity of energy transferred to or from the surroundings, when the reaction is carried out in an open container.

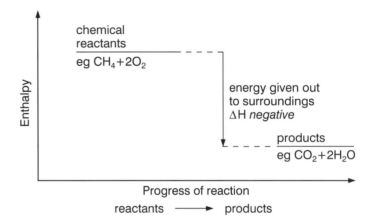

Figure 1 Energy level diagram for an exothermic reaction, eg burning methane, $CH_4 + 2O_2 \rightarrow CO_2 + 2H_2O$

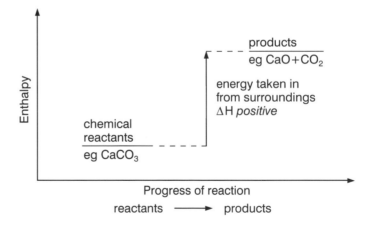

Figure 2 Energy level diagram for an endothermic reaction, eg decomposing calcium carbonate $CaCO_3 \rightarrow CaO + CO_2$

In an endothermic reaction, the reactants take in energy from the surroundings. This is shown in Figure 2.

How much?

As your study of chemistry becomes more advanced, you will find you need to put *numbers* to the features of chemical reactions – to make them quantitative. We measure the energy changes associated with chemical reactions by measuring enthalpy changes.

For an exothermic reaction ΔH is *negative*. This is because, from the point of view of the chemical reactants, energy has been *lost* to the surroundings. Conversely, for an endothermic reaction ΔH is *positive* – energy has been *gained* from the surroundings.

The units of ΔH

Enthalpy changes are measured in kilojoules per mole. For example, for the reaction of methane with oxygen, we write

$$CH_4(g) + 2O_2(g) \rightarrow CO_2(g) + 2H_2O(l); \quad \Delta H = -890\,\text{kJ mol}^{-1}$$

This means that for every mole of methane that reacts in this way, 890 kJ of energy are transferred to heat the surroundings. If we used 2 moles of methane, we should get $2 \times 890 = 1780$ kJ of energy transferred. This assumes that all the methane is converted into products, and that none is left unburned.

When calcium carbonate is heated, it decomposes. Energy is taken in – it is an endothermic reaction. We write

$$CaCO_3(s) \rightarrow CaO(s) + CO_2(g); \quad \Delta H = +572\,\text{kJ mol}^{-1}$$

For every mole of $CaCO_3$ that is decomposed, 572 kJ of energy is taken in. If we decomposed 0.1 mol of $CaCO_3$, 57.2 kJ would be taken in.

Standard conditions

Like most physical and chemical quantities, ΔH varies according to the conditions. In particular, ΔH is affected by temperature, pressure and concentration of solutions. So we choose certain *standard* conditions to refer to. We use
- a standard pressure of 1 atmosphere $(1.01 \times 10^5\,\text{N m}^{-2})$
- a standard concentration of $1\,\text{mol dm}^{-3}$.

The standard temperature is normally chosen as 298 K (25°C).

If ΔH refers to these standard conditions, it is written as $\Delta H^{\ominus}_{298}$, pronounced 'delta H standard, 298', or just 'delta H standard'. The values of ΔH given in the examples above are all standard values. You can use the Data Sheets to look up ΔH values.

Measuring enthalpy changes

Many enthalpy changes can be measured quite simply in the laboratory. We usually do it by arranging for the energy involved in the reaction to be exchanged with water. If it is an exothermic reaction, the water gets hotter; if it is endothermic, the water gets cooler. If we measure the temperature change of the water, and if we know its mass and specific heating capacity, we can work out how much energy was transferred to or from the water during the chemical reaction.

This method is explained further in **Activities F1.2 and F1.3**. Figure 3 shows apparatus that can be used to get a more accurate value for ΔH than the crude method used in these activities.

In an **exothermic** reaction, the enthalpy of the reacting system decreases.

ΔH is negative

In an **endothermic** reaction, the enthalpy of the reacting system increases.

ΔH is positive

System or surroundings?

When chemists talk about enthalpy changes they often refer to the **system**. This means the reactants and the products of the reaction they are interested in.

The system may lose or gain enthalpy as a result of the reaction. The **surroundings** means the rest of the world: the test tube, the air … and so on.

A word about temperature

There are two scales of temperature used in science and you should be familiar with both.

The kelvin (K) is the unit of **absolute temperature** and 0 K is called absolute zero. The kelvin is the SI unit of temperature and should always be used in *calculations* involving temperature.

However, you will usually *measure* temperature using a thermometer marked in degrees Celsius. You can convert temperatures from the Celsius scale to the absolute scale by adding 273. Thus, the boiling point of water = 100 + 273 = 373 K. Note that the unit does not have a degree symbol, °.

Note, too, that a *change* in temperature has the same numerical value on both scales.

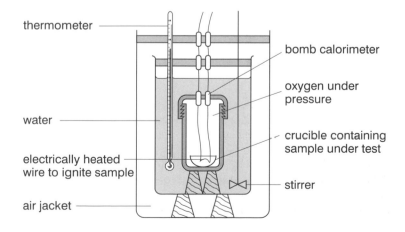

labels: thermometer, bomb calorimeter, oxygen under pressure, crucible containing sample under test, water, electrically heated wire to ignite sample, stirrer, air jacket

Figure 3 A bomb calorimeter for making accurate measurements of energy changes. The fuel is ignited electrically and burns in the oxygen inside the pressurised vessel. Energy is transferred to the surrounding water, whose temperature rise is measured. Note that the experiment is done at constant volume *in a* closed *container. Enthalpy changes are for reactions carried out at constant pressure, so the result needs to be modified accordingly*

For some reactions, measuring ΔH is very straightforward. Take the burning of methane, for example. You could make a rough measurement of ΔH for this reaction using ordinary kitchen equipment and a gas cooker (problem 3, **Section 4.1**).

For other reactions it is less simple. For example, decomposing $CaCO_3$ needs a temperature of over 800 °C, which makes it difficult to use water to measure the energy transferred. In cases like this, enthalpy changes are measured *indirectly*, using enthalpy cycles (see page 49).

Different kinds of enthalpy change

We can define the **standard enthalpy change for a reaction,** $\Delta H^{\ominus}_{r, 298}$, as the enthalpy change when molar quantities of reactants *as stated in the equation* react together under standard conditions. This means at 1 atmosphere pressure and 298 K, with all the substances in their standard states.

You must always give an equation when quoting an enthalpy change of reaction – otherwise things can get very muddled.

For example,

$$H_2(g) + \tfrac{1}{2}O_2(g) \rightarrow H_2O(l); \quad \Delta H^{\ominus}_{r, 298} = -286\,\text{kJ}\,\text{mol}^{-1}$$

but $2H_2(g) + O_2(g) \rightarrow 2H_2O(l); \quad \Delta H^{\ominus}_{r, 298} = -572\,\text{kJ}\,\text{mol}^{-1}$

The following kinds of enthalpy change are particularly important and are given special names:

Standard enthalpy change of combustion, $\Delta H^{\ominus}_{c, 298}$ is the enthalpy change that occurs when *1 mole* of a fuel is burned completely. In theory, the fuel needs to be burned under standard conditions – 1 atmosphere pressure and 298 K. In practice, that is impossible, so we burn the fuel in the normal way and then make adjustments to allow for non-standard conditions.

For example, the enthalpy change of combustion of octane, one of the alkanes found in petrol, is $-5470\,\text{kJ}\,\text{mol}^{-1}$. This is much bigger than the $\Delta H^{\ominus}_c$ for methane, ($-890\,\text{kJ}\,\text{mol}^{-1}$) because burning octane involves breaking and making more bonds than burning methane. (Note that if no temperature is given with $\Delta H^{\ominus}_c$, we assume the value refers to 298 K.)

Standard enthalpy change of formation $\Delta H^{\ominus}_{f, 298}$ is the enthalpy change when *1 mole* of a compound is formed from its elements – again with both the compound and its elements being in their standard states.

For example, the enthalpy change of formation of water, $H_2O(l)$, is $-286\,\text{kJ}\,\text{mol}^{-1}$. When you make a mole of water from hydrogen and oxygen, 286 kJ are transferred to the surroundings. This is summed up as

$$H_2(g) + \tfrac{1}{2}O_2(g) \rightarrow H_2O(l); \quad \Delta H^{\ominus}_{f, 298} = -286\,\text{kJ}\,\text{mol}^{-1}$$

Notice that the equation refers to 1 mole of H_2O, so only $\tfrac{1}{2}$ mole of oxygen is needed in the equation.

Remember too, that, by definition, the enthalpy change of formation of a pure element in its standard state is zero.

It is often impossible to measure enthalpy changes of formation directly. For example, the enthalpy change of formation of methane is $-75\,\text{kJ}\,\text{mol}^{-1}$. This refers to the reaction

$$C(s) + 2H_2(g) \rightarrow CH_4(g); \quad \Delta H^{\ominus}_{f, 298} = -75\,\text{kJ}\,\text{mol}^{-1}$$

but there is a problem. This reaction doesn't actually occur under normal conditions. So how did anyone manage to measure the value of $\Delta H^{\ominus}_{f, 298}$? It has to be done *indirectly*, using enthalpy cycles.

All combustion reactions are exothermic

$\therefore \Delta H_c$ is always negative.

Standard state

The standard state of a substance is its most stable state under standard conditions:

- a pressure of 1 atmosphere
- a stated temperature, usually 298 K

This may be a pure solid, liquid or gas.

ΔH_f may be positive or negative.

Enthalpy cycles

Figure 4 shows an enthalpy cycle, also known as an energy cycle. There is both a direct and an indirect way to turn graphite (C) and hydrogen (H_2) to methane (CH_4). We can't measure the enthalpy change for the *direct* route. The *indirect* route goes via CO_2 and H_2O, and involves two enthalpy changes which we *can* measure.

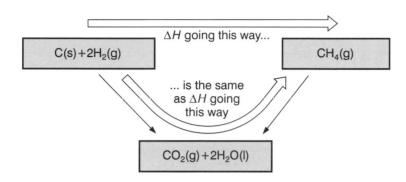

Figure 4 *An enthalpy cycle for finding the enthalpy change of formation of methane, CH_4*

The key idea is that the total enthalpy change for the indirect route is the same as the enthalpy change via the direct route. This makes sense if you think about it; energy cannot be created or destroyed – this is the law of Conservation of Energy. So *as long as your starting and finishing points are the same, the enthalpy change will always be the same no matter how you get from start to finish.* This is one way of stating **Hess's law**, and an enthalpy cycle like the one in Figure 4 is called a **Hess cycle** or a **thermochemical cycle**.

If you know the enthalpy changes involved in two parts of the cycle, you can work out the enthalpy change in the third (Figure 5).

> ### Hess's law
> The enthalpy change for any chemical reaction is independent of the intermediate stages, provided the initial and final conditions are the same for each route.

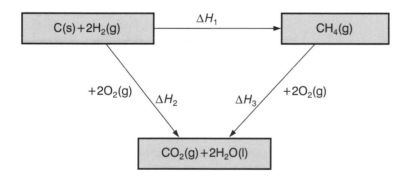

Figure 5 *Using an enthalpy cycle to find ΔH_1. $\Delta H_1 = \Delta H_2 - \Delta H_3$*

So if we can measure ΔH_2 and ΔH_3, we can find ΔH_1 – which is the enthalpy change that cannot be measured directly. Quite simply,

$$\Delta H_1 = \Delta H_2 - \Delta H_3$$

(It has to be minus ΔH_3 because the reaction to which ΔH_3 applies actually goes in the opposite direction to the way we want it to go in order to produce CH_4. To find the enthalpy change for the reverse reaction, we must reverse the sign.)

Now ΔH_2 is the sum of the enthalpy changes of combustion of 1 mole of carbon and 2 moles of hydrogen, and ΔH_3 is the enthalpy change of combustion of methane. ΔH_1 is the enthalpy change of formation of methane, $\Delta H_f^\ominus(CH_4)$, which is the quantity we are trying to find.

$$\Delta H_1 = \Delta H_f^\ominus(CH_4)$$
$$\Delta H_2 = \Delta H_c^\ominus(C) + 2\Delta H_c^\ominus(H_2)$$
$$\Delta H_3 = \Delta H_c^\ominus(CH_4)$$

ΔH_c° (C) $= -393\,kJ\,mol^{-1}$

ΔH_c° (H$_2$) $= -286\,kJ\,mol^{-1}$

ΔH_c° (CH$_4$) $= -890\,kJ\,mol^{-1}$

Calculations look cluttered if $\Delta H_{c,\,298}^\circ$ is written in full each time it occurs. So, we often use just ΔH_c° for the standard enthalpy change at 298 K.

Putting in the values of the enthalpies of combustion, which we can measure, we get

$$\Delta H_f^\circ (CH_4) = \Delta H_c^\circ (C) + 2\Delta H_c^\circ (H_2) - \Delta H_c^\circ (CH_4)$$
$$= -393\,kJ\,mol^{-1} + 2(-286)\,kJ\,mol^{-1} - (-890)\,kJ\,mol^{-1}$$
$$= -75\,kJ\,mol^{-1}$$

Using the enthalpy cycle has enabled us to find a value for an enthalpy change which we could not find directly. You will find these cycles very useful in other parts of your chemistry course.

PROBLEMS FOR 4.1

You will need to consult the Data Sheets to find values for standard enthalpy changes when doing these problems.

1 Explain what is meant by *standard enthalpy change of formation*.

2 Explain why standard enthalpy changes of formation may have a negative or a positive sign, but standard enthalpy changes of combustion are always negative.

3 Suppose you were asked to measure the enthalpy change of combination of methane, using a gas cooker and a saucepan.
  a What other equipment would you need?
  b What measurements would you make?
  c What would be the main sources of error?

4 The standard enthalpy change of combustion of heptane, C_7H_{16} is $-4817\,kJ\,mol^{-1}$.
  a What is the relative molecular mass of heptane?
  b Calculate the energy transferred when the following quantities of heptane are burned:
   i 10 g
   ii 10 kg.
   What assumptions have you made in this calculation?
  c What further information would you need in order to calculate the energy transferred when 1 *litre* of heptane is burned?

5 The standard enthalpy change of combustion of carbon is equal to the standard enthalpy change of formation of carbon dioxide. Explain why.

6 a Write an equation to represent the formation of 1 mole of water from its elements in their standard states.
  b Look up the standard enthalpy change of formation of water.
  c Calculate the enthalpy change when 1 g of hydrogen burns in oxygen. What assumption have you made?
  d What is the standard enthalpy change for the reaction
   $H_2O(l) \rightarrow H_2(g) + \tfrac{1}{2}O_2(g)$

7 a Look up the values of the standard enthalpy change of formation of each of the following compounds.
   i hydrogen chloride
   ii hydrogen iodide
  b Draw enthalpy level diagrams, similar to those in Figures 1 and 2, to represent the reactions that occur when each of these compounds is formed from its elements.

8 a Write an equation to represent the formation of propane, C_3H_8, from its elements in their standard states.
  b Draw an enthalpy cycle, similar to the one in Figure 5, to show the relationship between the formation of propane from carbon and hydrogen, and the combustion of these substances to give carbon dioxide and water.
  c Use your enthalpy cycle to calculate a value for the enthalpy change of formation of propane. You may use *only* the following data on enthalpy changes of combustion.
   ΔH_c° (C) $= -393\,kJ\,mol^{-1}$
   ΔH_c° (H$_2$) $= -286\,kJ\,mol^{-1}$
   ΔH_c° (C$_3$H$_8$) $= -2219\,kJ\,mol^{-1}$

9 Suppose you need to find the standard enthalpy change of formation of carbon monoxide. You cannot measure this directly, because you cannot normally convert carbon to carbon monoxide. However, you *can* convert both C and CO to CO_2, and you can use the enthalpy changes for these reactions to find the standard enthalpy change of formation of CO.
   $C(s) + O_2(g) \rightarrow CO_2(g)$;
    ΔH_c° (C) $= -393\,kJ\,mol^{-1}$
   $CO(g) + \tfrac{1}{2}O_2(g) \rightarrow CO_2(g)$;
    ΔH_c° (CO) $= -283\,kJ\,mol^{-1}$
   Draw an appropriate enthalpy cycle and use it to calculate the standard enthalpy change of formation of CO(g).

4.2 *Where does the energy come from?*

All chemical reactions involve breaking and making chemical bonds. Bonds break in the reactants and new bonds form in the products. The energy changes in chemical reactions come from the energy changes when bonds are broken and made.

> You can remind yourself of the basic ideas of chemical bonding by reading **Section 3.1**

Bond enthalpies

A chemical bond is basically an electrical attraction between atoms or ions. When you break a bond, you have to do work in order to overcome these attractive forces. To break the bond completely, you (theoretically) need to separate the atoms or ions so they are an infinite distance apart. Figure 6 illustrates this for the H—H bond in a molecule of hydrogen, H_2.

> Bond breaking is an *endothermic* process, so bond enthalpies are always *positive*.

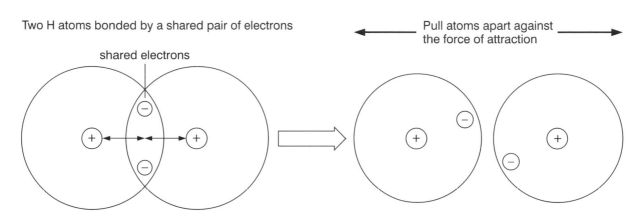

Two H atoms bonded by a shared pair of electrons

shared electrons

Both nuclei are attracted to the same shared pair of electrons. This holds the nuclei together.

Pull atoms apart against the force of attraction

Figure 6 Breaking a bond involves using energy to overcome the force of attraction

The quantity of energy needed to break a particular bond in a molecule is called the **bond dissociation enthalpy**, or **bond enthalpy** for short. For the bond shown in Figure 7, the process involved is

$$H_2(g) \rightarrow 2H(g); \quad \Delta H = +436\,\text{kJ mol}^{-1}$$

So the bond enthalpy of the H—H bond is $+436\,\text{kJ mol}^{-1}$. Notice that this is a *positive* ΔH value, because *breaking* a bond is an endothermic process – it needs energy. When you *make* a new bond, you get energy given out.

Bond enthalpies are very useful, because they tell you how strong bonds are. The stronger a bond, the more difficult it is to break, and the higher its bond enthalpy.

Bond enthalpy and bond length

When a bond like the one in Figure 6 forms, the atoms move together because of the attractive forces between nuclei and electrons. But there are also **repulsive** forces, between the nuclei of the two atoms, and these get bigger as the atoms stop moving together. The distance between them is now the equilibrium bond length (Figure 7). The shorter the bond length, the stronger the attraction between the atoms.

Table 1 on the next page gives some values for bond enthalpies and bond lengths. These are all *average* bond enthalpies, because the exact value of a bond enthalpy actually depends on the particular compound in which the bond is found. Notice these points:

- Double bonds have much higher bond enthalpies than single bonds. Triple bond enthalpies are even higher.

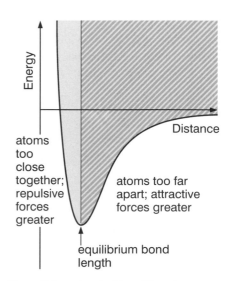

atoms too close together; repulsive forces greater

atoms too far apart; attractive forces greater

equilibrium bond length

Figure 7 In a chemical bond there is a balance between attractive and repulsive forces

Bond	Average bond enthalpy/kJ mol⁻¹	Bond length/nm
C—C	+347	0.154
C=C	+612	0.134
C≡C	+838	0.120
C—H	+413	0.108
O—H	+464	0.096
C—O	+336	0.143
C=O	+805	0.116
O=O	+498	0.121
N≡N	+945	0.110

Table 1 Average bond enthalpies and bond lengths

- In general, the higher the bond enthalpy, the shorter the bond. You can see this if you compare the lengths of the single, double and triple bonds between carbon atoms.

Bond enthalpies always refer to breaking a bond in the *gaseous* compound. This means we can make fair comparisons between different bonds.

Measuring bond enthalpies

It isn't easy to measure bond enthalpies, because usually there is more than one bond in a compound. Also, it is very difficult to make measurements when everything is in the gaseous state. For this reason, bond enthalpies are measured *indirectly*, using enthalpy cycles.

Breaking and making bonds in a chemical reaction

Let's look again at the reaction that occurs when methane burns:

$$CH_4(g) + 2O_2(g) \rightarrow CO_2(g) + 2H_2O(l); \quad \Delta H = -890\,kJ\,mol^{-1}$$

The reaction involves both breaking bonds and making new bonds. First you have to break four bonds between C and H in a methane molecule, and bonds between O and O in two O_2 molecules. This bond breaking requires energy. But once the bonds have been broken, the atoms can join together again to form new bonds: two C=O bonds in a CO_2 molecule and four O—H bonds in two H_2O molecules.

Figure 8 illustrates this – but note that you do not have to break *all* the old bonds before you can make new ones. New bonds start forming as soon as the first of the old bonds have broken.

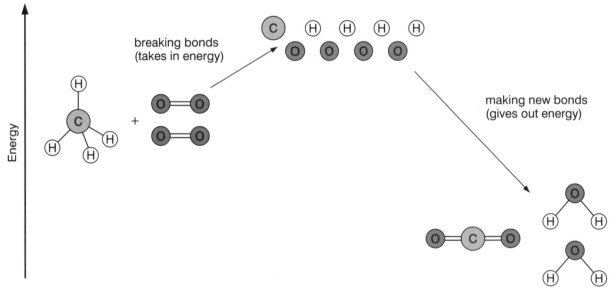

Figure 8 Breaking and making bonds in the reaction between methane and oxygen

Because you need to break bonds before product molecules can begin to form, many reactions need heating to get them started. All reactions need energy to stretch and break bonds and start them off: this is called the **activation enthalpy**. Some reactions need only a little energy, and there is enough energy available in the surroundings at room temperature to get them started. The reaction of acids with alkalis is one example. Other reactions have a higher activation enthalpy, and they need heating to get them started. The burning of fuels is an example. When you use a match to ignite a fuel, you are supplying the activation enthalpy that is needed to break bonds so new bonds can begin to form.

It isn't necessary for *all* the bonds to break before the reaction gets going. If it was, you would have to heat things to very high temperatures to make them react. Once one or two bonds have broken, new bonds can start to form and this usually gives out enough energy to keep the reaction going.

Chemists often represent the energy changes which occur during a reaction by drawing an **enthalpy** (or **energy**) **profile**. This is shown for the combustion of methane in Figure 9.

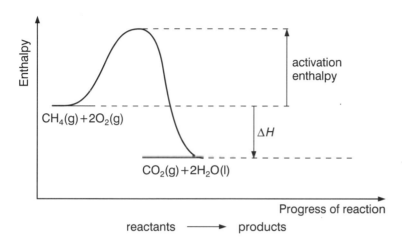

Figure 9 *Enthalpy profile for the combustion of methane*

Bonds and enthalpy cycles

We can represent bond-breaking and bond-making in an enthalpy cycle, like the one given in Figure 10.

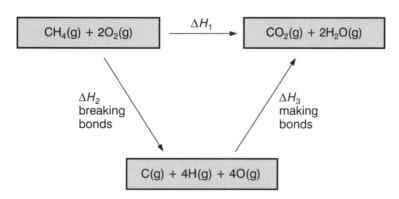

Figure 10 *An enthalpy cycle to show bond-breaking and bond-making in the combustion of methane*

We can use the enthalpy cycle in Figure 10 to work out a value for the enthalpy change of combustion of methane, represented by ΔH_1. The calculation looks like this:

ΔH_2 = enthalpy change when bonds are broken
= $4 \times E(C—H) + 2 \times E(O{=}O)$
= $+2648 \, \text{kJ mol}^{-1}$

ΔH_3 = enthalpy change when bonds are made
= $-[2 \times E(C{=}O) + 4 \times E(O—H)]$ (The minus sign occurs because energy is *released*, when the bonds are made.)
= $-3466 \, \text{kJ mol}^{-1}$

So the enthalpy change of combustion, ΔH_1, is given by:
$\Delta H_1 = \Delta H_2 + \Delta H_3 = +2648 \, \text{kJ mol}^{-1} - 3466 \, \text{kJ mol}^{-1}$
= $-818 \, \text{kJ mol}^{-1}$

Notice that this value is a little different from the $-890\,kJ\,mol^{-1}$ given for the *standard* enthalpy change of combustion of methane. The reason is that the value of ΔH_1 that we calculated here is not actually the standard value. In the equation we have been using, the water in the products is $H_2O(g)$, not $H_2O(l)$ as it would be under standard conditions. We used $H_2O(g)$ because when using bond enthalpies we have to work in the gaseous state.

PROBLEMS FOR 4.2

You will need to use the Data Sheets to look up values of bond enthalpies for some of these problems.

1 a Look up the bond enthalpies of the following bonds: C—F, C—Cl, C—Br, C—I. Comment on any trend you notice in terms of the positions of F, Cl, Br and I in the Periodic Table.

   b Chlorofluorocarbons (CFCs) are compounds containing carbon, chlorine and fluorine. An example is CCl_2F_2. Which of the bonds in this compound are likely to break most readily? What is the environmental significance of this?

2 a Draw a dot-cross diagram for a molecule of N_2.

   b Suggest an explanation for the unreactive nature of nitrogen gas in terms of bond enthalpy. (Look up any information you need).

3 Use bond enthalpies to calculate the energy needed to atomise 1 mole of methane, i.e. to turn 1 mole of methane to gaseous atoms:
$$CH_4(g) \rightarrow C(g) + 4H(g)$$

4 Use bond enthalpies to calculate the energy needed to atomise 1 mole of propane, C_3H_8.

5 a Would you expect a C=C bond to be twice as strong as a C—C bond?

   b Compare the bond enthalpy of C—C with that of C=C. Comment on the comparison in view of your answer to **a**.

6 a Construct an enthalpy cycle, similar to the one in Figure 10, to show bond-breaking and bond-making in the combustion of propane, C_3H_8.

   b Use your cycle, and the necessary bond enthalpies, to calculate a value for the enthalpy change of combustion of propane.

   c Look up the enthalpy change of combustion of propane in the Data Sheets. Account for any differences between this value and the one you have calculated.

7 Repeat question 6, using methanol, CH_3OH, instead of propane.

4.3 Entropy and the direction of change

Things happen by chance

If you spill some petrol in an enclosed space, such as a garage, you can soon smell it all over the place. The petrol vaporises, and the vapour **diffuses** (spreads out) to occupy all the available space. This is why petrol is such a serious fire risk; as the vapour spreads out it mixes with the air to make a highly flammable mixture.

But *why* does the vapour diffuse? Why doesn't it all stay in one part of the room? It is the laws of chance and probability that say it must diffuse.

Look at Figure 11. We have simplified the situation so that the petrol vapour starts off in one container and can diffuse into the other when the partition is removed. We have ignored the presence of air molecules and have shown just five 'petrol' molecules. (Of course, petrol is really a mixture of different hydrocarbons with many different kinds of molecules.)

Each molecule moves in a straight line, until it collides with another gas molecule or the wall of the container, when it changes direction.

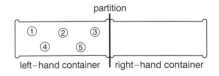

Figure 11 A simplified situation. 5 molecules in a container. What will happen when the partition is removed?

Figure 12 shows some of the things that can happen once the partition is removed. The molecules move around at random, and it is pure chance which container they end up in after a given length of time. Each of the molecules could end up in one of two places: the left-hand or the right-hand container. There are five molecules altogether, each with two places to be, so the total number of ways the molecules could arrange themselves once the partition is removed is $2 \times 2 \times 2 \times 2 \times 2 = 2^5 = 32$. Each of these ways is equally likely.

Only one of these 32 arrangements has all the molecules where they started, in the left-hand container. So the chance that they will all stay all in one container, instead of spreading out between the two, is 1 in 32. The molecules diffuse because there are more ways of being spread out than being all in one place.

Now consider the real-life situation, when there are billions of billions of molecules instead of just five. The number of ways all these molecules can spread out to fill the two containers is unimaginably large, so the chance that they will all remain in one is unimaginably tiny.

The idea of 'number of ways' is very important in chemistry (and in physics and biology for that matter) because it decides whether changes are likely to take place. The basic idea is that *the events that happen are the ones that are most likely to*. The more ways an event can occur, the more likely it is to happen.

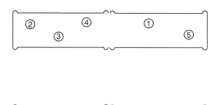

Figure 12 Possible arrangements of molecules after the partition has been removed

Why do liquids mix?

The mixing of liquids is another example of this rule. Figure 13 illustrates the point. If you have a jar half-full of roasted peanuts, and you carefully pour on to them half a jar of roasted cashew nuts, you get two layers of different nuts. If you shake the jar, they will mix, and you will have a jar of mixed nuts.

However much you shake them, the nuts will never unmix and give you two layers again. There are far more ways that the nuts can be mixed than unmixed. Each time you shake the jar, you produce another way of mixing them.

Unmixing could in theory happen – the different nuts could by chance get shaken to the top and bottom. But this is so unlikely that in practice it never happens. The nuts stay mixed.

If the different nuts represent molecules of different liquids, you have the situation when two liquids are mixed – for example when two different hydrocarbons are blended in petrol. The liquids mix because there are more ways of being mixed than unmixed.

It doesn't always happen like this. Some pairs of liquids don't mix; for example, petrol and water. This is because there is something to prevent the natural mixing process from happening. There are weak attractive forces between all molecules, but if the attractive forces between molecules of one liquid are stronger than those between the molecules in the other liquid (and stronger than the forces between the two different types of molecules) then mixing is unlikely. It is as if the peanuts in Figure 13 had a sticky coating that makes them stay attracting one another rather than getting mixed up with the cashew nuts.

Figure 13 Peanuts at the bottom, cashews on the top

The general rule about mixing is that *substances always tend to mix unless there is something stopping them*, such as strong attractive forces holding one set of molecules together so they cannot easily break away from each other and mix.

Entropy: measuring the number of ways

Clearly the 'number of ways' is very important in deciding how things change. It is useful to have a measure of the 'number of ways' a chemical

Substance	Standard molar entropy $S°/J\,K^{-1}\,mol^{-1}$
diamond, C(s)	2.4
hydrogen H_2(g)	130.6
iron, Fe(s)	27.2
sodium chloride, NaCl(s)	72.4
water (solid), H_2O(s)	48.0
water (liquid), H_2O(l)	70.0
water (gas), H_2O(g)	188.7
methane, CH_4(g)	186.2
ethane, C_2H_6(g)	229.5
propane, C_3H_8(g)	269.9

Table 2 Some values of standard molar entropy

system can be arranged, then we can tell how the system is likely to change. We use a quantity called **entropy** to measure 'number of ways'. The higher the number of ways, the higher the entropy. In general, the more spread-out, mixed-up or disordered a system, the higher its entropy will be.

We will see later that entropy is a precise physical quantity that can be measured and tabulated. The symbol for entropy is S. Table 2 gives some standard molar entropies for different substances. For the time being, you need not worry about how these numbers were obtained, but it is interesting to compare them with one another.

In general, gases have higher entropies than liquids, which have higher entropies than solids. You can see this by looking at the standard molar entropies of the three states of water. In gases, the molecules are arranged completely at random, so there are more ways they can be arranged than in liquids or solids. Liquids in turn have a more random arrangement than solids, and so more ways of arranging the molecules. The more regular and crystalline a solid is, the lower its entropy. Notice the very low entropy of diamond, which has one of the most perfectly regular structures of all.

Notice too that substances with more complex molecules, such as ethane, have higher entropies than substances with simpler molecules, such as methane.

It's not as simple as that …

In this treatment of entropy we have talked about the number of ways of arranging molecules of substances. But entropy does not simply measure arrangements of molecules: it also measures the number of ways that *energy* can be distributed among these molecules. This is a very important contribution to the entropy of a substance, and you will meet it later, in **Section 4.4**.

PROBLEMS FOR 4.3

1 For each of the following changes say whether the entropy of the system described would **A** increase **B** decrease **C** stay the same.
   a petrol vaporises
   b petrol vapour condenses
   c sugar dissolves in water
   d oil mixes with petrol
   e a suspension of oil in water separates into two layers.

2 Which of the following substances do you believe would have the higher standard molar entropy? Give your reasons.
   a solid wax or molten wax
   b Br_2(l) or Br_2(g)
   c separate samples of copper and zinc, or brass
   d pentane C_5H_{12}(l) or octane C_8H_{18}(l)

3 Look at Figure 11. Suppose there were eight molecules in the left-hand container instead of five. What would be the chance that they would all end up in the left-hand container after the partition had been removed?

4.4 *Energy, entropy and equilibrium*

Inside solids, liquids and gases

You should be familiar with the simple molecular-kinetic model for the structures of solids, liquids and gases. This is summarised in Figure 14.

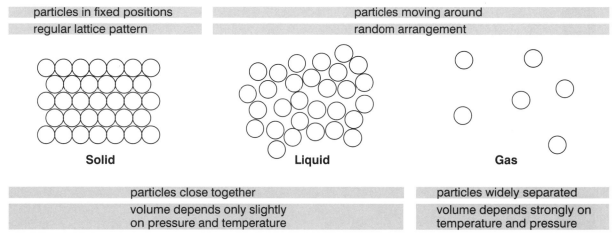

| particles in fixed positions | particles moving around |
| regular lattice pattern | random arrangement |

Solid   **Liquid**   **Gas**

| particles close together | particles widely separated |
| volume depends only slightly on pressure and temperature | volume depends strongly on temperature and pressure |

Figure 14 A comparison of solids, liquids and gases

Solids are *rigid* because the molecules are held together in a *lattice*. (We will use the word 'molecule' for the particles involved, though they could also be atoms or ions.) Because the molecules in liquids and gases are moving around, we say these substances are *fluids*. They can take up any shape; usually it is the shape of the container they are in.

Solids and liquids do not expand much when you heat them but, at a given pressure, gases expand a lot on heating. What is more, gases can be easily compressed by increasing the pressure. The **Ideal Gas law**

$$pV = nRT$$

tells us how p, V and T arc related for a fixed amount in moles, n, of gas. Solids and liquids, in contrast, have low *compressibilities*. **Section 1.5** tells you more about the Ideal Gas law.

If we heat a solid, liquid or gas, its temperature increases. There are only *two* exceptions to this rule. If the solid happens to be at its melting point, or the liquid at its boiling point, the temperature does not increase – it stays constant while the solid melts or the liquid boils.

Stacks of energy

We use **specific heating capacity** (symbol, c_p) as a measure of how much energy is required to warm something up. It tells us how many joules are needed to raise the temperature of 1 g of substance by 1 K. So c_p has units of $J\,g^{-1}K^{-1}$.

The temperature of a substance is related to the *kinetic energy* of the molecules. A substance feels hotter when its molecules are moving more energetically. That is because movement energy can easily be passed on to the molecules in you, or the thermometer, or whatever.

There are three forms of kinetic energy which molecules can possess:

- **translation**   movement of the whole molecule from one place to another
- **rotation**   spinning around
- **vibration**   stretching and compressing bonds.

You can read more about these forms of energy in **Section 6.2**.

If we continue putting energy into a substance we may increase the **electronic energy** of the molecules, or break bonds between them or even within them. But none of these processes alter the temperature because changes in electronic energy and **bonding energy** do not affect the *motion* of the molecules.

Energy is quantised, and molecules are restricted to particular levels of electronic, vibrational, rotational and translational energy. The size of the quanta (i.e. the gaps between the energy levels) increase in the order

translation < rotation < vibration < electronic

If we think of just one electronic level of a simple molecule such as H_2, the energy levels can be illustrated as shown in Figure 15.

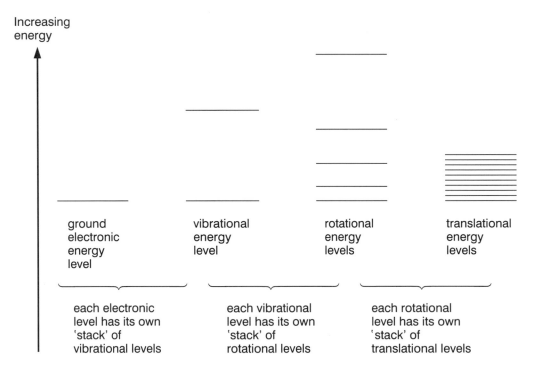

Figure 15 Each electronic energy level has within it several vibrational, rotational and translational energy levels. Note that the levels are not *to scale*

Overall, there is a large number of closely spaced energy levels for the molecules to be in. There are even more for the more complicated molecules.

Entropy matters

Molecules don't all have the same energy: they are spread out among the energy levels. As molecules collide and exchange energy with each other, their energies change and they move up and down the energy level stack. The molecules are distributed among the energy levels in the way that gives the greatest *entropy*.

Entropy (symbol S) was introduced in **Section 4.3**. We can think of entropy as measuring the number of ways in which something can be arranged. In **Section 4.3** we discussed entropy in terms of the number of ways that *molecules* can be arranged. Entropy is also a measure of the number of ways that *quanta of energy* can be arranged.

We can explain diffusion and mixing by saying there are more ways of arranging the positions of the molecules if they are spread out in space; the entropy is higher when the molecules are spread out. Similarly, there are more ways of arranging the quanta of energy in molecules if they are spread out in an energy stack.

Imagine 2 molecules with 4 quanta of energy between them. They could each have 2 quanta, but there are several other possibilities too. In fact, there are five ways in which the energy can be divided between the molecules:

Notice that, although the energies are spread out and the molecules can change from one level to another when they collide, the *total energy* must always equal 4 quanta. Energy cannot be created or destroyed. That's one way of expressing the **First law of Thermodynamics** – the law you make use of whenever you do Hess's law calculations.

What happens if we give the molecules 6 quanta instead of 4? There are now *seven* ways of distributing the energy:

So when we heat something, we increase the total number of quanta, and we increase the entropy. This idea will be very useful – we shall come back to it later.

The entropy also increases if we increase the number of molecules sharing the energy. For example, there are 15 ways in which 3 particles can share 4 quanta. See if you can work out why.

If the energy levels are closely spaced, the quanta are smaller, but there are more of them at a given temperature. The more closely spaced the energy levels are, the greater the number of quanta, and the higher the entropy.

Generally, solids have lower entropies than liquids and gases. This is because solids have a regular lattice structure, whereas the molecules in liquids and gases are arranged randomly. Gases tend to have greater entropies than liquids because gas molecules occupy a greater volume and so there is more randomness in their distribution.

Energy levels are more closely spaced for molecules which contain heavier atoms. The number of energy levels tends to increase with the number of atoms in a molecule, thus making adjacent levels closer together in larger molecules. So molecules with heavier atoms, and molecules with larger numbers of atoms, lead to higher entropies.

Quanta possessed by molecule 1	Quanta possessed by molecule 2
0	4
1	3
2	2
3	1
4	0

Quanta possessed by molecule 1	Quanta possessed by molecule 2
0	6
1	5
2	4
3	3
4	2
5	1
6	0

Here is a summary of the main points about entropy which have just been covered.

- Entropy is a measure of the number of ways of arranging molecules and distributing their quanta of energy.

- A collection of molecules has greater entropy if the molecules are spread out as much as possible.

- There are more ways of arranging the energy of a collection of molecules if they spread out among the energy levels available to them.

- The entropy is increased if the energy is shared among more molecules

- The entropy depends on the number of quanta of energy available. This is turn depends on
    the temperature
    the spacing of the energy levels

- Substances have higher entropies if their molecules contain
    heavier atoms
    larger numbers of atoms .

- In general, gases have higher entropies than liquids, and liquids have higher entropies than solids.

... *the whole universe if necessary*

Let's look at melting and freezing in more detail. Why should 0 °C mark the change from water to ice?

$$H_2O(l) \rightarrow H_2O(s); \quad \Delta S = -22.0 \, J \, K^{-1} mol^{-1}$$
$$\Delta H = -6.01 \, kJ \, mol^{-1}$$

ΔS is the *entropy change* for this process. It is negative, because the entropy *decreases* when liquid water becomes solid. ΔH is also negative; the change is exothermic, and energy is transferred from the water to the surroundings by heating.

So far we have only considered the molecules in the *system* we are studying – the water and the ice, but changes to the system also affect the *surroundings*. These changes in the surroundings are very important in determining what happens.

Freezing is an exothermic process. Energy is transferred from the chemical system to the surroundings by heating. When ice freezes on a window pane, energy is transferred to the glass, but it soon spreads out – by conduction, convection and radiation – into things around the glass. Lots of things, such as the frame, the wall, the air and the plant beside the window, get hotter, and therefore increase in *entropy*.

We can't work out how much each individual substance in the surroundings has increased in entropy because it is impossible to say exactly how the energy has been shared out. Fortunately, we don't need to know this; there is a very simple relationship which allows us to think just in terms of *surroundings* – whatever they are.

The relationship tells us that 'the entropy of the surroundings increases by a quantity equal to the energy they gain divided by the temperature'. Since the energy *gained* by the *surroundings* is the same as the energy *lost* by the *chemical system*, it is equal to $-\Delta H$. So the **entropy change in the surroundings, ΔS_{surr},** is given by

$$\Delta S_{surr} = \frac{-\Delta H}{T}$$

The powerful thing about this relationship is that it allows us to make a prediction about the surroundings of a chemical process, whatever they are (the rest of the universe if necessary), using measurements which we can make on the chemical system in the laboratory.

Will it or won't it?

To find the **total entropy change, ΔS_{total},** for a process, we need to combine the entropy change for the chemical system, ΔS_{sys}, with the entropy change in the surroundings, ΔS_{surr}.

$$\Delta S_{total} = \Delta S_{sys} + \Delta S_{surr}$$

We can use this to find the total entropy change when water freezes at, say, $-10\,°C$ (263 K).

$$\Delta S_{sys} = +22.0\,J\,K^{-1}\,mol^{-1}$$

$$\Delta S_{surr} = \frac{-\Delta H}{T} = \frac{-(-6010\,J\,mol^{-1})}{263\,K} = +22.9\,J\,K^{-1}\,mol^{-1}$$

$$\Delta S_{total} = \Delta S_{sys} + \Delta S_{surr} = (-22.0\,kJ\,mol^{-1}) + (+22.9\,kJ\,mol^{-1})$$
$$= +0.9\,J\,K^{-1}\,mol^{-1}$$

Overall, the process leads to an increase in entropy.

Now let's work at a higher temperature and see what happens if ice *melts* at $10\,°C$ (283 K). We will assume that ΔS and ΔH do not change with temperature.

$$\Delta S_{sys} = +22.0\,J\,K^{-1}\,mol^{-1}$$
(Notice that the sign of ΔS_{sys} has changed because we are now producing a liquid from a solid)

$$\Delta S_{surr} = \frac{-\Delta H}{T} = \frac{-(+6010\,J\,mol^{-1})}{283\,K} = -21.2\,J\,K^{-1}\,mol^{-1}$$

(We are melting ice this time, so notice another sign change: ΔH is *endothermic.*)

$$\Delta S_{\text{total}} = \Delta S_{\text{sys}} + \Delta S_{\text{surr}} = (+22.0\,\text{kJ mol}^{-1}) + (-21.2\,\text{kJ mol}^{-1})$$
$$= +0.8\,\text{J K}^{-1}\text{mol}^{-1}$$

Again, the process leads to an increase in entropy.

The entropy data are summarised in Figure 16.

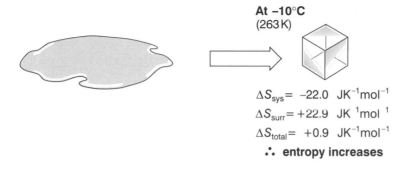

At −10°C
(263 K)

$\Delta S_{\text{sys}} = -22.0$ JK^{-1}mol^{-1}
$\Delta S_{\text{surr}} = +22.9$ JK^{-1}mol^{-1}
$\Delta S_{\text{total}} = +0.9$ JK^{-1}mol^{-1}
∴ **entropy increases**

Figure 16 Entropy changes for water freezing at −10°C and ice melting at 10°C

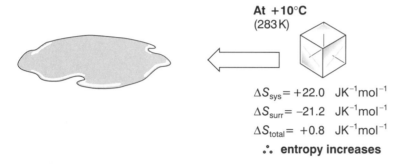

At +10°C
(283 K)

$\Delta S_{\text{sys}} = +22.0$ JK^{-1}mol^{-1}
$\Delta S_{\text{surr}} = -21.2$ JK^{-1}mol^{-1}
$\Delta S_{\text{total}} = +0.8$ JK^{-1}mol^{-1}
∴ **entropy increases**

Both these changes are spontaneous: ice melts at 10°C and water freezes at −10°C, of their own accord and without any help from us. Both processes result in an *increase* in ΔS_{total}. This is part of a general rule; we can tell if a process will be spontaneous or not by finding the value of ΔS_{total}.

ΔS_{total} is positive for a spontaneous process

This is one way of expressing the **Second law of Thermodynamics**. It is a simple but powerful law which allows us to predict whether or not something *should* happen. Chemistry is all about making things happen and explaining why they happen, so the 'Second law' is one of the foundations of chemistry. Indeed, the Second law of Thermodynamics is fundamental to the whole of science; it has been said that ignorance of the Second law is equivalent to never having read a word of Shakespeare.

Here are some other ways in which famous scientists have expressed the Second law

'You can never restore everything to its original condition after a change has occurred.' (based on *G.N. Lewis*)

'Entropy is time's arrow.' (*A. Eddington*)

'Gain in information is loss in entropy.' (*G.N. Lewis*)

Freezing seawater

Let's now put the Second law of Thermodynamics to work to explain why seawater freezes at a lower temperature than pure water, and why putting salt on the roads helps to keep them clear of ice in winter.

Remember that for the process

$$H_2O(l) \rightarrow H_2O(s); \quad \Delta S = -22.0\,\text{J K}^{-1}\text{mol}^{-1}$$
$$\Delta H = -6.01\,\text{kJ mol}^{-1}$$

> You have learned a lot more about entropy in the last few pages. Here is a summary of the key points which have been discussed.
>
> - When changes take place in a chemical *system* there are nearly always accompanying changes occurring in the *surroundings*.
> - To predict whether or not a change will take place, we need to take account of the *entropy changes* in the system *and* its surroundings.
> - The total entropy change for a process is given by
> $$\Delta S_{\text{total}} = \Delta S_{\text{sys}} + \Delta S_{\text{surr}}.$$
> - $\Delta S_{\text{surroundings}}$ can be found using the relationship
> $$\Delta S_{\text{surr}} = \frac{-\Delta H_{\text{sys}}}{T}$$
> - The Second law of Thermodynamics tell us that for a spontaneous chemical change, the total entropy must increase
> $$\Delta S_{\text{total}} > 0$$

We will once again assume that ΔH and ΔS do not vary with temperature, though in fact they do vary a bit.

When seawater freezes, pure ice is produced. The energy released will be the same as when molecules from pure water fit together to form a lattice in an ice crystal; ΔH will still be $-6.01\,\text{kJ}\,\text{mol}^{-1}$.

However, ΔS for the system is different. The entropy of salt solution is greater than the entropy of pure water because the ions are spread out in the solution. There are more ways of arranging water molecules *and* the ions in the salt than water molecules alone. Therefore ΔS is *more negative* when salt solution freezes (Figure 17).

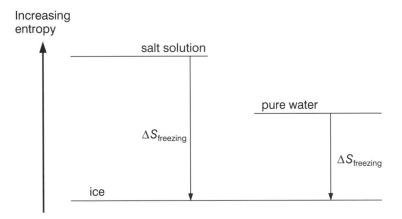

Figure 17 The entropy change when salt water freezes is greater than when pure water freezes

When seawater freezes, ΔS_{sys} is more *negative* than $-22.0\,\text{J}\,\text{K}^{-1}\,\text{mol}^{-1}$. Before freezing can occur, ΔS_{surr} has to be *more positive* than $+22.0\,\text{kJ}\,\text{mol}^{-1}$.

Because $\Delta S_{\text{surr}} = \dfrac{-\Delta H}{T}$, this can only happen if we divide ΔH by a smaller value of T. So salt solution will *not* freeze at $0\,^\circ\text{C}$; a *lower* temperature is required.

Taking it further ...

Melting and freezing do not involve changing one compound into another; but the ΔS_{total} rule applies just as much to the production of new chemicals. We can see this by looking at a chemical reaction which is described in **The Oceans** storyline in the section on Perrier water.

Limestone rocks have resisted for a long time. At the temperatures on the Earth's surface, the calcium carbonate has not broken down into calcium oxide and carbon dioxide; but this is exactly what happens when limestone gets very hot, for example when it is next to lava which has forced its way into the Earth's crust. The process we need to consider is

$$CaCO_3(s) \rightarrow CaO(s) + CO_2(g); \quad \Delta S = +160.4\,\text{J}\,\text{K}^{-1}\,\text{mol}^{-1}$$
$$\Delta H = +117.9\,\text{kJ}\,\text{mol}^{-1}$$

What are the total entropy changes at, say $298\,\text{K}$ ($25\,^\circ\text{C}$) and $1273\,\text{K}$ ($1000\,^\circ\text{C}$)?

At 298 K

$\Delta S_{\text{sys}} = +160.4\,\text{J}\,\text{K}^{-1}\,\text{mol}^{-1}$

$\Delta S_{\text{surr}} = \dfrac{-(+\,117900\,\text{J}\,\text{mol}^{-1})}{298\,\text{K}} = -395.6\,\text{J}\,\text{K}^{-1}\,\text{mol}^{-1}$

$\begin{aligned}\Delta S_{\text{total}} &= (+160.4\,\text{J}\,\text{K}^{-1}\,\text{mol}^{-1}) + (-395.6\,\text{J}\,\text{K}^{-1}\,\text{mol}^{-1}) \\ &= -235.2\,\text{J}\,\text{K}^{-1}\,\text{mol}^{-1}\end{aligned}$

At 1273 K

$\Delta S_{\text{sys}} = +160.4\,\text{J}\,\text{K}^{-1}\,\text{mol}^{-1}$

$\Delta S_{\text{surr}} = \dfrac{-(+\,117900\,\text{J}\,\text{mol}^{-1})}{1273\,\text{K}} = -92.6\,\text{J}\,\text{K}^{-1}\,\text{mol}^{-1}$

$\begin{aligned}\Delta S_{\text{total}} &= (+160.4\,\text{J}\,\text{K}^{-1}\,\text{mol}^{-1}) + (-92.6\,\text{J}\,\text{K}^{-1}\,\text{mol}^{-1}) \\ &= +67.8\,\text{J}\,\text{K}^{-1}\,\text{mol}^{-1}\end{aligned}$

At the Earth's surface, at 298 K, the reaction would correspond to a large entropy *decrease*. Therefore it does not take place. At higher temperatures, near the hot lava, the total entropy change is positive and the reaction is favourable. It is the temperature which has made the difference. When we divide ΔH by a small temperature, we get a large negative value for ΔS_{surr}, which dominates the situation and makes it unfavourable. Dividing ΔH by a much bigger value of T doesn't change the sign of ΔS_{surr}, but it does reduce the significance of ΔS_{surr} by making it smaller.

The same relationship between ΔS_{sys} and ΔS_{surr} allows us to explain dissolving and crystallisation. The arguments are slightly different this time because the temperature remains constant. Why should salt dissolve to produce a solution with a concentration like that of seawater, but crystallise from solution in the salt works when most of the water has been evaporated?

The process involved is

$$NaCl(s) + aq \rightarrow Na^+(aq) + Cl^-(aq)$$

When a $1\,mol\,dm^{-3}$ salt solution is produced, $\Delta S = +39.2\,J\,K^{-1}\,mol^{-1}$ and $\Delta H = +3.9\,kJ\,mol^{-1}$.

At 298 K

$$\Delta S_{sys} = +39.2\,J\,K^{-1}\,mol^{-1}$$

$$\Delta S_{surr} = \frac{-\Delta H}{T} = \frac{-(+3900\,J\,mol^{-1})}{298\,K} = -13.1\,J\,K^{-1}\,mol^{-1}$$

$$\Delta S_{total} = (+39.2\,J\,K^{-1}\,mol^{-1}) + (-13.1\,J\,K^{-1}\,mol^{-1}) = +26.1\,J\,K^{-1}\,mol^{-1}$$

ΔS_{total} is positive, so salt dissolves to produce a $1\,mol\,dm^{-3}$ solution at 298 K.

When the solution is more concentrated, ΔH may be very slightly different. However, we can neglect this and assume that ΔH and, therefore ΔS_{surr} remain the same. We cannot, however, neglect the change to ΔS_{sys}. This changes because, among other things, there is a smaller volume of water for the ions to be dispersed into, thus there are fewer ways of arranging their positions.

This causes ΔS_{sys} to be less positive for the formation of a more concentrated solution. Eventually ΔS_{sys} becomes less than $+13.1\,J\,K^{-1}\,mol^{-1}$, which leads to a *negative* value of ΔS_{total} for the dissolving process. Dissolving becomes unfavourable, so salt crystallises from the solution, rather than dissolving.

What about equilibrium?

A change always takes place in the direction which corresponds to an increase in ΔS_{total}. This is the case for $H_2O(s) \rightarrow H_2O(l)$ at $+10\,°C$.

At $-10\,°C$, however, ΔS_{total} for the melting of ice is negative. Therefore the reverse process occurs, and water freezes.

What happens when ΔS_{total} is equal to zero? There is no net change in either direction – liquid and solid are in **equilibrium**. We can check this by looking at the figures again: this time at $0\,°C$ (273 K).

At 273 K

$$H_2O(s) \rightleftharpoons H_2O(l); \qquad \Delta S_{sys} = +22.0\,J\,K^{-1}\,mol^{-1}$$
$$\Delta H = +6.01\,kJ\,mol^{-1}$$

$$\Delta S_{surr} = \frac{-\Delta H}{T} = \frac{-(+6010\,J\,mol^{-1})}{273\,K} = -22.0\,J\,K^{-1}\,mol^{-1}$$

$$\Delta S_{total} = (+22.0\,J\,K^{-1}\,mol^{-1}) + (-22.0\,J\,K^{-1}\,mol^{-1}) = 0\,J\,K^{-1}\,mol^{-1}$$

Liquid water and ice exist together at $0\,°C$ – they are in equilibrium – and at this temperature $\Delta S_{total} = 0$.

The requirement for equilibrium is that ΔS_{total} must be zero.

The very important point is that it is the total entropy change *under the actual experimental conditions* which must be zero at equilibrium. If conditions happen to be standard, the standard entropy change will be zero; but the standard entropy change need not be zero at equilibrium, in fact it rarely is.

PROBLEMS FOR 4.4

1 Explain the differences between the entropies of the following pairs of substances at 298 K.

	Substance	Entropy/J K^{-1} mol^{-1}
a	$F_2(g)$	101.4
	$Cl_2(g)$	111.5
b	$Br_2(l)$	75.8
	$I_2(s)$	58.4
c	$O_2(g)$	103.0
	$H_2S(g)$	206.0

2 Explain the pattern in the entropies of the first five alkanes at 298 K, given in the following table.

Alkane	$CH_4(g)$	$C_2H_6(g)$	$C_3H_8(g)$	$C_4H_{10}(g)$	$C_5H_{12}(l)$
Entropy/ J K^{-1} mol^{-1}	186	230	270	310	261

3 For each of the following reactions state whether you would expect the entropy of the products to be greater or less than that of the reactants. In each case explain your answer.

a $N_2(g) + 3H_2(g) \rightarrow 2NH_3(g)$

b $C(s) + H_2O(g) \rightarrow CO(g) + H_2(g)$

c $NH_3(g) + HCl(g) \rightarrow NH_4Cl(s)$

d $4P(s) + 5O_2(g) \rightarrow P_4O_{10}(s)$

e $(NH_4)_2Cr_2O_7(s) \rightarrow Cr_2O_3(s) + N_2(g) + 4H_2O(g)$

4 The values of ΔH and ΔS which follow refer to changes at 298 K under standard conditions. Predict whether or not the following changes will be spontaneous at 298 K. Explain your answer in each case.

a $Ca^{2+}(aq) + CO_3^{2-}(aq) \rightarrow CaCO_3(s)$;

$\Delta S = +205\,J\,K^{-1}\,mol^{-1}$

$\Delta H = +13\,kJ\,mol^{-1}$

b $H_2O_2(l) \rightarrow H_2O(l) + \frac{1}{2}O_2(g)$;

$\Delta S = +62\,J\,K^{-1}\,mol^{-1}$

$\Delta H = -98\,kJ\,mol^{-1}$

c $N_2(g) + O_2(g) \rightarrow 2NO(g)$;

$\Delta S = +25\,J\,K^{-1}\,mol^{-1}$

$\Delta H = +180\,kJ\,mol^{-1}$

d $NH_4NO_3(s) \rightarrow N_2O(g) + 2H_2(l)$;

$\Delta S = -231\,J\,K^{-1}\,mol^{-1}$

$\Delta H = -288\,kJ\,mol^{-1}$

e $C(graphite) \rightarrow C(diamond)$;

$\Delta S = -3.3\,J\,K^{-1}\,mol^{-1}$

$\Delta H = +2.0\,kJ\,mol^{-1}$

5 Use the data given for the process in question **4e** to explain why you would not expect to be able to make diamonds from graphite at any temperature at atmospheric pressure.

4.5 *Energy changes in solutions*

Some ionic substances dissolve in water, but others do not. What decides whether an ionic substance will dissolve? We can find the answer by looking at the energy changes that are involved.

Lattice enthalpy

Before an ionic solid can dissolve, we need to separate the ions from the lattice so they can spread out in the solution (Figure 18). This means supplying energy to overcome the electrical attraction between the oppositely-charged ions.

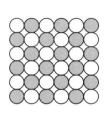 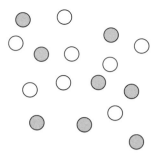

ions in solid lattice   ions spread out in solution

Figure 18 When an ionic solid dissolves, the ions in the lattice get spread out through the solution

We measure the strength of the ionic attractions in a lattice by the **lattice enthalpy** of the solid. The lattice enthalpy, ΔH_{LE}, is the enthalpy change when 1 mole of solid is formed by the coming together of the separate ions. When ions are separated from one another we can think of them as being in the gaseous state. When they are together in the lattice they are in the solid state. So we define lattice enthalpy as the enthalpy change involved in processes such as

$Na^+(g) + Cl^-(g) \rightarrow NaCl(s)$ for NaCl, $\Delta H_{LE} = -780\,kJ\,mol^{-1}$,
$Mg^{2+}(g) + 2Br(g) \rightarrow MgBr_2(s)$ for $MgBr_2$, $\Delta H_{LE} = -2440\,kJ\,mol^{-1}$.

All lattice enthalpies are large *negative* quantities. If we want to break down a lattice, we have to put in energy equal to $-\Delta H_{LE}$ (the minus sign is there because we are putting energy *in*). This tends to stop substances dissolving – unless the energy is 'paid back' later. Table 3 sets out some lattice enthalpy values for some simple ionic compounds.

Compound	ΔH_{LE} /kJ mol^{-1}	Compound	ΔH_{LE} /kJ mol^{-1}	Compound	ΔH_{LE} /kJ mol^{-1}
Li_2O	−2814	MgO	−3792	Al_2O_3	−15916
Na_2O	−2478	CaO	−3406		
K_2O	−2232	SrO	−3220		
LiF	−1033	MgF_2	−2957		
NaF	−915	CaF_2	−2630		
KF	−816				

Table 3 Lattice enthalpies for some ionic compounds

Table 4 lists values for the radii of the cations present in the compounds in Table 3. Notice how lattice enthalpy depends on the *sizes* and *charges* of the ions. Lattice enthalpy become more negative (i.e. more energy is given out) when
• the ionic charges increase
• the ionic radii decrease.

Ion	Radius/nm	Ion	Radius/nm	Ion	Radius/nm
Li^+	0.074	Mg^{2+}	0.072	Al^{3+}	0.053
Na^+	0.102	Ca^{2+}	0.100		
K^+	0.138	Sr^{2+}	0.113		
Rb^+	0.149				

Table 4 Ionic radii for some cations

This is because ions with a higher charge attract one another more strongly. They also attract more strongly if they are closer together. Stronger attractions mean higher lattice enthalpies (Figure 19). Substances with very high lattice enthalpies, such as Al_2O_3, are usually insoluble.

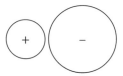

Small, highly charged ions get close together and attract strongly: ΔH_{LE} more negative, more energy given out.

Large, singly charged ions are further apart and attract less strongly: ΔH_{LE} less negative, less energy given out.

Figure 19 The factors deciding the size of lattice enthalpies

Hydration and solvation

Figure 20 Water molecules possess a dipole

Despite the need to supply lattice enthalpy, many ionic substances *do* dissolve. Something else must happen to supply the lattice enthalpy.

For the moment, let us restrict our thinking to aqueous solutions where the solvent is water. Water molecules possess a dipole (Figure 20)

The tiny charges on the water molecules are attracted to the charges on ions. This happens on the surface of an ionic solid which is placed in water, so the ions are pulled into solution (Figure 21).

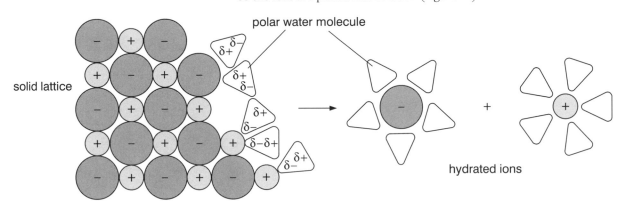

Figure 21 Polar water molecules attract the ions in a solid lattice

The ions in solution are **hydrated** – they have water molecules bound to them. Table 5 shows approximately how many water molecules are likely to be attached to particular cations in solution. You can see that the smaller and more highly charged ions attract larger numbers of water molecules. There can be several layers of water molecules around the ion – this is how Al^{3+} gets 26 water molecules fitted around it. (There is more about the sizes of ions in **Sections 3.2** and **5.1**.)

Ion	Average number of attached water molecules	Ion	Average number of attached water molecules
Li^+	5	Mg^{2+}	15
Na^+	5	Ca^{2+}	13
K^+	4	Al^{3+}	26

Table 5 Approximate extent of hydration for some cations

Water molecules bind weakly to some ions, with the result that they are not extensively hydrated. Other ions are extensively hydrated with some water molecules bound very strongly to them. When these ions crystallise out of solution the strongly bound water molecules crystallise with them to give hydrated crystals, such as blue hydrated copper(II) sulphate, $CuSO_4.5H_2O$.

The strength of the attractions between ions and water molecules is measured by the **enthalpy of hydration** or ΔH_{hyd}. This is the enthalpy change for the production of a solution of ions from 1 mol of gaseous ions. For example,

$$Na^+(g) + aq \rightarrow Na^+(aq); \quad \Delta H_{hyd} = -390 \, kJ \, mol^{-1}$$
$$Br(g) + aq \rightarrow Br(aq); \quad \Delta H_{hyd} = -337 \, kJ \, mol^{-1}$$

(The symbol aq is used in an equation to represent water when it is acting as a solvent.)

Enthalpies of hydration depend on the concentration of the solution produced. Values quoted refer to a $1 \, mol \, dm^{-3}$ concentration.

Enthalpies of hydration are always negative, i.e. hydration is *exothermic* and energy is given out. Some values are listed in Table 6. If you compare Table 6 with Table 4 you will see that the most exothermic values occur for the ions with

- the greatest charge
- the smallest radii.

The reasons are very similar to those used earlier to explain the variation in ΔH_{LE}. Small, highly charged ions can get close to water molecules and attract them strongly (Figure 22).

Figure 22 The magnitude of hydration enthalpy depends on the size and charge of the ion

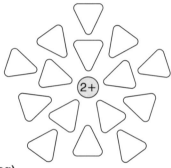

Mg²⁺(aq)

Small, doubly-charged Mg^{2+} ion attracts H_2O molecules strongly. $\Delta H_{hyd} = -1891 \, kJmol^{-1}$

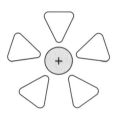

Na⁺(aq)

Larger Na^+ ion with smaller charge attracts H_2O molecules less strongly. $\Delta H_{hyd} = -390 \, kJmol^{-1}$

Ion	ΔH_{hyd} /kJ mol⁻¹	Ion	ΔH_{hyd} /kJ mol⁻¹	Ion	ΔH_{hyd} /kJ mol⁻¹
Li^+	−499	Mg^{2+}	−1891	Al^{3+}	−4613
Na^+	−390	Ca^{2+}	−1562		
K^+	−305	Sr^{2+}	−1413		
Rb^+	−281				

Table 6 Enthalpies of hydration of some ions

Molecules of some other solvents, such as ethanol, are also polar and can bind to ions. When we are dealing with solvents other than water we talk more generally about the **enthalpy of solvation** or ΔH_{solv}, rather than enthalpy of hydration.

Enthalpies of solution

Hydration of ions favours dissolving and helps to supply the energy needed to separate the ions from the lattice. The difference between the enthalpies of hydration of the ions and the lattice enthalpy gives the **enthalpy of solution**, or $\Delta H_{solution}$. $\Delta H_{solution}$ can be measured experimentally; it is the enthalpy change when one mole of a solute dissolves to form a solution of concentration $1 \, mol \, dm^{-3}$.

$$\Delta H_{solution} = \Delta H_{hyd(cation)} + \Delta H_{hyd(anion)} - \Delta H_{LE}$$

We can represent the enthalpy changes involved using an enthalpy cycle like the ones you met in **Section 4.1** (Figure 23). Notice that the enthalpy change for breaking up the ionic lattice is *minus* ΔH_{LE}, because ΔH_{LE} is defined as the enthalpy change when the lattice is *created* – the opposite of breaking it up.

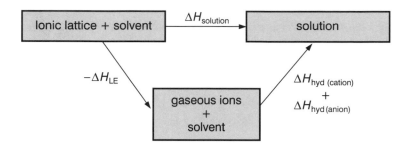

Figure 23 An enthalpy cycle to show the dissolving of an ionic solid

Figures 24, 25 and 26 show the same enthalpy cycle, but this time in the form of an enthalpy level diagram. This makes it easier to compare the sizes of the different enthalpy changes involved.

Figure 24 represents a solute for which $\Delta H_{solution}$ is slightly *negative*. The hydration of the ions provides slightly more energy than is needed to break up the lattice. This solute dissolves, giving out a little energy in the process.

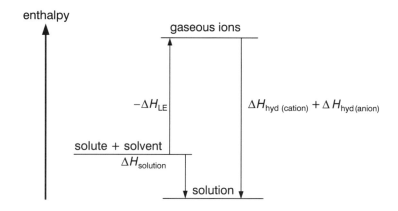

Figure 24 An enthalpy level diagram for a solute with $\Delta H_{solution}$ slightly negative. This solute will dissolve

Figure 25 represents a solute for which $\Delta H_{solution}$ has a large *positive* value. The hydration of the ions does not provide as much energy as is needed to break up the lattice. This solute does not dissolve.

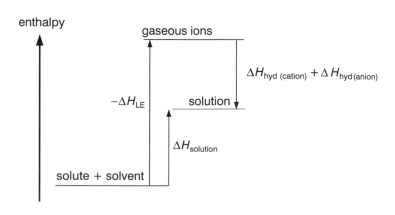

Figure 25 An enthalpy level diagram for a solute with $\Delta H_{solution}$ positive. The solute will not dissolve because too much energy is needed

Figure 26 represents a solute for which $\Delta H_{solution}$ is slightly positive, *but which dissolves nevertheless*. Many ionic solutes are like this; they dissolve even though they need a little energy from the surroundings to do so. The reason this can happen concerns *entropy*.

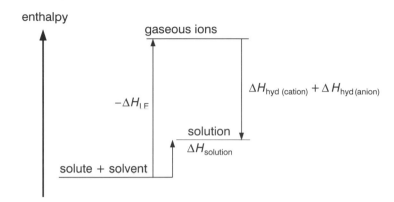

Figure 26 An enthalpy level diagram for a solute with $\Delta H_{solution}$ slightly positive. If the entropy increase is favourable, this solution may dissolve despite needing energy from the surroundings

Entropy and dissolving

When a solute dissolves there is normally an entropy increase, because the solute becomes more disordered as it spreads out through the solvent. (You can read more about entropy in **Section 4.3**.) This increase in entropy favours dissolving – even if a little energy is needed. So substances with a small positive $\Delta H_{solution}$ can still dissolve, provided there is a favourable increase in entropy. But if $\Delta H_{solution}$ is *very* large and positive, the substance will not dissolve even if the entropy change is favourable. This is why salt dissolves in the sea but the White Cliffs of Dover do not; NaCl has a small, slightly negative $\Delta H_{solution}$ while $CaCO_3$ has a large, positive one.

Non-polar solvents

Ionic solids, like NaCl, are insoluble in **non-polar solvents** like hexane. The molecules in non-polar solvents have no regions of slight positive and negative charge, so they are unable to interact strongly with ions. The enthalpy level diagram for dissolving an ionic solid in a non-polar solvent would look like Figure 27. The large positive value of $\Delta H_{solution}$ prevents the solid dissolving.

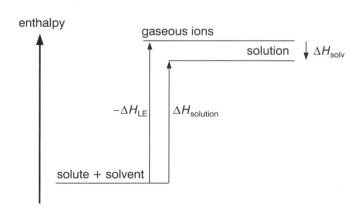

Figure 27 The situation when you try to dissolve an ionic solid in a non-polar solvent such as hexane. The enthalpy change of solvation is so small that $\Delta H_{solution}$ is large and positive, and dissolving is unlikely to occur

Ionic precipitation

$\Delta H_{solution}$ for silver chloride, $AgCl(s)$, is $+95\,kJ\,mol^{-1}$, so silver chloride is insoluble. Very few ions will be present in solution when silver chloride is added to water. What happens if we add a solution containing a high concentration of chloride ions to a solution with a high concentration of silver ions?

You can find out by mixing $0.1\,mol\,dm^{-3}$ silver nitrate solution and $0.1\,mol^{-3}$ potassium chloride solution. Solid silver chloride precipitates out. Ag^+ and Cl^- ions come together to form a lattice. The combination of enthalpy and entropy factors favours the solid in this case and an **ionic precipitation** reaction takes place,

$$Ag^+(aq) + Cl^-(aq) \rightarrow AgCl(s)$$

Ionic precipitation is another important class of chemical reaction – as are redox and acid-base reaction. It is involved, for example, in the formation of chalcopyrite in mineral lodes,

$$Cu^{2+}(aq) + Fe^{2+}(aq) + 2S^{2-}(aq) \rightarrow CuFeS_2(s)$$

and in the production of rust (iron(III) hydroxide),

$$Fe^{3+}(aq) + 3OH^-(aq) \rightarrow Fe(OH)_3(s)$$

PROBLEMS FOR 4.5

1 Explain the difference between the lattice enthalpies of the following pairs of compounds.
   a LiF and NaF
   b Rb_2O and K_2O
   c MgO and Na_2O
   d Al_2O_3 and MgO.

2 Which of the following pairs of compounds will have the more negative (i.e. more exothermic) lattice enthalpy? In each case explain your answer.
   a RbF and SrF_2
   b Cs_2O and BaO
   c CuO and Cu_2O.

3 Explain the difference between the hydration enthalpies of the following pairs of ions.
   a Li^+ and Na^+
   b Mg^{2+} and Ca^{2+}
   c Na^+ and Ca^{2+}

4 Lattice enthalpy values ($kJ\,mol^{-1}$) for the silver halides are

	AgF	AgCl
	−958	−905

The enthalpies of hydration of the ions in these compounds ($kJ\,mol^{-1}$) are

	Ag^+	F^-	Cl^-
	−446	−506	−364

   a Explain the changes in
     i the lattice enthalpies of the two silver halides
     ii the enthalpies of hydration of the ions.

   b Calculate values for the enthalpies of solution of the silver halides.

   c Use your $\Delta H_{solution}$ values to comment on the solubilities of the silver halides in water.

5 a Use the following table of lattice enthalpies and hydration enthalpies to draw enthalpy level diagrams (to scale) to determine the enthalpy of solution of
     i magnesium sulphate
     ii barium sulphate.

   b Use the enthalpies of solution that you have calculated in part a to predict the relative solubilities of magnesium sulphate and barium sulphate.

Compound	Lattice enthalpy $\Delta H_{LE}/kJ\,mol^{-1}$	Ion	Hydration enthalpy $\Delta H_{LE}/kJ\,mol^{-1}$
$MgSO_4$	−2833	Mg^{2+}	−1891
$BaSO_4$	−2374	Ba^{2+}	−1360
		SO_4^{2-}	−1004

5.1 *Ions in solids and solutions*

Ionic solids

In ionic solids, ions are held together by their opposite electrical charges. Each **cation** (+ ion) attracts several **anions** (– ions), and vice versa. The ions build up into a giant ionic lattice, in which very large numbers of ions are arranged in fixed positions.

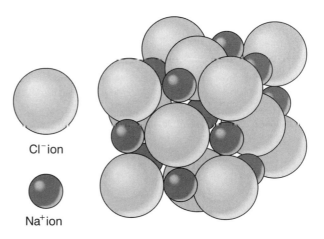

Cl⁻ ion

Na⁺ ion

Figure 1 The sodium chloride lattice

One of the simplest examples is sodium chloride, $Na^+Cl^-(s)$ (Figure 1). In the sodium chloride lattice, each Na^+ ion is surrounded by six Cl^- ions, and each Cl^- ion is surrounded by six Na^+ ions. Each Na^+ is attracted to the six Cl^- round it, but repelled by other Na^+ ions which are a bit further away (there are 12 of these), and attracted to the next lot of Cl^- ions (eight of them) which are further away still … and so on. It adds up to an infinite series of attractions and repulsions, but overall the attractions are stronger than the repulsions, which is why the lattice holds together. Indeed, it holds together very strongly, which is why ionic solids are hard and have high melting and boiling points.

It is important to realise that an ionic solid like Na^+Cl^- isn't just pairs of Na^+ and Cl^- ions, but a huge lattice. You can see from Figure 1 that the lattice has a cubic shape which accounts for the fact that sodium chloride crystals are cubic.

The simple cubic lattice structure of sodium chloride is also found in some other Group 1 halides, such as potassium fluoride and lithium chloride. However, other ionic solids may have more complicated lattices, depending on the number and size of the different ions present.

Modifying ionic lattices

Double salts

You can think of the sodium chloride lattice as being made from a lattice of Cl^- anions with the smaller Na^+ ions fitted into it. It is possible to fit *two different* sorts of cations into an anion lattice, provided the + and – charges still balance so that the lattice is electrically neutral overall.

An example is potassium magnesium chloride (carnallite), $KCl.MgCl_2.H_2O$. Both K^+ and Mg^{2+} are fitted into the lattice of Cl^- ions. KCl and $MgCl_2$ can both exist as separate salts, but carnallite is not a mixture of the two salts – it is one single substance called a **double salt**. Substances like this are not just mixtures of the two salts because the different cations are arranged throughout the lattice in a regular pattern.

Hydrated crystals

Notice that the formula of carnallite includes a molecule of water, H_2O. The water is not just mixed with the carnallite crystals – that would make them damp. Instead, the H_2O molecules are *fitted in the lattice* in the same regular way as the ions.

The lattice of carnallite is, therefore, an interweaving regular pattern of four different particles: K^+ ions, Mg^{2+} ions, Cl^- ions and H_2O molecules. The water in compounds like this is called **water of crystallisation**, and the crystals are called **hydrated crystals**.

Ionic substances in solution

Many ionic substances dissolve readily in water. When they do so, the ions become surrounded by water molecules: *they are hydrated* (see **Section 4.5**) The hydrated ions spread out through the water. Figure 2 illustrates this for sodium chloride. The hydrated ions, $Na^+(aq)$ and $Cl^-(aq)$, are no longer regularly arranged; they are scattered through the water at random. what's more, now the Na^+ and Cl^- are separated, they behave independently of each other. It is as if each has forgotten the other exists.

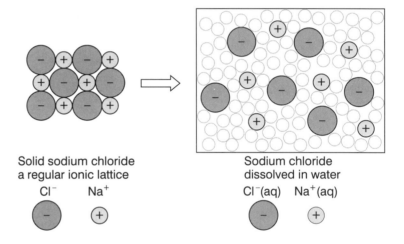

Solid sodium chloride
a regular ionic lattice
Cl^- Na^+

Sodium chloride
dissolved in water
$Cl^-(aq)$ $Na^+(aq)$

Figure 2 What happens when an ionic substance like sodium chloride dissolves

This applies to all ionic substances. As soon as they are dissolved, the positive and negative ions separate and behave independently.

Ionic equations

Ions in solution behave independently – and this includes their chemical reactions. The reactions of an ionic substance like sodium chloride quite often involve only one of the two types of ion. The other ion does not get involved in the reaction.

For example, if you add a solution of sodium chloride to silver nitrate(V) solution, you get a white precipitate of silver chloride. Silver ions, Ag^+, and chloride ions, Cl^-, have come together to form insoluble silver chloride, which precipitates out (Figure 3). We can write an equation for this reaction. Sodium chloride contains $Na^+(aq)$ and $Cl^-(aq)$, and silver nitrate contains $Ag^+(aq)$ and $NO_3^-(aq)$, so we write

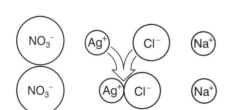

Figure 3 Making a precipitate of silver chloride by mixing silver nitrate(V) and sodium chloride solutions. Only the $Ag^+(aq)$ and $Cl^-(aq)$ ions react

$$Ag^+(aq) + NO_3^-(aq) \quad + \quad Na^+(aq) + Cl^-(aq)$$
silver nitrate(V) solution   sodium chloride solution

$$\rightarrow \quad AgCl(s) \quad + Na^+(aq) + NO_3^-(aq)$$
silver chloride
precipitate

Notice that $Na^+(aq)$ and $NO_3^-(aq)$ ions are on both sides of the equation. They do not take part in the reaction – it is as if they are on the sidelines, watching the action going on between $Ag^+(aq)$ and $Cl^-(aq)$ ions without getting involved. They are described as **spectator ions**.

Because $Na^+(aq)$ and $NO_3^-(aq)$ are not involved in the reaction, we can leave them out of the equation altogether. This simplifies the equation so that it only shows the ions that are actually taking part in the reaction.

$$Ag^+(aq) + Cl^-(aq) \rightarrow AgCl(s)$$

This type of equation, showing only the ions that take part in the reaction and excluding the spectator ions, is called an **ionic equation**.

PROBLEMS FOR 5.I

1 a Many of the compounds of Group 2 elements are insoluble in water. Write ionic equation (with state symbols) for the precipitates formed when the following solutions are mixed.

Aqueous solutions mixed	Precipitate (colour)
i barium chloride, $BaCl_2$, and sodium sulphate, Na_2SO_4	barium sulphate (white)
ii magnesium sulphate, $MgSO_4$, and sodium hydroxide, NaOH	magnesium hydroxide (white)
iii calcium chloride, $CaCl_2$, and sodium carbonate, Na_2CO_3	calcium carbonate (white)
iv barium nitrate(V), $Ba(NO_3)_2$, and potassium chromate(VI), K_2CrO_4	barium chromate(VI) (yellow)

b i When a saturated solution of aqueous sodium thiosulphate is shaken it crystallises rapidly. A mass of hydrated crystals are produced with the formula $Na_2S_2O_3.5H_2O$. Write an ionic equation, with state symbols, for its formation from the separate ions in the solution.

ii Large, deep purple crystals of chrome alum (potassium chromium(III) sulphate) may be obtained from its solution. The formula of this double salt may be written $KCr(SO_4)_2.12H_2O$. Write an ionic equation, with state symbols, for its formation from the separate ions in the solution.

2 Figure 4 shows a section of a lattice structure for caesium chloride.

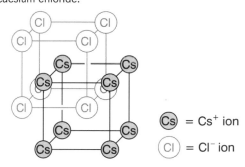

Figure 4 The lattice structure of caesium chloride. A caesium ion lies at the centre of a cube of chloride ions, and a chloride ion lies at the centre of a cube of caesium ions

a i Predict the shape of caesium chloride crystals.

ii How many chloride ions closely surround one caesium ion?

iii How many caesium ions closely surround one chloride ion?

b Explain why caesium chloride is hard and has a high melting point (918 K).

Element	Boiling point/K
He	4
Ne	27
Ar	87
Kr	121
Xe	166

Table 1 Boiling points for the noble gases

Alkane	Boiling point/K
hexane	342
3-methylpentane	336
2-methylpentane	333
2,3-dimethylbutane	331
2,2-dimethylbutane	323

Table 2 Boiling points of the isomers of C_6H_{14}

5.2 *Forces between molecules: induced dipoles*

The boiling points of noble gases and alkanes

Look at the boiling points given in Table 1 for the noble gases.

When the noble gases change from liquid to gas, energy is required to separate the atoms. There are no covalent bonds in the noble gases but there must be some forces holding the atoms together.

These forces must be much weaker than covalent bonds because the enthalpies of evaporation are much smaller than bond enthalpies: for example,

$$\Delta H_{vap}(Xe) = +15\,kJ\,mol^{-1},$$
whereas bond enthalpy of (Cl—Cl) $= +242\,kJ\,mol^{-1}$.

Now look at the boiling points of the isomeric alkanes with the formula C_6H_{14}. These are given in Table 2.

There is a pattern to these boiling points; the isomer with the unbranched carbon chain has the highest boiling point, and the more branched the chain the lower the boiling point. There must be stronger forces between the straight-chain molecules than between the branched-chain molecules. More energy and higher temperatures are needed to overcome these stronger forces and set the alkane molecules free. To understand how the **intermolecular forces** between molecules such as those of alkanes arise, and what affects their relative sizes, you need to know about the charges on molecules, called dipoles.

Dipoles

You have already come across polar bonds in **Section 3.1**: a **dipole** is simply a molecule (or part of a molecule) with a positive end and a negative end. Hydrogen chloride has a dipole; we show the molecule as

δ+ δ–
H—Cl

When a molecule has a dipole we say it is **polarised**. There are several ways a molecule can become polarised.

Permanent dipoles

Permanent dipoles occur when a molecule has two atoms bonded together which have substantially different electronegativity, so that one atom attracts the shared electrons much more than the other. Hydrogen chloride has a permanent dipole, because chlorine is much more electronegative than hydrogen, and so attracts the shared electrons more.

Instantaneous dipoles

Some molecules do not possess a permanent dipole, because the atoms that are bonded together have the same, or very similar, electronegativity, so that the electrons are evenly shared. The chlorine molecule, Cl_2, is an example.

However, even though the molecule does not have a permanent dipole, a **temporary**, or **instantaneous dipole** can arise. We can think of the electrons in the chlorine molecule as forming a negatively-charged cloud. The electrons in this cloud are in constant motion, and at a particular instant they may not be evenly distributed over the two atoms. This means that one end of the chlorine molecule has a greater negative charge than the other end; instantaneously, the molecule has developed a dipole (Figure 5).

Electron cloud evenly distributed; no dipole.

At some instant, more of the electron cloud happens to be at one end of the molecule than the other; molecule has an instantaneous dipole.

Figure 5 How a dipole forms in a chlorine molecule

Left on its own, this dipole only lasts for an instant before the swirling electron cloud changes its position, cancelling out or even reversing the dipole. However, if there are other molecules nearby, the instantaneous dipole may affect them and produce **induced dipoles**.

Induced dipoles

If an unpolarised molecule finds itself next to a dipole, the unpolarised molecule may get a dipole induced in it. The dipole attracts or repels electrons in the charge cloud of the unpolarised molecule, **inducing** a dipole in it. Figure 6 illustrates this by showing what happens if an unpolarised chlorine molecule finds itself next to an HCl dipole.

In Figure 6, the dipole has been induced by the effect of a *permanent* dipole. A dipole can also be induced by the effect of an *instantaneous* dipole. This makes it possible for a whole series of dipoles to be set up in a substance that contains no permanent dipoles. How this occurs in a noble gas like xenon is explained later in this section.

An unpolarised Cl_2 molecule finds itself next to an HCl molecule with a permanent dipole.

Electrons get attracted to the positive end of the HCl dipole, inducing a dipole in the Cl_2 molecule.

Figure 6 How a dipole can be induced in a chlorine molecule.

Dipoles and intermolecular forces

If a molecular substance contains dipoles, they can attract each other (Figure 7).

$$----\ \overset{\delta+}{H}\!-\!\overset{\delta-}{Cl}\ -----\ \overset{\delta+}{H}\!-\!\overset{\delta-}{Cl}\ -----\ \overset{\delta+}{H}\!-\!\overset{\delta-}{Cl}\ -----\ \overset{\delta+}{H}\!-\!\overset{\delta-}{Cl}\ ----$$

Figure 7 Intermolecular forces arise from the attractions between dipoles. (The dotted lines represent the attractive forces between HCl molecules; the solid lines represent the stronger covalent bonds inside each molecule.)

All intermolecular forces arise from the attractive forces between dipoles. There are three kinds of attraction:

- **Permanent dipole–permanent dipole**, where two or more permanent dipoles attract one another. This kind of attraction occurs between the permanent dipoles in HCl.
- **Permanent dipole–induced dipole**, in which a permanent dipole induces a dipole in another molecule, then the two attract one another. This kind of attraction occurs between HCl and Cl_2 molecules.
- **Instantaneous dipole–induced dipole**, in which an instantaneous dipole induces a dipole in another molecule, then attracts it. This kind of attraction occurs between Cl_2 molecules.

In this section we will look in more detail at instantaneous dipole-induced dipole forces, which are particularly important in polymers such as poly(ethene). **Section 5.3** deals with the other kinds of dipole attractions.

Instantaneous dipole-induced dipole attractions

These forces act between *all* molecules, because instantaneous dipoles can arise in molecules which already have a permanent dipole. However, you notice them most easily in substances like the noble gases and the alkanes because there are no other intermolecular forces present. Figure 8 shows how an instantaneous dipole in a xenon atom induces polarisation in a neighbouring atom.

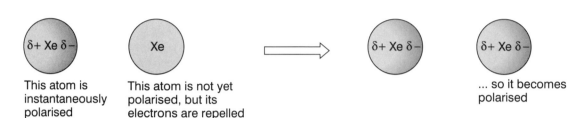

This atom is instantaneously polarised

This atom is not yet polarised, but its electrons are repelled by the dipole next to it ...

... so it becomes polarised

Figure 8 How an induced dipole is formed in an Xe atom

The polarisation illustrated here is not permanent. Electrons are always moving around and the original polarity will be lost and replaced by another which will be different – possibly the other way round. In the gaseous or liquid state, the atoms are always moving around. As they move past each other the dipoles constantly change position, although they are always lined up so that they attract rather than repel one another.

The forces between atoms in xenon are stronger than those between helium atoms. The xenon atom is bigger, and has more electrons. The outer electrons are less strongly held by the positive nucleus and more easily pushed around from outside. We say that xenon atoms are more **polarisable** than helium atoms. This leads to bigger induced dipoles which in turn lead to bigger forces of attraction, which is why xenon has a higher boiling point than helium.

The importance of molecular shape

In the straight-chain and branched alkanes in Table 2, the atoms in the different isomers are identical, but in straight chain alkanes there are more contacts between atoms of different molecules. There are, therefore, more opportunities for induced dipole forces to operate. That is why straight chain alkanes have higher boiling points than their branched isomers.

Polymers are molecules with very long chains. If the chains can line up neatly, the intermolecular forces are quite strong, making the polymer material strong too.

If you are not sure about the amount of contact which is possible between molecules of different shape, building models will help. Try building two models each of hexane and 2,2-dimethylbutane and compare how closely they fit together.

To sum up ...

Instantaneous dipole-induced dipole forces arise because electron movements in molecules cause instantaneous dipoles which then induce polarity in neighbouring molecules. The attractive forces between molecules are the result of electrostatic interaction between the dipoles. These interactions are normally strongest for large atoms, which have more electrons and are more polarisable than small atoms. For molecules, this type of induced dipole force is greatest in large molecules and where the molecular structure allows the molecules to have extensive contact with each other.

Remember that these forces are present in *all* substances.

PROBLEMS FOR 5.2

1 For each pair of chemicals given below, arrange the formulae in the order in which the strength of the instantaneous dipole-induced dipole forces increases.

a CCl_4 and CH_4
b $SiCl_4$ and SiH_4
c $CH_3CH_2CH_2CH_2CH_2CH_3$
and $CH_3CH(CH_3)CH(CH_3)CH_3$
d $CH_3CH_2CH_2CH_3$ and $CH_3CH(CH_3)CH_3$
e

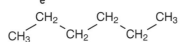 and

2 The hydrides of silicon and sulphur have the following formulae and boiling points:
SiH_4 161 K; H_2S 212 K

a **i** How many electrons are present in each molecule?
   ii How will the strengths of instantaneous dipole-induced dipole attractions compare in the two compounds?
   iii Does either molecule possess a permanent dipole? Explain your answer.

b Use your answers in part **a** to explain the relative sizes of the boiling points of these compounds.

5.3 *Forces between molecules: permanent dipoles and hydrogen bonding*

There are attractive forces between all atoms and molecules. In **Section 5.2** you were introduced to forces between instantaneous dipoles and induced dipoles. These arise because electron movements in molecules produce *temporary dipoles* which induce dipoles in neighbouring molecules. These forces exist in *all* substances.

In this section, we will look in more detail at the additional attractive forces which can arise when molecules have **permanent dipoles.**

Bond polarity and dipoles

Section 3.1 covers bond polarity, and describes how you can use electronegativity values to decide how polar a particular bond will be.

Let's look at an important example of a polar molecule: H_2O. Oxygen is much more electronegative than hydrogen, so the O—H bond is polar:

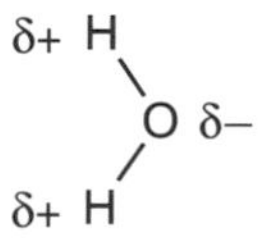

$$\delta+ \ H$$
$$O \ \delta-$$
$$\delta+ \ H$$

One side of the molecule is positive, the other is negative. The water molecule has a dipole with the positive charge centred between the two hydrogen atoms. We could also write it like this:

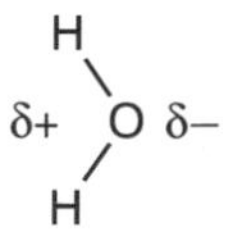

$$H$$
$$\delta+ \ O \ \delta-$$
$$H$$

An example of a more complicated dipolar molecule is 1,1,1-trichloroethane. The three chlorine atoms make one end of the molecule negatively charged; the three hydrogen atoms at the other end are positively charged. The dipole can be seen more clearly if we draw the structure as

$$\overset{\delta+ \quad \delta-}{H_3C - CCl_3}$$

When the atoms which form a molecule have no electronegativity difference, or when the electronegativity differences are small (in other words, when the bonds are non-polar), any dipole will be very small. It is also possible for a molecule to have no overall dipole even though the bonds are polar. CCl_4 is an example of such a molecule. Each Cl atom in CCl_4 carries a small negative charge and the central carbon is positive.

The Cl atoms are distributed symmetrically around the carbon atom, and CCl_4 has a tetrahedral shape, like that of methane.

Because the Cl atoms are distributed tetrahedrally around the carbon, the centre of negative charge is midway between all the chlorines. It is at the centre of the molecule and is superimposed on the positive charge on the carbon. The molecule has no overall dipole.

So, *bond* polarity depends on electronegativity differences. A *molecular* dipole depends on electronegativity differences *and* the shape of the molecule. You may need to build a model to help you decide if a molecule has a dipole.

You need a table of electronegativity values before you can do problems and make decisions about whether or not molecules contain dipoles. Several ways of estimating electronegativities have been used. Each method gives different numbers but they place elements in a similar order – they lead to the same *relative* electronegativity values. The figures in Table 3 for some common elements are derived from a method suggested by Linus Pauling, a US chemist and winner of two Nobel Prizes.

If you exclude the noble gases, the elements are more electronegative at the top of a Group and at the right hand side of the Periodic Table. The highest electronegativity values correspond to non-metal elements with small atomic radii.

Notice that C and H have similar electronegativities (2.5 and 2.1), so for practical purposes we can think of the C—H bond as non-polar.

Atom	Electronegativity
F	4.0
O	3.5
Cl	3.0
N	3.0
Br	2.8
I	2.5
S	2.5
C	2.5
H	2.1

Table 3 Pauling electronegativity values for some common elements

Dipole-dipole interactions

Let's look at the forces between molecular dipoles like those in 1,1,1,-trichloroethane. This substance is a liquid at room temperature, and in the liquid phase the molecules are constantly moving and tumbling around. Sometimes the negative end of one molecule will be lined up with the positive end of another, making them attract, but often there will be two negative ends or two positive ends together, in which case they will repel (Figure 9). Overall, there will be more attraction than repulsion between the permanent dipoles, which is one reasons why the molecules stay together as a liquid.

In addition to the permanent dipoles, there will also be some *induced* dipoles in the molecules in the liquid, and these will give rise to permanent dipole-induced dipole forces. These forces are always attractive, and they play an important part in holding the molecules in the liquid together.

In the liquid state, the molecular dipoles are constantly tumbling around.

Figure 9 Molecular dipoles in the liquid state

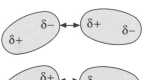

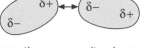

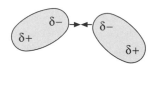

Sometimes opposite charges are next to one another, causing attraction ...

... sometimes like charges are next to one another, causing repulsion.

In the solid state, things are rather different. In a crystalline solid the molecules are in fixed positions rather than tumbling around, and it is possible for the permanent dipoles to line up with opposite charges next to one another. The attractive forces between these aligned dipoles help to hold the molecules together in the solid state.

Permanent dipole forces are present in many molecular substances, for example: propanone (CH_3COCH_3), ethyl ethanoate ($C_2H_5OCOCH_3$), poly(chloroethene) (pvc). The attractive forces are stronger if the sizes of the dipoles increase, or if the interacting dipoles are able to approach each other closely.

Take the two factors separately. First, what leads to particularly large dipoles? A large dipole comes about when a very electronegative atom is present, or several electronegative atoms in close proximity. Thus, the $CHCl_3$ molecule has a relatively large dipole because there are three chlorine atoms pulling electrons towards themselves.

Second, when can the dipoles approach closely? They can do this when the molecules are small.

Hydrogen bonding

What is special about hydrogen bonding?

Hydrogen bonding is a particularly strong intermolecular force that involves three features:

- a *large dipole* between an H atom and a highly electronegative atom such as O, N or F
- the *small H atom* which can get very close to other atoms
- a *lone pair of electrons* on another O, N or F atom, with which the positively charged H atom can line up.

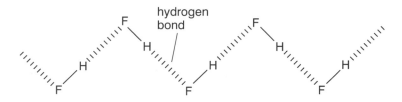

Figure 10 Hydrogen bonding in liquid hydrogen fluoride

Figure 10 illustrates hydrogen bonding in liquid hydrogen fluoride, HF. The H atoms have a strong positive charge because of the highly electronegative F to which they are bonded. These positive charges line up with a lone pair on another F atom, which provides a region of concentrated negative charge (Figure 11). The H and F atoms can get very close, and therefore attract very strongly, because the H atom is so small. The lining up of HF molecules is important because it means that positive and negative

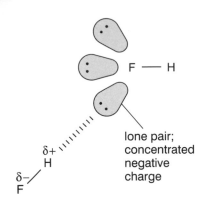

Figure 11 The positively charged H atom lines up with the lone pair on an F atom

Compound	Boiling point/K
HF	293
HCl	188
HBr	206
HI	222

Table 4 Boiling points of the hydrogen halides

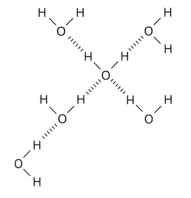

Figure 12 Hydrogen bonding between water molecules

charges are always lined up with one another, so the forces are always attractive. As a result, the HF molecules are held together in the molecular chains you see in Figure 10.

Hydrogen bonds are much stronger than the other kinds of intermolecular forces that we have met.

Dipole-dipole forces are present in all the hydrogen halides, but the greater strength of the hydrogen bonds in HF can be seen from a comparison of the boiling points. Look at Table 4.

HF has the highest boiling pint despite having the lowest relative molecular mass.

Intermolecular attractive forces affect the melting point and boiling point of a covalent substance because energy is needed to overcome these forces and to separate molecules when the substance melts and then boils. If the attractive forces *between molecules* are strong, then the melting point and boiling point of the substance tend to be high – whereas if forces are weak, the melting point and boiling point will tend to be low. In the case of hydrogen fluoride, the strong hydrogen bonds give it an exceptionally high boiling point compared with the other hydrogen halides.

Hydrogen bonding is also present in water (Figure 12).

Notice that water molecules can form twice as many hydrogen bonds as hydrogen fluoride. The oxygen atom possesses two lone pairs of electrons with which positive hydrogen atoms can interact. There are twice as many hydrogens as oxygens – just the right number to maximise the bonding to the lone pairs.

Water is unique in this respect. In HF the fluorine has three lone pairs but there are only as many H atoms as F atoms. So only one-third of the available lone pairs can be used. In NH_3, another hydrogen-bonded substance, there is only one lone pair on the N so, on average, only one of the three hydrogen atoms can form hydrogen bonds.

Section 5.5 explains how the hydrogen bonding in water gives it unique properties.

The effects of hydrogen bonding

Hydrogen bonds are responsible for the strong intermolecular forces between the polymer chains of many fibres. In nylon, for example, there are interactions between the H atoms of the N—H groups and the O atoms of the C=O groups. The importance of hydrogen bonding in nylon and Kevlar is described in the **Polymer Revolution** storyline.

Hydrogen bonding also helps fibres to absorb water. The protein chains in wool contain lots of O and N atoms that can hydrogen bond to water, which means that wool can absorb sweat and doesn't feel sticky to wear. Polythene, on the other hand, cannot form hydrogen bonds at all, so polythene clothes would feel very damp and sticky.

In summary, for hydrogen bonding to occur there needs to be a hydrogen atom made positively charged by being attached to a highly electronegative atom or group of atoms. There also needs to be a small, highly electronegative atom with at least one lone pair of electrons for the H atom to interact with. N, O and F atoms satisfy these conditions (Figure 13).

$$\text{Y} \longrightarrow \text{H} \mid\mid\mid\mid\mid\mid\mid\mid\mid \text{X} \longrightarrow$$

X = F, O or N
Y = F, O or N

hydrogen bond

Figure 13 Hydrogen bonds only form between particular kinds of atoms

Hydrogen bonds are stronger than other intermolecular forces, but much weaker than covalent bonds. Some typical bond strengths are shown in Table 5.

Type of bond	Bond enthalpy/kJ mol⁻¹
O—H covalent bond	464
hydrogen bond	10–20
instantaneous dipole-induced dipole forces	<10

Table 5 Relative strengths of covalent bonds and some intermolecular forces

PROBLEMS FOR 5.3

1 Which of the covalent bonds in the following list will be significantly polar? In cases where there is a polar bond, show which atom is positive and which is negative, using the $\delta+/\delta-$ convention.

a   C—F   e   H—N
b   C—H   f   S—Br
c   C—S   g   C   O
d   H—Cl

2 Which of the following molecules will possess a dipole? You may need to build models in some cases to help you answer the question.

a CO_2 f   benzene
b $CHCl_3$ g   *cis*-1,2-difluoroethene
c C_6H_{12}(cyclohexane)   h   *trans*-1,2-difluoroethene
d CH_3OH i   benzene-1,2-dicarboxylic acid
e $(CH_3)_2CO$ j   benzene-1,4-dicarboxylic acid

3 List the types of intermolecular forces which will exist in the following compounds. The types of forces to choose from are

- instantaneous dipole-induced dipole
- permanent dipole-induced dipole
- permanent dipole-permanent dipole
- hydrogen bonding

a $CH_3CCH_2CH_3$
        ‖
        O

c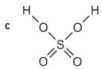

b $CH_3C—OH$
      ‖
      O

d   S=C=S

e Br_2

(Remember, more than one type of intermolecular force may operate for each molecule.)

4 Hydrogen bonding can occur between *different* molecules *in mixtures*. Draw diagrams to show where the hydrogen bonds form in the following mixtures.

a NH_3 and H_2O
b H_2SO_4 and H_2O

5 a   Figure 14 shows the boiling points for the hydrides of Group 6 elements (H_2O, H_2S, H_2Se, H_2Te). Explain why the boiling point of water appears to be abnormally high.

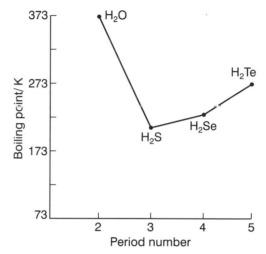

Figure 14 Boiling points for Group 6 hydrides

b The boiling points of the hydrides of Group 4 elements are

CH_4 112 K; SiH_4 161 K; GeH_4 185 K; SnH_4 221 K

Plot these on a graph similar to that in Figure 14. Explain the trend in boiling points for these compounds.

Note on problems 6 to 9. In these problems you are asked to explain the differences in T_g and T_m values for different polymers in terms of the intermolecular forces which act between the polymer chains. These problems should therefore be attempted after you have studied **Section 5.4** which is where T_g and T_m are introduced.

6 Poly(propene), poly(chloroethene) and poly(etheneitrile) all have side chains of similar size.

$$\left(\!\!\begin{array}{c} CH-CH_2 \\ | \\ X \end{array}\!\!\right)_n \qquad \text{where X} = CH_3, Cl \text{ or } C\equiv N$$

Explain the differences in their T_g values in terms of the intermolecular forces in the polymers.

	T_g/K
poly(propene) (X = CH_3)	258
poly(chloroethene) (X = Cl)	353
poly(ethenenitrile) (X = C≡N)	378

7   **a**   Which of the polymers below would you expect to have the higher T_g value?

$$\text{i} \quad \left(\!\!-CH_2-CH_2\!-\!\right)_{\!n} \qquad \text{ii} \quad \left(\!\!-CH_2O\!-\!\right)_{\!n}$$

   b   Explain your answer in terms of the intermolecular forces in the polymers.

8   Use your knowledge of intermolecular forces to explain the difference in T_g values for poly(caprolactone) and poly(caprolactam), better known as nylon-6.

		T_g/K
poly(caprolactone)	$\left(\!\!-(CH_2)_5COO\!-\!\right)_{\!n}$	213
poly(caprolactam)	$\left(\!\!-(CH_2)_5CONH\!-\!\right)_{\!n}$	333

9   You may find it helpful to make models to assist you in answering this question

   a   Explain the difference in T_m values for the following two nylons by referring to their structures and the forces between the polymer chains.

		T_m/K
i	nylon-6	486
ii	nylon-11	457

   b   Explain why nylon-6,10 is more flexible than nylon-6,6.

5.4 The structure and properties of polymers

What is a polymer?

A **polymer** molecule is a long molecule made up from lots of small molecules called **monomers**. If all the monomer molecules are the same, and they are represented by the letter A, an A–A polymer forms:

... A + A + A + A ... → ... –A–A–A–A– ...

Poly(ethene) and pvc are examples of A–A polymers.

If two different monomers are used, an A–B polymer is formed, in which A and B monomers alternate along the chain:

... A + B + A + B ... → ... –A–B–A–B– ...

Nylon-6,6 and polyesters are examples of this type of A–B polymer.

Writing out the long chain in a polymer molecule is very time consuming – we need a short-hand version. See how this is done for poly(propene),

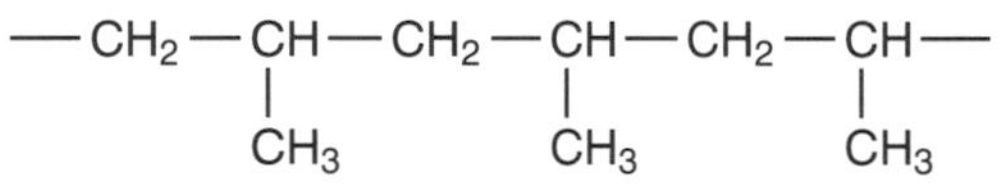

$$-CH_2-\underset{\underset{CH_3}{|}}{CH}-CH_2-\underset{\underset{CH_3}{|}}{CH}-CH_2-\underset{\underset{CH_3}{|}}{CH}-$$

In the chain, the same basic unit is continually repeated, so the chain can be abbreviated to

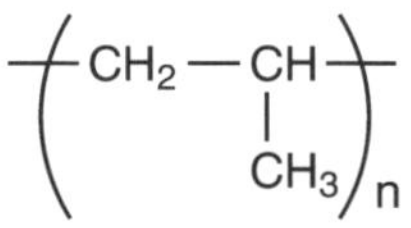

$$\left(\!\!-CH_2-\underset{\underset{CH_3}{|}}{CH}-\!\!\right)_{\!n}$$

where n is a very large number.

Elastomers, plastics and fibres

Polymer properties vary widely. Polymers which are soft and springy, which can be deformed and then go back to their original shape, are called **elastomers**. Rubber is an elastomer.

Poly(ethene) is not so springy and when it is deformed it tends to stay out of shape, undergoing permanent or plastic deformation. Substances like this are called **plastics**.

Stronger polymers, which do not deform easily, are just what you want for making clothing materials. Some can be made into strong, thin threads which can then be woven together. These polymers, such as nylon, are called **fibres**.

Poly(propene) is on the edge of the plastic/fibre boundary. It can be used as a plastic like poly(ethene), but it can also be made into a fibre for use in carpets.

What decides the properties of a polymer?

The physical properties of a polymer, such as its strength and flexibility, are decided by the characteristics of its molecules. The particularly important characteristics are:

- *chain length*: in general, the longer the chains the stronger the polymer
- *side groups*: polar side groups give stronger attraction between polymer chains, making the polymer stronger
- *branching*: straight, unbranched chains can pack together more closely than highly branched chains, giving polymers that are more crystalline and therefore stronger
- *cross-linking*: if polymer chains are linked together extensively by covalent bonds, the polymer is harder and more difficult to melt. Thermosetting polymers have extensive cross-linking.

Thermoplastics and thermosets

We can put polymers into two groups, depending on how they behave when they are heated.

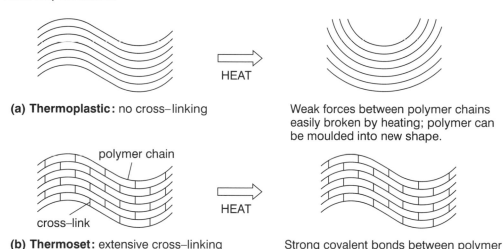

(a) **Thermoplastic**: no cross–linking

Weak forces between polymer chains easily broken by heating; polymer can be moulded into new shape.

(b) **Thermoset**: extensive cross–linking

Strong covalent bonds between polymer chains cannot be easily broken; polymer keeps shape on heating.

Figure 15 Thermoplastics and thermosets

Thermoplastics are polymers without cross-links between the chains. The intermolecular forces between the chains are much weaker than the covalent cross-links in a thermoset and so the attractive forces can be broken by warming. The chains can then move relative to one another (Figure 15a).

When this happens the polymer changes its shape – it is deformed – and thermoplastics can be moulded. When the polymer is cool the intermolecular forces reform, but between different atoms.

Over 80% of all the plastics produced are thermoplastics. This includes all poly(alkenes) such as poly(ethene), and polyamides like nylon.

Thermosetting polymers or **thermosets** have extensive cross-linking. The cross-links prevent the chains moving very much relative to one another, so the polymer stays set in the same shape when it is heated (Figure 15b).

Bakelite is an example of a thermosetting polymer. It is made by heating together phenol and methanal, to form a network which is cross-linked by strong covalent bonds (Figure 16).

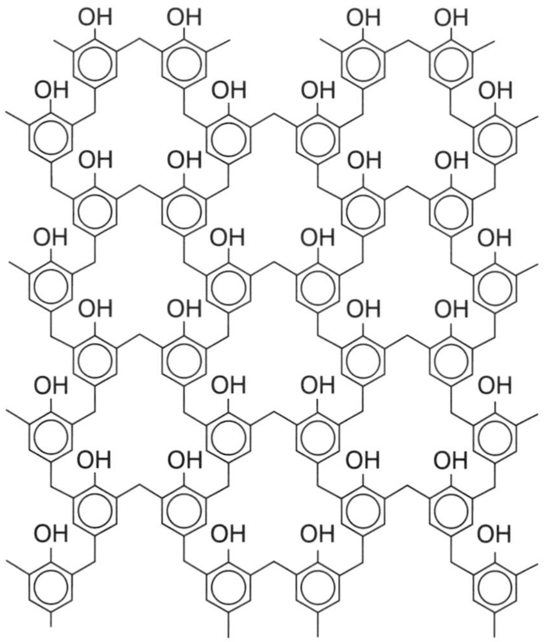

Figure 16 The structure of Bakelite, a thermosetting polymer. (This is only one possible arrangement)

This huge cross-linked structure would require a massive amount of energy to break it down. Instead of melting when it is heated, Bakelite chars and decomposes.

The structure would also have to be broken down into much smaller pieces before it could dissolve. Bakelite is a very insoluble substance because the energy needed to do this is not normally available.

Bakelite was the first synthetic polymer to be made. It is still used occasionally, for example in the brown plastic used in some electrical insulators.

Longer chains make stronger polymers

In general, the longer the chains, the stronger the polymer. However, it is not quite as simple as this. A critical length has to be reached before strength increases. This length is different for different polymers. For hydrocarbon polymers like poly(ethene), an average of at least 100 repeating units is necessary. Polymers like nylon may need only 40 repeating units.

Tensile strength is a measure of how much you have to pull on a sample of polymer before it snaps. Figure 17 shows how tensile strength and chain length are related for a typical polymer whose chains are arranged in a random way. You get a reminder of poly(ethene)'s rather limited tensile strength whenever you overload your polythene carrier bag with heavy shopping.

There are two factors which lead to this increase in tensile strength of a polymer with increasing chain length:
- longer chains are more tangled together
- when chains are longer they have more points of contact with chains of neighbouring polymer molecules. There are, therefore, more **intermolecular forces** to hold the chains to one another. (These attractive forces *between molecules* are quite weak for poly(ethene) but are much stronger for some other polymers, such as nylon.)

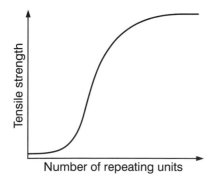

Figure 17 The relationship between tensile strength and chain length for a polymer

There is so much tangling, and there are so many intermolecular forces between long polymer molecules, that it is difficult for each molecule to move. It's like trying to pull a single piece of spaghetti out of a bowlfull.

Crystalline polymers

You probably think of *crystals* as regular flat sided solids, such as copper(II) sulphate and sodium chloride crystals. These crystals have a regular shape externally because, inside, the ions are packed in a very ordered way.

Polymer chemists use the term **crystalline** to describe *the areas in a polymer* where the chains are closely packed in a regular way.

Many polymers contain mixtures of both crystalline (ordered) regions and amorphous (random) regions, where the chains are further apart and have more freedom of movement. Any one polymer chain may be involved in both crystalline and amorphous regions along its length. Figure 18 shows the crystalline and amorphous regions in a polymer.

The polymers most likely to contain crystalline regions are the ones with regular chain structures, such as isotactic poly(propene), and without bulky side groups or extensive chain branching, such as high density poly(ethene).

The percentage of crystallinity in a polymer is very important in determining its properties. The more crystalline the polymer, the stronger and less flexible it becomes.

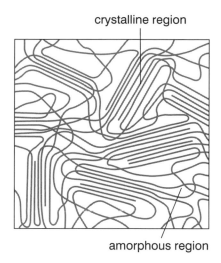

Figure 18 *Crystalline and amorphous regions of a polymer*

Cold-drawing

All polymers contain some crystalline regions. When the polymer is stretched (cold-drawn) a *neck* forms (see Figure 19). In the neck the polymer chains are lined up to form a more crystalline region. The process carries on until all the polymer, except the ends which are being pulled, has aligned chains.

Cold-drawing leads to a large increase in the polymer's strength. It is used to produce tough fibres.

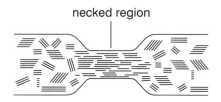

Figure 19 *When a polymer is cold-drawn, a neck forms, in which the polymer chains become aligned*

How are polymers affected by temperature changes?

When you heat solids made of small molecules they simply melt to form free flowing liquids. Carry on heating and eventually the liquid boils.

With polymers it is not so simple. You may have seen a piece of rubber tubing cooled in liquid nitrogen and then hit with a hammer. The rubber shatters – it has become brittle or **glassy**. Or, you may have used a margarine tub for storing food in the freezer. Some of these tubs are made of poly(propene) which becomes brittle at around –10 °C. When you try to get the frozen food out, the tub splits apart; quite unlike a similar tub from the fridge.

The structure of many polymers is a mixture of ordered (crystalline) regions and random (amorphous regions. In the glassy state the tangled polymer chains in the amorphous regions are 'frozen' so that easy movement of the chains is not possible. If the polymer has to change shape it does so by breaking.

If you heat up the glassy material, the polymer chains will reach a temperature at which they can move relative to each other. This temperature is called the **glass transition temperature (T_g)**. When the polymer is warmer than this, it becomes flexible and shows the typical plastic properties we expect of polymers.

On further heating, the **melting temperature (T_m)** is reached, the crystalline areas break down and the polymer becomes a viscous fluid.

These processes are reversible for thermoplastics. A graph of strength against temperature is shown in Figure 20 on the next page.

Figure 20 A graph of strength against temperature for a typical part-crystalline polymer

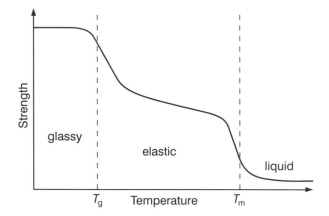

Today's polymers are designed to have T_g and T_m values which are suitable for the particular properties needed by a manufacturer.

Matching polymer properties to needs

Different uses require polymers with different glass transition temperatures (T_g). Two important ways of changing T_g involve *copolymerisation* and *plasticisers*.

Copolymerisation

Poly(propene) has a T_g of about –10 °C. For some applications a polymer is needed which has the same general properties as poly(propene) but with a lower T_g. One way to make this material is to add some ethene to the propene during the polymerisation process. This is called **copolymerisation**. Both monomers become incorporated into the final polymer.

A section of the copolymer chain could look like this:

$$-CH_2-CH-CH_2-CH-CH_2-CH_2-CH_2-CH-CH_2-CH-$$
$$\quad\quad\;\; CH_3\quad\quad\; CH_3\quad\quad\quad\quad\quad\quad CH_3\quad\quad CH_3$$

Plasticisers

Pure poly(chloroethene) – commonly called pvc – has a T_g of about 80 °C. It is rigid and quite brittle at room temperature. This is the form used to make items such as drain pipes. It is sometimes called *unplasticised* pvc, or upvc.

If it is to be made more flexible, the T_g value must be lowered. One way of doing this is to copolymerise the chloroethene with ethenyl ethanoate.

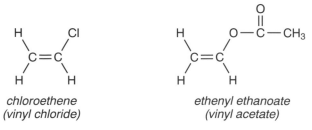

chloroethene   *ethenyl ethanoate*
(vinyl chloride)   *(vinyl acetate)*

Another way is to use a kind of 'molecular lubricant'. This allows the pvc chains to slide around on one another more easily. Such a substance is called a **plasticiser**. Figure 21 shows a plasticiser in place between polymer chains.

Plasticisers have to be chosen carefully so that they are compatible with the polymer. Di-(2-ethylhexyl)hexanedioate is commonly used as a plasticiser for pvc. However, there is now evidence that some plasticisers added to pvc film used for wrapping food can dissolve into fatty food and may be harmful to health.

(a) Unplasticised
Pvc chains attract strongly and cannot slide past each other. Polymer is rigid.

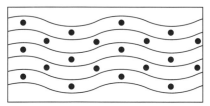

(b) Plasticised
Plasticiser molecules push pvc chains apart and help them slide. Polymer is flexible.

Figure 21

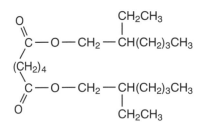

di-(2-ethylhexyl)hexanedioate

PROBLEMS FOR 5.4

1 In the section of the copolymer chain shown on page 86, how many monomer units are there of
  a propene?
  b ethene?

2 Look back at the section on plasticisers on page 86. Draw a section of the polymer chain for the chloroethene/ethenyl ethanoate copolymer.

3 Try stretching a rectangle (approx 10 cm by 5 cm) of poly(ethene) cut from a thin, clean food bag. Stretch the film slowly and place it against your lips. You should find that the film becomes warm. You will also see that the film becomes narrower. You may see opaque regions appearing in the film and it will gradually become more difficult to stretch.

    Explain your observations in terms of what is happening to the poly(ethene) molecules.

4 The range of polymers available to designers and manufacturers is now enormous. Select from those listed below, the most appropriate polymer for each of the following:
  a cassette recording tape
  b cycling shorts or leotard
  c food container for use from freezer to microwave
  d compact disc
  e climbing rope
  f body armour for the police or security services.

Polymer	Properties
poly(ester)	tough, resistant to stretching and tearing, not affected by low temperatures, higher melting point than many polymers, thermoplastic
poly(aramid)	weight for weight has a much higher tensile strength than steel, can be moulded and made into fibres, very resistant to high temperatures and combustion.
nylon	a thermoplastic with a high tensile strength which is capable of absorbing energy by stretching under stress.
poly(carbonate)	can be rapidly moulded with microscopic precision and to a very high degree of flatness without built-in stress.
poly(urethane)	very elastic, versatile polymer which can be made into fibres

5 Flexibility occurs when polymer chains are not tightly packed and some freedom of movement is possible. Bulky side groups, for example, reduce flexibility by preventing chains moving past each other. They also limit the amount of rotation which can take place about chain bonds. Strong intermolecular forces between chains limit movement and reduce flexibility. So too does the presence of groups in the polymer which make the chain rigid.

  a Explain why poly(ethene) has a lower T_g value than polymers with side groups.

  b Explain the difference between the T_g values of the pairs of polymers in **i**, **ii** and **iii** below.

T_g/K

		T_g/K
i	A	302
	B	355
ii	A	243
	B	378
iii	A	373
	B	408

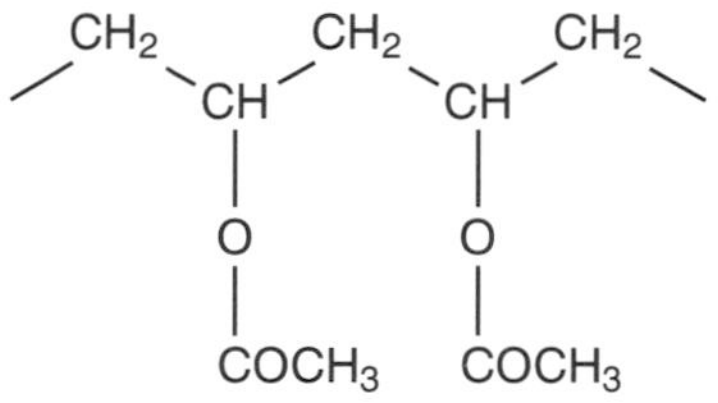

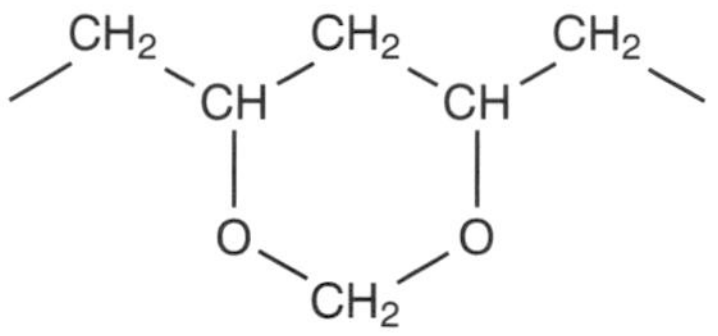

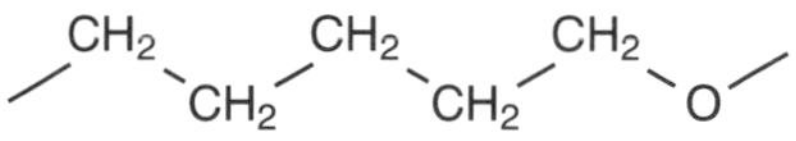

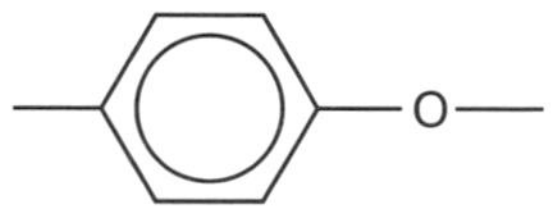

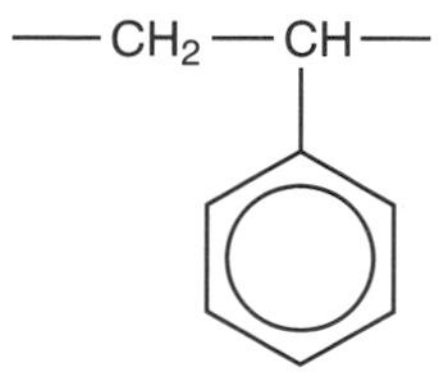

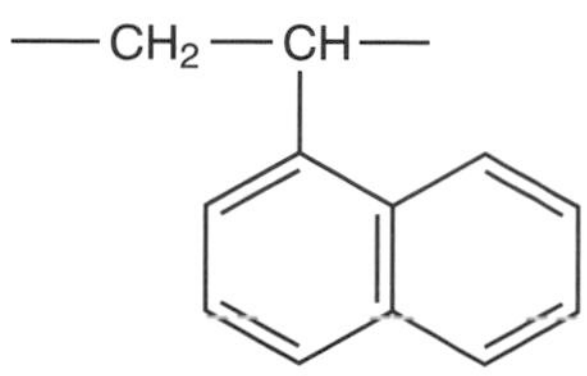

5.5 *Water: hydrogen bonding in action*

Unusual properties

Many of the properties of water are unusual when we compare them with those of other similar compounds of low molecular mass. In particular water shows

- unusually high values for boiling point and enthalpy change of vaporisation
- a greater specific heating capacity than almost any other liquid
- a *decrease* in density when it freezes.

The graph in Figure 22 shows how the boiling points of hydrides change as you go down a Group in the Periodic Table.

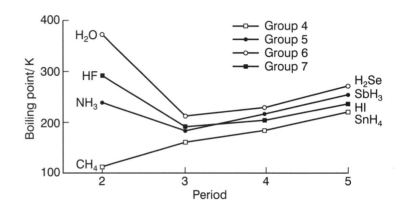

Figure 22 Variation in the boiling points of the hydrides of some Group 4, 5, 6 and 7 elements

The data for Group 4 (CH_4 to SnH_4) show how we might *expect* the boiling points to behave. The pattern corresponds to what we know about instantaneous dipole – induced dipole intermolecular forces and the polarisability of atoms. (These were discussed in **Section 5.2**) The points for H_2O, NH_3 and HF are clearly out of line.

There is a similar situation for the enthalpy changes of vaporisation as shown in Figure 23.

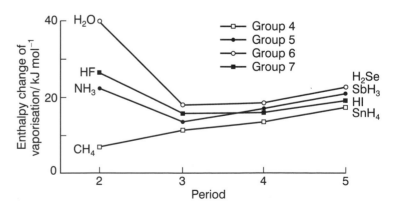

Figure 23 Variation in the enthalpy changes of vaporisation of some Group 4, 5, 6 and 7 elements

The behaviour of water, ammonia and hydrogen fluoride can be explained in terms of **hydrogen bonding.** Hydrogen bonding holds molecules together more strongly than other types of intermolecular forces. The hydrogen bonding in water is particularly strong because there are two lone pairs of electrons *and* two positively charged hydrogen atoms per oxygen atom, so intermolecular interactions are maximised. This is shown in Figure 24; note that the four groups around each O atom are actually arranged tetrahedrally in three dimensions.

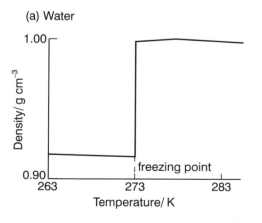

Figure 24 Hydrogen bonding between water molecules

Water molecules are close together in the liquid state, but they become widely separated in the gaseous state. For this to happen, the strong intermolecular hydrogen bonding in the liquid has to be overcome. The **enthalpy change of vaporisation** measures the energy we have to put into the liquid to overcome the intermolecular forces and turn one mole of molecules from liquid to vapour.

Specific heating capacities

Water has a high specific heating capacity; it takes a lot of energy to raise its temperature. Its value, together with those for some other substances, is given in Table 6.

To raise the temperature of 1 g of copper by 1 K requires 0.39 J. To do the same for water requires over ten times this quantity of energy. Water is an excellent substance for absorbing and storing energy, which is just as well for life in and around the oceans – and for the manufacturers of hot water bottles!

Water's unusually high specific heating capacity is another consequence of its hydrogen bonding. A lot of energy is used in overcoming hydrogen bonding within clusters of water molecules. This energy is not available for increasing the molecules' kinetic energies; it is not available for raising the temperature. For liquids where there is no hydrogen bonding between molecules, a greater proportion of the energy can be used for increasing the molecules' kinetic energies, and so raising the temperature; thus such liquids have a lower heating capacities.

Substance		Specific heating capacity/$Jg^{-1}K^{-1}$
water,	$H_2O(l)$	4.18
ethanol,	$C_2H_5OH(l)$	2.43
heptane,	C_7H_{16}	2.05
copper,	$Cu(s)$	0.39
mercury,	$Hg(l)$	0.14

Table 6 Some specific heating capacities

The density of water and ice

If we take a tube of water at room temperature and cool it to 4 °C (277 K), the water contracts. Most substances contract on cooling, so this is not unexpected. What is unusual is what happens next; the water *expands* as its temperature approaches 0 °C (273 K), and there is a further expansion as water freezes.

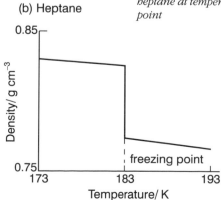

Figure 25 Density changes in water and heptane at temperatures around the freezing point

Figure 25a shows how the *density* of water changes with temperature around the freezing point. A similar graph for heptane is shown in Figure 25b. The changes in the density of heptane are typical of most solids and liquids. The changes for water are very unusual, and are due to the effects of hydrogen bonding.

The structure of ice is shown in Figure 26. The arrangement of water molecules maximises the hydrogen bonding between them, but leads to a very 'open' structure with large spaces in it. Therefore, the density of ice at 273 K is less than the density of water at the same temperature. When ice melts, the open structure collapses and water molecules fall into some of the open spaces, thus giving water a greater density than the ice it came from.

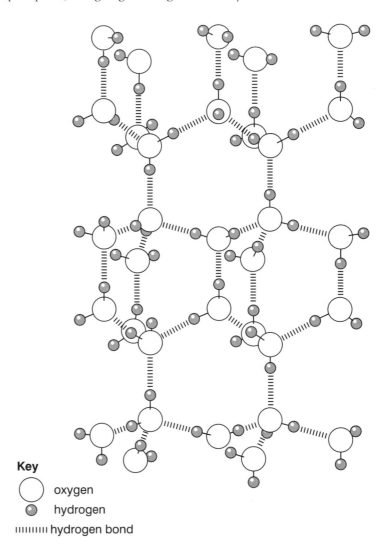

Key

◯ oxygen

⬤ hydrogen

⁞⁞⁞⁞⁞⁞⁞ hydrogen bond

Figure 26 The arrangement of water molecules in ice

When water is cooled, it seems that regions in the liquid begin to adopt the open structure of ice before the freezing point at 273 K is reached. This explains why the fall in density begins as the liquid is cooled below 277 K.

The lower density of ice is bad news for people who fail to insulate exposed water pipes in cold weather or who forget to add anti-freeze to the coolant system of their car. But it is good news for fish. Icebergs float and ponds freeze from the top down. The layer of surface ice insulates the water underneath. If ice sank there would be no insulation; ponds and lakes would be more likely to freeze solid, killing fish and other creatures. An occasional burst pipe is a small price to pay for the environmental advantages of the expansion of freezing water.

1 a Why does the density of most solids or liquids decrease with increasing temperature?

b Explain the following properties of water in terms of hydrogen bonding:

  i the greater density of liquid water compared with ice;

  ii its relatively high boiling point;

  iii its relatively high specific heating capacity;

  iv the fact that a needle can float on the surface of water. (You can try this for yourself. Float a piece of tissue on the surface of some water in a bowl. Place a needle carefully on top of the tissue. After a few minutes the tissue should sink, leaving the needle floating on the surface until it is disturbed.)

5.6 *Bonding, structure and properties: a summary*

The properties of substances are decided by their bonding and structure. Bonding means the way atoms are held together. Structure means the way the atoms are arranged relative to one another. You have already met the major types of structure in different parts of this course. They are summarised in Figure 27.

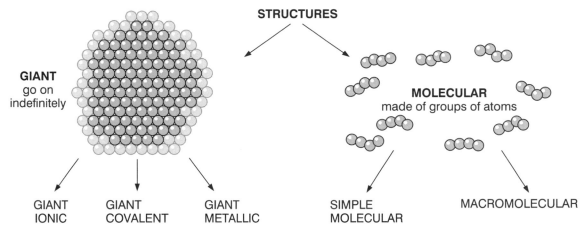

Figure 27 A summary of the structures of substances

How do bonding and structure decide properties?

The properties of a solid substance are decided by three main factors:

- *the type of particles it contains*; these may be atoms, ions or molecules. For example, if the substance contains ions (as in giant ionic substances like sodium chloride), it will conduct electricity when molten or dissolved. If a substance contains ions or polar molecules, it may dissolve in water.
- *the way the particles are bonded together*; this may be ionic, covalent, metallic or weak intermolecular bonds. The stronger the bonds, the higher the melting and boiling point of the substance, and the greater its hardness. For example, silica, SiO_2, has strong covalent bonds linking every atom to several others in a giant structure that goes on indefinitely. This makes the atoms difficult to separate; silica is hard, and very difficult to melt. On the other hand, although carbon dioxide, CO_2, has strong covalent bonds between C and O atoms – so it doesn't decompose into carbon and oxygen – it has only weak intermolecular forces between each CO_2 molecule. This makes the molecules easy to separate, so CO_2 has a low melting and boiling point and is a gas at room temperature.

- *the way the particles are arranged relative to one another* (the structure); the particles may be arranged in 1-dimensional chains (like poly(ethene)), 2-dimensional sheets (like clays) and many different kinds of 3-dimensional arrangements. For example, clays with their 2-dimensional sheets are flaky and slippery, while silica with its 3-dimensional giant structure is very hard and gritty.

Table 7 on page 93 summarises the different types of structures and their main properties. It also tells you where you can find out more details about each type.

Structure and the Periodic Table

As you go across a Period of the Periodic Table, there is a trend in the structures of the elements. There are similar trends in the structures of oxides and chlorides. These trends are summarized in Table 8 for Period 3 (sodium to argon).

Group	1	2	3	4	5	6	7	0
Element	Na	Mg	Al	Si	P	S	Cl	Ar
Structure of element	←——giant metallic ——→			giant covalent	←——————simple molecular ——————→			
Structure of oxide	←———— giant ionic ————→			giant covalent	←——simple molecular ——→			
Structure of chloride	←————giant ionic ————→		←——————————simple molecular ——————————→					

Table 8 Trends in structure in Period 3

The trends in *structure* in the oxides and chlorides are closely linked to trends in their *properties*. Here are two examples.

Acid-base properties of oxides

Table 9 shows the trend in the acid-base character of the oxides of the elements of Period 3.

Group	1	2	3	4	5	6	7	0
Formula of oxide	Na_2O	MgO	Al_2O_3	SiO_2	P_4O_{10} P_4O_6	SO_3 SO_2	Cl_2O_7 Cl_2O	—
Acid-base character of oxide	←——basic ——→		ampho- teric	←——————————— acidic ———————————→				

Table 9 The acid-base character of oxides of Period 3 elements

If you compare Table 9 with Table 8, you can see how the acid-base character of oxides is linked to their structure. The general pattern is that oxides with giant ionic structures are basic, whereas oxides with simple covalent structures are acidic. The bonding in aluminium oxide has some covalent character, so Al_2O_3 can behave as an acid or as a base – it is *amphoteric*.

	GIANT			MOLECULAR	
	Ionic	**Covalent**	**Metallic**	**Macromolecular**	**Simple molecular**
What substances have this type of structure?	compounds of metals with non-metals	some elements in Group 4 and some of their compounds	metals	polymers	some non-metal elements and some non-metal/non-metal compounds
Examples	sodium chloride, NaCl; calcium oxide, CaO	silicon(IV) oxide, SiO_2; diamond C; graphite, C	copper, Cu; iron, Fe	polythene, proteins	carbon dioxide CO_2, chlorine Cl_2, water H_2O
What type of particles does it contain?	ions	atoms	positive ions surrounded by delocalised electrons	long-chain molecules	small molecules
How are the particles bonded together?	strong ionic bonds; attraction between oppositely charged ions	strong covalent bonds; attraction of atoms' nuclei for shared electrons	strong metallic bonds; attraction of atoms' nuclei for delocalised electrons	weak intermolecular bonds between molecules; strong covalent bonds between the atoms within each molecule	weak intermolecular bonds between molecules; strong covalent bonds between the atoms within each molecule
What are the typical properties?					
Melting point and boiling point	high	very high	generally high	moderate (often decompose on heating)	low
Hardness	hard but brittle	very hard	hard but malleable	many are soft but often flexible	soft
Electrical conductivity	conduct when molten or dissolved in water; electrolytes	do not normally conduct	conduct when solid or liquid	do not normally conduct	do not conduct
Solubility in water	often soluble	insoluble	insoluble (but some react)	usually insoluble	usually insoluble, unless molecules contain groups which can hydrogen bond with water
Solubility in non-polar solvents (e.g. hexane)	insoluble	insoluble	insoluble	sometimes soluble	usually soluble
Where can you find out more?	Section 5.1	Section 11.3	Section 11.6	Section 5.4	Sections 5.2 and 5.3

Table 7 A summary of types of structure

Behaviour of chlorides in water

Table 10 shows how the chlorides of the elements of Period 3 behave when they are put into water.

Group	1	2	3	4	5	6	7	0
Formula of chloride	NaCl	$MgCl_2$	$AlCl_3$	$SiCl_4$	PCl_5 PCl_3	S_2Cl_2	Cl_2	—
What happens when the chloride is put into water?	←— dissolves —→		←——————— reacts with water ———————→ producing fumes of hydrogen chloride and acidic solution				some chlorine reacts with water	

Table 10 The behaviour in water of the chlorides of Period 3 elements

If you compare Table 10 with Table 8, you can see how the behaviour of chlorides in water is linked to their structure. The general pattern is for chlorides with giant ionic structures simply to dissolve in water, with no chemical reaction, whereas chlorides with simple covalent structures react, producing fumes of hydrogen chloride and forming acidic solutions.

The covalent chlorides are hydrolysed in water, forming hydrogen chloride and a hydroxy compound. For example, with aluminium chloride

$$AlCl_3(s) + 3H_2O(l) \rightarrow Al(OH)_3(aq) + 3HCl(aq)$$
$$\textit{aluminium hydroxide}$$

With silicon chloride
$$SiCl_4(s) + 4H_2O(l) \rightarrow Si(OH)_4(aq) + 4HCl(aq)$$
$$\textit{silicon hydroxide}$$
$$\textit{(silicic acid)}$$

PROBLEMS FOR 5.6

1 Which type of structure would you expect each of substances A to E to have?
  A A white solid which starts to soften at 200 °C and can be drawn into fibres.
  B A liquid which conducts electricity and solidifies at −39 °C.
  C A white solid which melts at 770 °C and conducts electricity when molten.
  D A hard grey solid which conducts electricity and melts at 3410 °C.
  E A white solid which melts at −190 °C.

2 Explain the following in terms of particles, structure and bonding.
  a Ionic substances do not conduct electricity when they are solid, but metallic substances do.
  b Molten ionic substances are decomposed when they conduct electricity;
  c Substances with giant covalent structures are insoluble in all solvents.
  d Substances that are gases at room temperature always contain simple molecules.

3 What type of structures would you expect each of the following substances to have in the solid state?
  a argon
  b an alloy of nickel and cobalt
  c silicon carbide
  d silk
  e rubidium bromide
  f xenon tetrafluoride

4 Table 8 shows the trends in types of structure across the Periodic Table.
  a Draw up a similar table showing the structures you would expect for Period 4 (potassium to krypton). Use a copy of the Periodic Table to help you.
  b Suggest explanations for the following:
    i silicon has a giant covalent oxide, but a molecular chloride
    ii silicon, phosphorus, sulphur and chlorine all form covalent bonds, but only silicon has a giant covalent structure
    iii aluminium has an ionic oxide, but a covalent chloride.

6

RADIATION AND MATTER

6.1 Light and electrons

Most of our understanding of the electronic structure of atoms has come from the area of science known as **spectroscopy** – the study of how light and matter interact. Before you go on to learn about this, you need to know more about the nature of light.

Chemists use two models to describe the behaviour of light; the **wave model** and the **particle model**. Neither theory fully explains all the properties of light. Some are best described by the wave model; the particle model is better for others. We choose the theory which is most appropriate to the situation.

The wave theory of light

Light is one form of electromagnetic radiation. Like all electromagnetic radiation, it behaves like a wave, with a characteristic wavelength and frequency.

A wave of light will travel the distance between two points in a certain time. It doesn't matter what kind of light it is, the time is always the same. The speed at which the wave moves, the *speed of light* (symbolised by c), is the same for all kinds of light, and indeed for all kinds of electromagnetic radiation. It has a value of $2.99 \times 10^8 \, \mathrm{m \, s^{-1}}$ when the light is travelling in a vacuum.

Like all waves, the light wave has a **wavelength** (symbol: λ) and a **frequency** (symbol: ν). Different colours of light have different wavelengths. Figure 1 illustrates this for two different waves.

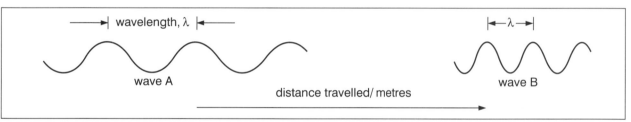

Figure 1 Wavelength measures the distance (in metres) travelled by the wave during one **cycle**. *Wave A has twice the wavelength of wave B.*

The two waves in Figure 1 also have different frequencies. Frequency tells us how many *cycles* a wave goes through every second and is measured in units of $\mathrm{s^{-1}}$ or hertz (Hz). Both waves in Figure 1 travel at the same speed, so they both travel the same *distance* per second. Wave B therefore has twice the frequency of Wave A.

Frequency and wavelength are very simply related. Wave B has twice the frequency but half the wavelength of wave A. If you multiply wavelength and frequency together you get a constant. This constant is the speed of light. So c, λ and ν are related by the very simple equation

$$c = \lambda \nu$$

When we use the term 'light' we normally mean the visible light to which our eyes respond. But visible light is only a small part of the **electromagnetic spectrum**, which includes all the different forms of electromagnetic radiation. There are other regions, for example: radio waves, ultra-violet, infra-red and γ-rays. A full version of the electromagnetic spectrum is shown in Figure 2.

Figure 2 The electromagnetic spectrum (the spectrum continues above $10^{20} \, \mathrm{Hz}$ and below $10^5 \, \mathrm{Hz}$)

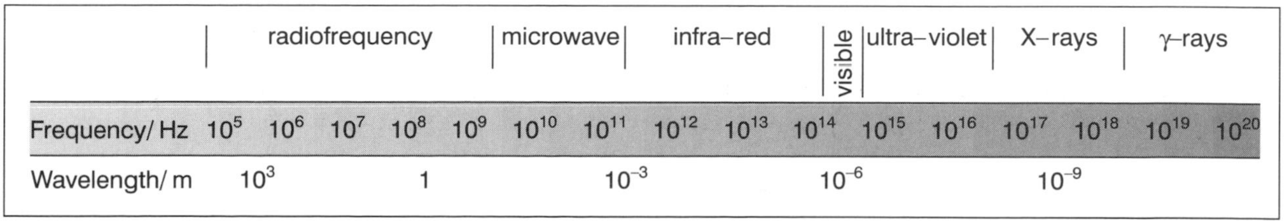

	radiofrequency		microwave	infra-red	visible	ultra-violet	X-rays	γ-rays
Frequency/Hz 10^5 10^6 10^7 10^8 10^9	10^{10} 10^{11} 10^{12} 10^{13} 10^{14} 10^{15} 10^{16} 10^{17} 10^{18} 10^{19} 10^{20}
Wavelength/m 10^3	1	10^{-3}	10^{-6}	10^{-9}

The particle theory of light

In some situations, the behaviour of light is easier to explain by thinking of it not as waves but as particles. This idea, first proposed by Albert Einstein in 1905, regards light as a stream of tiny 'packets' of energy called **photons**. The energy of the photons is related to the position of the light in the electromagnetic spectrum. For example, photons with energy 3×10^{-19} J would correspond to red light.

The two theories of light – the wave and photon models – are linked by a relationship which was proposed by Max Planck around the turn of the century:

$$E = h\nu$$

E, the energy of a photon, is equal to the frequency of the light on the wave model multiplied by a constant, h. This is known as the **Planck constant** and has a value of 6.63×10^{-34} J Hz^{-1}.

For a photon of red light with an energy of 3×10^{-19} J, the frequency would be given by

$$E = h\nu$$
$$\text{or } 3 \times 10^{-19} \text{J} = (6.63 \times 10^{-34} \text{J Hz}^{-1})\nu$$
$$\text{so, } \nu = 4.5 \times 10^{14} \text{Hz}$$

Emission spectra

Atoms can become **excited** by absorbing energy, for example, from flames, from an electric discharge or from radiation in the stratosphere or in outer space. When atoms lose energy, it is often emitted as electromagnetic radiation. This radiation is usually in the infra-red visible or ultra-violet region. The emitted light can be split up into an **emission spectrum** by passing it through a prism or a diffraction grating (see Figure 3).

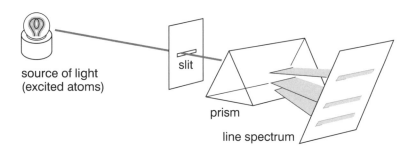

source of light
(excited atoms)

slit

prism

line spectrum

Figure 3 Obtaining a line emission spectrum

The spectrum consists of a series of lines. A *continuous spectrum* (like one which shows all the colours of the rainbow) is *not* produced because the atoms can only emit certain precise frequencies.

The sequence of lines is characteristic of each element and, like a human finger print, can be used to identify the element. The composition of stars, for example, can be determined from vast distances away by using this technique.

The spectrum of hydrogen atoms

The characteristic emission spectrum of hydrogen atoms in the ultra-violet region is shown in Figure 4. This series of lines is called the **Lyman series** after the scientist who first observed it.

The full spectrum contains other series of lines, one in the visible and several in the infra-red region. The spectrum was interpreted in 1913 by the Danish scientist, Niels Bohr. Bohr's theory explained why the hydrogen atom only emits a limited number of specific frequencies. The frequencies predicted by Bohr's theory matched extremely well with the observed lines.

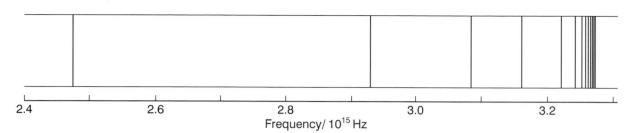

Figure 4 The Lyman series in the hydrogen atom emission spectrum

How Bohr's theory explained emission spectra

The basic idea behind Neils Bohr's theory was this. Atomic emission spectra are caused by electrons in atoms moving between different energy levels (we now call these energy levels *shells* and *sub-shells* – see **Sections 2.3** and **2.4**). When an atom is excited, electrons jump into higher energy levels. Later, they drop back into lower levels again – and emit the extra energy as electromagnetic radiation, which gives an emission spectrum.

Bohr's theory not only explained how we get emission spectra, it also gave scientists a model for the electronic structure of atoms.

However, Bohr's theory was controversial. It made use of the idea of **quantisation of energy**. This idea was new at the time, and at odds with much of what was thought about energy. But Bohr's theory predicted experimental observations so well that it lent great support to the radical new *quantum theory*.

The main points of Bohr's theory were

- the electron in the H atom is only allowed to exist in certain definite **energy levels** (Figure 5)
- a photon of light is emitted or absorbed when the electron changes from one energy level to another
- the energy of the photon is equal to the difference between the two energy levels (ΔE)
- the frequency of the emitted or absorbed light is related to ΔE by: $\Delta E = h\nu$.

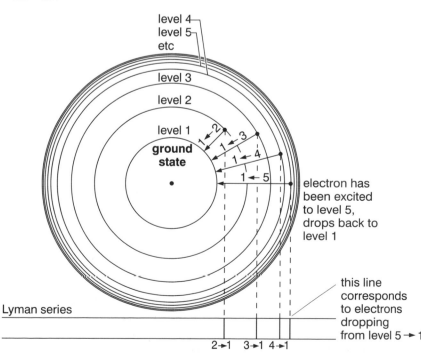

Figure 5 How the Lyman series is related to energy levels in the H atom

The first of these points is the one about quantisation of energy: the electron can only possess definite quantities of energy, or **quanta**. The electron's energy cannot change continuously – it is not able to change to any value, only those values that are allowed.

Figure 5 shows how Bohr's ideas explained the lines of the Lyman series. The rings represent the energy levels of the electron in the hydrogen atom. The further away from the nucleus, the higher the energy. Levels are labelled with numbers starting at 1 for the lowest level – the **ground state**.

The lines of the Lyman series correspond to changes in electronic energy from various upper levels to one common lower level, level 1. Each line corresponds to a particular energy level change, such as level 4 to level 1.

The series of lines which lies in the visible region, the Balmer series, arises from changes to level 2 from levels 3, 4, 5 …. etc.

Ionisation energy

Notice that as energy increases the levels become more closely spaced, until they converge. After this point, which corresponds to the electron breaking away from the atom, the electron is free to move around with any energy. The H atom has lost its electron and become an H^+ ion. This is **ionisation** and the energy difference between this point and the ground state is called the **ionisation energy**.

Ionisation can be represented by the equation

$$X(g) \rightarrow X^+(g) + e^-$$

where X stands for an atom of any element and e^- for an electron. Notice that X is shown as X(g), indicating that the atoms are separated from one another, in the gaseous state. In the case of hydrogen, we can work out the ionisation energy from the point where the lines of the Lyman series converge together.

Ionisation energies are useful in deciding which ions a metal will form: there is more about this in **Section 11.6**.

_____ **PROBLEMS FOR 6.1** _____

1 a A radio station in the UK transmits at a frequency of 1089 kHz. What is the wavelength of the transmission waves?
      (1 kHz = 1 kilohertz = 1000 hertz)

   b Dental X-rays have a wavelength of 3 nm. What is their frequency?
      (1 nm = 1 nanometre = 10^{-9} metres)

   c How many times more energy is possessed by the photons of dental X-rays compared with the photons of the radio station waves?

2 a Copy out the diagram of the H atom energy levels from Figure 5 and draw arrows on it to represent the energy changes which give rise to the four _lowest energy_ lines of the Balmer series.

   b Underneath your diagram draw a sketch (like the one in Figure 5 for the Lyman series) to show how these first four lines of the Balmer series would be arranged.

3 Figure 5 shows the point in the Lyman series where the lines converge. The frequency of this convergence limit may be used to calculate the ionisation energy for hydrogen. This point corresponds to the energy required to move an electron from the $n = 1$ level to the point where it is no longer attracted by the nucleus of the atom.

   a What is the name used for the energy state in which the electrons of an atom are in their lowest energy levels?

   b For hydrogen, the convergence limit occurs at a frequency of 3.27×10^{15} Hz.
      i Calculate the ionisation energy for hydrogen.
      ii Is the value that you have calculated the ionisation energy for a single atom or for a mole of atoms? Justify your answer.
      iii If necessary, convert your value to that for one mole of atoms and compare your figure with the figure from a data book.

6.2 What happens when radiation interacts with matter?

Energy interacts with matter

Electromagnetic radiation can interact with matter, transferring energy to the chemicals involved. The chemicals absorb energy, and the absorbed energy can make changes happen in the chemicals. Just what changes occur depend on

- the chemical involved
- the amount of energy involved.

For the moment, let's look at chemicals with molecular structures.

Molecules are doing energetic things all the time. They move around, they rotate, and the bonds in the molecule vibrate. The electrons in the molecule have energy too, and they can move between the different electronic energy levels.

A molecule has energy associated with several different aspects of its behaviour, including:

- energy associated with **translation** (the molecule moving around as a whole)
- energy associated with **rotation** (of the molecule as a whole)
- energy associated with **vibration** of the bonds
- energy associated with **electrons**.

These different kinds of energetic activities involve different amounts of energy – for example, making the bonds in a molecule vibrate generally involves more energy than making the molecule as a whole rotate. The energy needed increases in the general order shown in Figure 6.

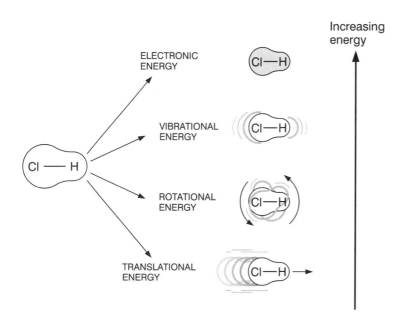

Figure 6 *An HCl molecule has energy associated with different aspects of its behaviour*

You should be familiar with the idea that electrons can occupy definite energy levels, from **Section 6.1**. The electronic energy of an atom or molecule changes when an electron moves from one level to another. We say that electronic energy is **quantised**, with fixed levels.

Now here is a crucial point: *all these other types of energy* (translational, rotational and vibrational) *are quantised too*. For example, the HCl molecule can only occupy certain fixed levels of vibrational energy (see Figure 7). When its vibrational energy changes, it moves to a new, fixed energy level, in the same way that changes in electronic energy involve moves to new energy levels.

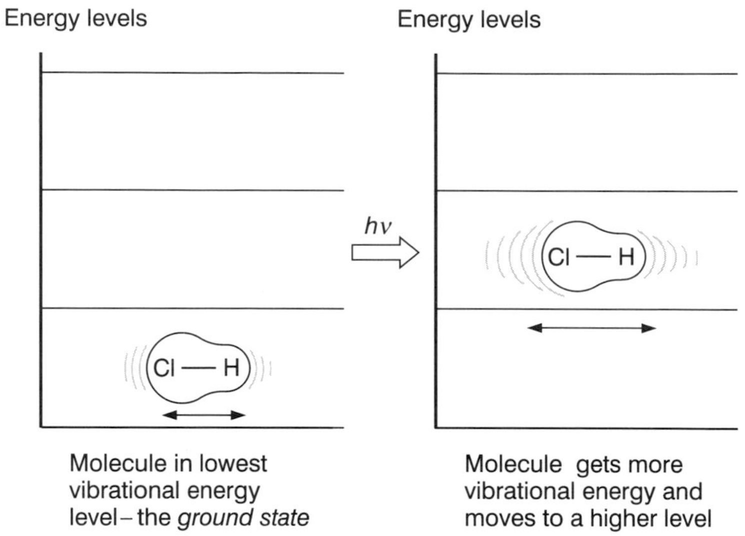

Energy levels   Energy levels

hv

Molecule in lowest
vibrational energy
level – the *ground state*

Molecule gets more
vibrational energy and
moves to a higher level

Figure 7 Vibrational energy changes in an HCl molecule

It happens that the gap between vibrational energy levels for HCl corresponds to the energy of a photon of infra-red radiation. So infra-red radiation is able to make HCl molecules – and other molecules too – increase their vibrational energy and move from one vibrational energy level to a higher one. As a result, we find that HCl absorbs infra-red radiation of a certain frequency; the frequency needed to make it vibrate in a particular way. There is more about this in the section on infra-red spectroscopy, **Section 6.7**.

Different energy changes for different parts of the spectrum

The spacing between **vibrational energy levels** corresponds to the infra-red part of the spectrum. We sense infra-red radiation as heat: when you feel the warmth radiated by a fire, you are sensing infra-red radiation. The radiation makes bonds in the chemicals in your skin vibrate more energetically, and this is why you feel warmer.

To rotate molecules requires less energy than to make their bonds vibrate. Therefore, changes in **rotational energy** correspond to a lower energy, lower frequency part of the electromagnetic spectrum, namely the microwave region. Microwaves are the basis of microwave cookery (as you will soon see).

The spacings between **translational energy levels** are even smaller. So small, in fact, that we can treat translational energy as being continuous.

On the other hand, to make **electronic** changes occur in a molecule requires *higher* energy than for vibrational changes. Electronic changes involve electrons jumping to higher electronic energy levels within the molecule, and to make this happen, energy corresponding to the **visible and ultra-violet** parts of the spectrum is needed.

Table 1 summarises the changes mentioned above. Other changes can occur with energy of different frequencies. For example, γ-rays cause changes in the energy levels within the nuclei of atoms. However, the three changes shown in the table are the only ones that need concern us here.

How does a microwave cooker work?

Microwave cookers can cook food very quickly. They do this by using microwave radiation which can penetrate deep into the food.

Microwave radiation corresponds to the energy of molecular rotations. The microwave radiation used in cookers is chosen to have a precise frequency, 2.45×10^9 Hz. This corresponds to the energy needed to rotate *water* molecules. The radiation is therefore absorbed by water molecules in the food, making them move to a higher rotational energy level. They rotate faster, and when they bump into each other they do so more energetically. In this way the extra energy gets shared around and the water generally becomes more energetic. In other words, its temperature rises.

Since most food has a high water content, microwaves provide an excellent way of heating food quickly. However, like most electromagnetic radiation, microwaves are reflected by metal, so the food must be heated in a non-metallic container – usually plastic. Figure 8 shows the inside of a typical microwave cooker.

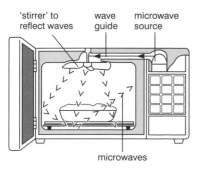

'stirrer' to   wave   microwave
reflect waves   guide   source

microwaves

Figure 8 How a microwave cooker works

Change occurring	Size of energy change/J	Type of radiation absorbed
change of rotational energy level	1×10^{-22} to 1×10^{-20}	microwave
change of vibrational energy level	1×10^{-20} to 1×10^{-19}	infra-red
change of electronic energy level	1×10^{-19} to 1×10^{-16}	visible and ultra-violet

Table 1 Summary of molecular energy changes

Notice that the table gives *ranges* of energy. The particular value of the energy change depends on the substance involved. Different substances have different chemical structures, and this gives them different energy levels. For example, the C—F bond is stronger than the C—Br bond, so it takes more energy to make a C—F bond vibrate than to make a C—Br bond vibrate. So compounds containing C—F bonds absorb infra-red of a higher energy than compounds containing C—Br bonds.

What kind of electronic changes occur when substances absorb ultra-violet radiation?

Electrons in atoms occupy definite energy levels. The gaps between these electronic energy levels correspond to the energy of visible and ultra-violet light. So when an atom absorbs photons of visible or ultra-violet radiation, electrons jump to higher energy levels.

The same kind of thing happens with molecules. The electrons in a molecule like Cl_2 occupy definite energy levels. This is true of the electrons that bond the atoms together, and also of the other, non-bonding electrons. The outer shell electrons are in the highest energy levels, and so can move most easily to higher levels still.

When the molecule absorbs radiation, one of three things can happen, depending on the amount of energy involved:

1   Electrons may be **excited** to a higher energy level. Later the electrons fall back to a lower energy level, and nothing permanent has happened to the molecule. Chlorine owes its green colour to this kind of excitation: it happens that Cl_2 absorbs visible light of such a frequency that the remaining, unabsorbed light looks green.

2   If higher energy radiation is used, then the molecule may absorb so much energy that the bonding electrons can no longer act to bond the atoms together. The two atoms in the molecule break apart. This is **dissociation**, or more precisely **photodissociation** because it is caused by light. As a result, radicals (**Section 6.3**) are formed. Radicals are very reactive, and their formation may lead to further chemical reactions.

3   With very high energy photons, the molecules may acquire so much energy that an electron is able to leave it: the molecule has **ionised**.
Figure 9 illustrates these three possibilities:

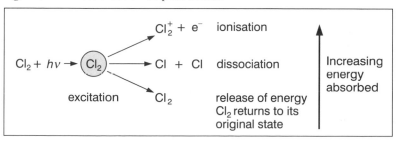

Figure 9 When Cl_2 molecules absorb radiation, they become excited. The excited molecules may then ionise, or dissociate, or just release the energy

_____ **PROBLEMS FOR 6.2** _____

1 To change one mole of molecular HCl from the lowest vibrational energy level (ground state) to the next vibrational level requires 32.7 kJ.

a How much energy, in joules, would be needed to change *one molecule* of HCl from the ground state to the next vibrational level?

b What frequency of radiation would be absorbed when HCl absorbs energy in this way? What *type* of radiation is this?

2 You want to heat up a cup of coffee in a microwave cooker. The cooker uses radiation of frequency 2.45×10^9 Hz. The cup contains $150\,cm^3$ of coffee, which is mainly water.

a How much energy is transferred to the water by each photon of microwave radiation? ($h = 6.63 \times 10^{-34}$ J Hz^{-1})

b How much energy is transferred to the water by each mole of photons?

c How many moles of photons would be needed to raise the temperature of the water by 30 °C?

(Specific heating capacity of water = 4.2 J g^{-1} K^{-1}. This is the energy needed to raise the temperature of 1 g of water by 1 °C.)

3 Ozone in the troposphere is a secondary atmospheric pollutant produced by the reaction of oxygen radicals with dioxygen.
$$O(g) + O_2(g) \rightarrow O_3(g)$$
The oxygen radicals are produced by the photodissociation of nitrogen dioxide in sunlight.
$$NO_2(g) \rightarrow NO^\bullet(g) + O(g) \quad \Delta H = +214 \text{ kJ mol}^{-1}$$
Ozone production continues for as long as the light is sufficient because the nitrogen monoxide, NO, formed can react with dioxygen to generate more nitrogen dioxide. The build up of ozone and other secondary pollutants leads to the formation of photochemical smog.

a Calculate the minimum energy required to break one N—O bond in a single molecule of NO_2.

b What is the minimum frequency of radiation that would break this bond?

c What type of radiation is this?

6.3 *Radiation and radicals*

Ways of breaking bonds

All reactions involve the breaking and remaking of bonds. Breaking bonds is sometimes called **bond fission**. The way bonds break has an important influence on reactions.

In a covalent bond, a pair of electrons is shared between two atoms. For example, in the HCl molecule:

H **:** Cl

When the bond breaks, these electrons get redistributed between the two atoms. There are two ways this can happen.

Heterolytic fission

In this type of fission, when the bond breaks both of the shared electrons go to just one of the atoms. This atom becomes negatively charged, because it has one more electron than it has protons. The other atom becomes positively charged.

In the case of HCl

H **:** Cl $\longrightarrow$ H$^+$ + **:** Cl$^-$

Heterolytic fission is more common where a bond is already *polar* (**Section 3.1**). For example, bromomethane contains a polar C—Br bond, and under certain conditions this can break heterolytically:

$$\begin{array}{ccc} H & & H \\ | & & | \\ H-C-Br & \longrightarrow & H-C^+ \quad + \quad Br^- \\ | & & | \\ H & & H \end{array}$$

Homolytic fission

In this type of bond fission, one of the two shared electrons goes to each atom. In the case of Br_2

$$Br : Br \longrightarrow Br^{\bullet} + Br^{\bullet}$$

The dot beside each atom shows the unpaired electron that the atom has gained from the shared pair in the bond. The atoms have no overall electric charge because they have gone back to the electronic structure they had before they shared their electrons to form the bond.

The unpaired electron has a strong tendency to pair up again with another electron from another substance. These highly reactive atoms or groups of atoms with unpaired electrons are called **radicals**.

Another example of radical formation is when a C—H bond in methane is broken:

$$H - \underset{\underset{H}{|}}{\overset{\overset{H}{|}}{C}} - H \longrightarrow H - \underset{\underset{H}{|}}{\overset{\overset{H}{|}}{C}}^{\bullet} + H^{\bullet}$$

Radicals are most commonly formed when the bond being broken has electrons that are more or less equally shared, but many polar bonds can break this way too – particularly when the reaction is taking place in the gas phase and in the presence of light.

More about radicals

The key feature of a radical is its unshared electron, which makes it particularly reactive. The unpaired electron is shown as a dot, although of course it is just one of the outer shell electrons.

$: \overset{\bullet\bullet}{\underset{\bullet\bullet}{Cl}} \bullet$	$Cl \bullet$	Cl
showing all outer shell electrons	showing unpaired electron only	showing none of the electrons

Sometimes, the dot is omitted altogether and a chlorine radical is simply represented by the symbol for the atom. We have used this less cluttered approach in **Chemical Storylines**.

Some radicals are a bit more subdued in their reactivity. This allows them to live long enough to behave as ordinary molecules. Nitrogen monoxide, NO, is an example. One way of showing the arrangement of electrons in NO is

$$\bullet \overset{\bullet\bullet}{\underset{\times}{N}} \overset{\bullet\,\times\times}{\underset{\times\,\times\times}{O}}$$

Notice that the N atom breaks the normal 'full outer shell' rule by having only seven electrons in its outer shell. The odd, unpaired electron makes NO a radical.

Some radicals have more than one unpaired electron. for example, the oxygen atom, O, has *two* unpaired electrons and is a **biradical**:

$\bullet \overset{\bullet\bullet}{\underset{\bullet\bullet}{O}} \bullet$	$\bullet O \bullet$
showing all outer shell electrons	showing unpaired electrons only

You will be able to explain why the two electrons in oxygen prefer not to pair with each other when you have studied **The Steel Story**.

Another, important biradical is the dioxygen molecule, O_2. It has two unpaired electrons, one on each O atom. We can represent it as

$$\bullet O - O \bullet$$

The line between the atoms just represents some bonding: it does not imply a single bond in this case.

Like NO, dioxygen is an example of a relatively stable molecular radical. But unlike NO, you cannot explain why dioxygen is a biradical from writing out its dot-cross diagram. You would need a more sophisticated treatment of bonding than we can give here.

Radicals are reactive

Filled outer electron shells are more stable than unfilled ones. Radicals are reactive because they tend to try and fill their outer shells by grabbing an electron from another atom or molecule. For example, when a chlorine radical collides with a hydrogen molecule, the chlorine grabs an electron from the pair of electrons in the bond between the H atoms. The effect is to make a new bond between the Cl and H atoms.

$$Cl \cdot \qquad H \overset{x}{\underset{x}{\cdot}} H$$

chlorine radical   *hydrogen molecule*

The curved arrow indicates the movement of an electron. The 'tail' of the arrow shows where the electron starts and the 'head' shows where it finishes. Look carefully at the head of the arrow. It is drawn this way to show the movement of a *single* electron. (You will also meet full-headed arrows which indicate the movement of *pairs* of electrons.)

Notice that a hydrogen radical is also formed in this reaction. This, too, is highly reactive and it will combine with another molecule, once again creating a new radical – and so it goes on. It is a **chain reaction**.

The reaction between chlorine and hydrogen is a typical **photochemical** radical chain reaction. If you mix these two gases in the dark, nothing much happens: as soon as you shine ultra-violet light on them, they react explosively.

Like all radical chain reactions this reaction has three key stages: *initiation*, *propagation* and *termination*.

Initiation   Chlorine radicals are initially formed by the photodissociation of chlorine molecules
$$Cl_2 + h\nu \rightarrow Cl\bullet + Cl\bullet$$
Only a few chlorine radicals are formed, but they are so reactive that they soon react with something else. They *initiate* the reaction.

Propagation   Chlorine radicals react with hydrogen molecules. This produces hydrogen chloride molecules and hydrogen radicals. The hydrogen radicals can then go on to react with chlorine molecules.
$$Cl\bullet + H_2 \rightarrow HCl + H\bullet$$
$$H\bullet + Cl_2 \rightarrow HCl + Cl\bullet$$
These two reactions produce new radicals which keep the reaction going – they *propagate* the reaction.

Termination   Every now and then, two radicals collide with each other. This isn't common, because very few radicals are present at any one time. When it does happen the reaction chain is *terminated*, because the radicals have been taken out of circulation.
$$H\bullet + H\bullet \rightarrow H_2$$
$$Cl\bullet + Cl\bullet \rightarrow Cl_2$$
$$H\bullet + Cl\bullet \rightarrow HCl$$

The overall effect of these three stages is to convert hydrogen and chlorine into hydrogen chloride

$$H_2 + Cl_2 \rightarrow 2HCl$$

Radical chain reactions have particular features:
- they often occur in the gas phase or in a non-polar solvent

- they are often initiated by heating or by light
- they usually go very fast.

Radical chain reactions are very common. Combustion involves radical reactions, and so do most explosions. Many of the reactions that occur in the troposphere and stratosphere also involve radicals.

PROBLEMS FOR 6.3

1 Decide whether or not each of the following is a radical. You may need to draw dot-cross diagrams to help you decide. (They all obey simple bonding rules.)

a	F	c	H_2O	e	NO_2
b	Ar	d	OH	f	CH_3

2 The hydroxyl radical (OH•) is an important species in atmospheric chemistry. Reaction A shows one process in which OH• is produced. The reaction is brought about by radiation with a wavelength below 190 nm.

Reaction A $H_2O + h\nu \rightarrow H• + OH•$

OH• is very reactive and scavenges many other species which are present in the atmosphere. One set of reactions which involve stratospheric ozone is:

Reaction B $OH• + O_3 \rightarrow HO_2• + O_2$
Reaction C $HO_2• + O_3 \rightarrow OH• + 2O_2$

a   i What term is used to describe a process like Reaction A in which a molecule is split up by light?
   ii Is this splitting a homolytic or a heterolytic process?

b Which of the terms: initiation, propagation and termination are most appropriate for each of the reactions A, B and C?

c Which species other than OH• in the mechanism are radicals?

d   i Write an equation which represents the overall result of reactions B and C.
   ii What is the role of OH• in this process?

e Explain why it is important for chemists to have an accurate idea of the concentration of radicals like OH• when trying to understand the details of atmospheric chemistry.

f One process which removes OH• from the atmosphere is:
$OH• + HO_2• \rightarrow H_2O + O_2$
What name is given to a reaction like this one which removes the radicals in a reaction?

3 The creation of significant amounts of nitrogen monoxide from human activities is of concern because it is thought to lead to a loss of ozone from the atmosphere. Reactions D and E show how this loss can occur.

Reaction D $NO• + O_3 \rightarrow NO_2• + O_2$
$\Delta H = -100 \, kJ \, mol^{-1}$
Reaction E $NO_2• + O \rightarrow NO• + O_2$
$\Delta H = -192 \, kJ \, mol^{-1}$

a Name one human activity which leads to the production of a significant amount of NO.

b Are reactions D and E endothermic or exothermic?

c   i What is the overall effect of reactions D and E?
   ii What is the role of NO in this process?
   iii Calculate the value of ΔH for the overall process.

d Explain why we should be worried about the effects which might arise from the emission of small amounts of NO into the atmosphere.

4 In **Developing Fuels**, you learned about the importance of cracking and reforming reactions to modify a hydrocarbon's performance in a car engine. The details of these reactions are complex, but the effect of strongly heating ethane is well understood. Bringing a reaction about in this way is called **pyrolysis**.

The following reactions occur when ethane is pyrolysed at 620 °C:

1   $C_2H_6 \rightarrow 2CH_3•$
2   $CH_3• + C_2H_6 \rightarrow CH_4 + C_2H_5•$
3   $C_2H_5• \rightarrow C_2H_4 + H•$
4   $H• + C_2H_6 \rightarrow H_2 + C_2H_5•$
5   $2C_2H_5• \rightarrow C_2H_4 + C_2H_6$
6   $2C_2H_5• \rightarrow C_4H_{10}$
(C_2H_4 is ethene, $CH_2 = CH_2$)

a Classify these reactions under the headings: initiation, propagation and termination.

b Which of the reactions do you expect will be
   i endothermic,
   ii exothermic?
   Explain your decision.

c For which reactions are you unable to make a decision about whether they will be exothermic or endothermic? What further information would you need before you could decide?

d Which of the species in the reaction sequence are radicals? Give their names and formulae.

5 The reaction of methane (CH_4) with chlorine in the presence of sunlight proceeds via a radical chain reaction. The overall equation for the reaction is:
$CH_4 + Cl_2 \rightarrow CH_3Cl + HCl$

a Write out a possible mechanism for the reaction showing clearly which reactions correspond to the three stages: initiation, propagation and termination.

b Explain why the reaction product also contains some CH_3Cl, CH_2Cl_2, $CHCl_3$ and CCl_4.

6.4 *Where does colour come from?*

You 'see' an object because light reflected from it enters your eyes. If all the incident sunlight is reflected, the object appears white. Some objects are transparent, like glass, and allow light to pass through them. Figure 10 shows what happens if all wavelengths of visible light are transmitted or reflected.

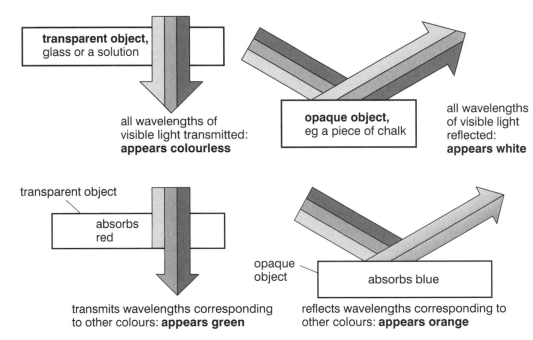

Figure 10 How light behaves with transparent and opaque objects

transparent object, glass or a solution

all wavelengths of visible light transmitted: **appears colourless**

opaque object, eg a piece of chalk

all wavelengths of visible light reflected: **appears white**

transparent object

absorbs red

transmits wavelengths corresponding to other colours: **appears green**

opaque object

absorbs blue

reflects wavelengths corresponding to other colours: **appears orange**

Figure 11 How colours arise from absorption of light

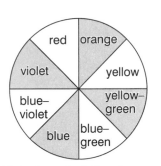

Figure 12 A colour wheel. Complementary colours are opposite one another

However, most objects appear coloured. This is because wavelengths corresponding to particular colours are being absorbed. Two examples are given in Figure 11.

When absorption occurs, wavelengths corresponding to one colour are removed from the white light and you see the **complementary** colour. The relationship is sometimes shown as a colour wheel in which complementary colours are opposite one another. Thus an object absorbing violet appears yellow-green.

It is common for people working with visible and ultra-violet radiation to use wavelength, rather than frequency, as a unit. The relationship between them is explained in **Section 6.1.**

Figure 13 is a chart showing the frequency, wavelength and colour of light.

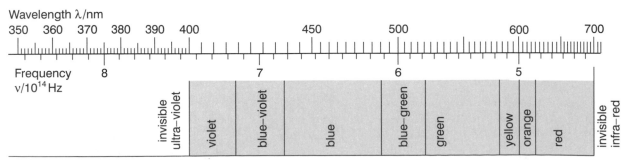

Figure 13 Wavelength, frequency and colour of light

Fluorescence

What happens when molecules absorb ultra-violet radiation?

When an *atom* absorbs ultra-violet radiation, an electron moves to a higher energy level. We often represent an atom in an excited (higher energy) state as X*. The same thing happens when *molecules* absorb ultra-violet radiation:

$$X_2 \rightarrow X_2{}^*$$

where $X_2{}^*$ represents a molecule in an electronically excited state.

Molecules can also change their energy in other ways (**Section 6.1**). The bonds in the molecule can vibrate. The energy gaps between vibrational energy levels are about 100 times smaller than the energy gaps between electronic energy levels. In any one electronic level a molecule has many possible vibrational energies. This is shown for a simple diatomic molecule in Figure 14.

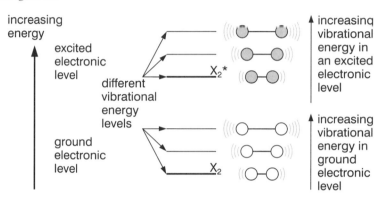

Figure 14 The electronic and vibrational energy levels of a diatomic molecule X_2. $X_2{}^*$ represents the excited state of X_2

When a molecule absorbs ultra-violet radiation, it goes into one of the several possible excited electronic levels.

What happens when molecules fluoresce?

Fluorescent substances are able to absorb invisible ultra-violet radiation and re-emit it as visible radiation. A molecule of the substance first absorbs ultra-violet radiation which gives it both a high electronic and a high vibrational energy. The relatively small quantities of vibrational energy are rapidly lost to other molecules during collisions, and so the molecule drops to a lower vibrational level. The electronic energy change is larger, and this energy is re-emitted as electromagnetic radiation. One example of how this might happen is shown in Figure 15.

Figure 15 Energy changes during fluorescence

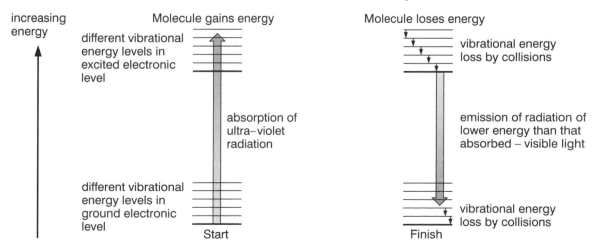

The radiation emitted has a lower energy than the radiation absorbed. Thus some molecules can absorb higher energy invisible ultra-violet radiation, and emit lower energy visible radiation. The whole process is very rapid and emission stops almost as soon as the ultra-violet radiation is switched off.

This explains why clothes which have been washed in soap powders containing fluorescent compounds glow only when the ultra-violet disco lights are on.

PROBLEMS FOR 6.4

1   What range of wavelengths do we perceive as
    a   green?
    b   red?

2   A solution appears orange-red. What range of wavelengths of visible light is being absorbed by the solution?

3   A substance absorbs both red and violet light. What colour is it likely to appear? Sketch a possible absorption spectrum for this substance showing intensity of absorption against wavelength.

4   Calculate the frequency of light corresponding to wavelengths of
    a   430 nm
    b   350 nm
    c   700 nm.

5   With the aid of an energy level diagram, explain why a solution of chlorophyll transmits green light but fluoresces red when light is reflected from its surface.

6.5 *Ultra-violet and visible spectroscopy*

Coloured compounds

Why are carrots orange? If a substance absorbs radiation in the visible region of the spectrum, then the light that reaches your eye from the substance will be lacking in certain colours. It will no longer appear white. For example, carrots contain the pigment *carotene*,

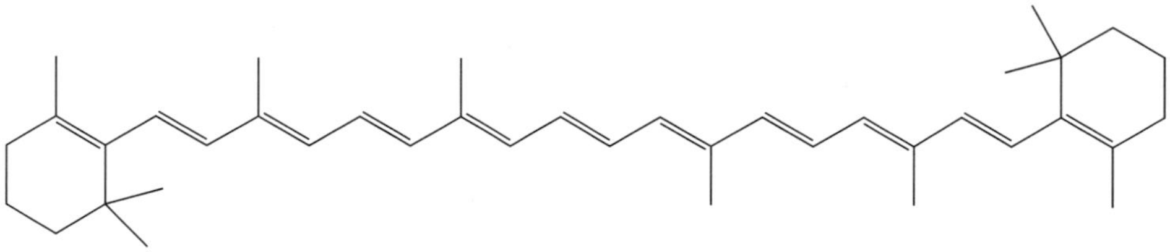

carotene (all-*trans* isomer)

Carotene absorbs blue light strongly, so the light reaching your eye is lacking in blue light and the carrot appears orange-red.

If you make a solution of carotene in a suitable solvent, you can use a **spectrometer** to measure the quantity of light absorbed by the solution at each wavelength. The recorder plots out the intensity of absorption against the wavelength. The result is an **absorption spectrum** of carotene. You can see what this looks like in Figure 16.

Absorption spectra

An **ultra-violet and visible spectrometer** works on much the same principle as the infra-red spectrometer (see **Section 6.7**). In this case, the source is ultra-violet and visible radiation rather than infra-red. Light from the source is split into two identical beams. One beam passes through the sample solution. The other passes through pure solvent. The light in the two emerging beams is compared to give the absorption spectrum of the sample.

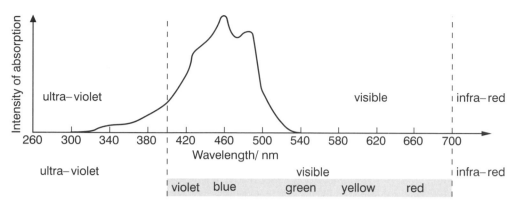

Figure 16 The absorption spectrum of carotene (in solution in hexane)

Most spectrometers scan wavelengths in the ultra-violet as well as the visible region to give a continuous spectrum like the one in Figure 16.

Our eyes cannot detect ultra-violet light so if a compound absorbs in this region, the absorption does not affect its colour. This means that a compound like benzene which absorbs only in the ultra-violet region (see Figure 17) appears colourless because it transmits all visible radiation.

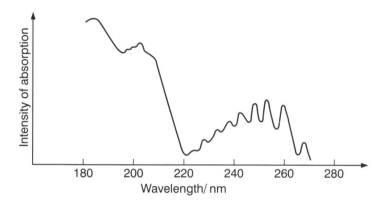

Figure 17 The absorption spectrum of benzene

Ultra-violet and visible spectra look different from the infra-red spectra in **Section 6.7**. The most obvious difference is that the peaks in an ultra-violet and visible spectrum rise from a base line and show the radiation *absorbed* by the sample. The absorption bands in an infra-red spectrum hang down from a 'base line' at the top of the spectrum, because the spectrometer is plotting the radiation *transmitted* by the sample. It is just two different ways of recording the same effect.

Another difference is in the units used to measure the radiation absorbed. In an infra-red spectrum it is the *wavenumber $1/\lambda$* (cm^{-1}) which is plotted on the horizontal axis. In an ultra-violet and visible spectrum, it is much more usual to plot the *wavelength* λ (nm). Don't let this worry you. *Always work in the units of the data you are given, unless you are specifically asked to convert from one set of units to another.*

When you interpret an infra-red spectrum you can assign specific absorption bands to particular groups in the molecule. For example, a $C{=}O$ group is usually responsible for an absorption around $1710\,cm^{-1}$ and so on. In contrast, an ultra-violet and visible spectrum is often a broad absorption band which is characteristic of general structural features of the molecule rather than individual functional groups.

A **colorimeter** is a simple type of visible spectrophotomer. Colorimetry is used to measure the intensity of absorption of coloured compounds over a narrow range of frequencies: it provides a useful way of finding the *concentration* of a coloured compound.

Interpreting the spectrum

Colour chemists are interested in three main features of the spectrum:

- the wavelength of the radiation absorbed (remember, for a compound to be coloured, at least part of the absorption band must be in the visible region)
- the intensity of the absorption
- the shape of the absorption band.

When recording the spectrum, chemists often give the wavelength of the maximum absorption (λ_{max}). For carotene (see Figure 16), λ_{max} occurs at 453 nm in the blue region of the spectrum.

The intensity of the absorption depends on the concentration of the solution and on the distance the light travels through the solution. Standard molar values are quoted so that values for different compounds can be compared. (You don't need to know about the units used to measure intensity for this course. It is enough to know that the higher the peak, the more intense the absorption.) The intensity of the absorption is important commercially because it determines the amount of pigment or dye needed to produce a good colour.

The shape and width of the absorption band is important because it governs the shade and purity of the colour seen.

You will find out more about the relationship between the structure of a compound and its electronic spectrum in **Section 6.6**.

Reflectance spectra

To measure an absorption spectrum, you need to make a solution of the coloured substance. However, it is not always possible to do this – for example, when the coloured substance is a pigment on the surface of a painting. In cases like this, chemists can use a different type of visible spectrum called a **reflectance spectrum**.

Here they shine light onto the paint surface and examine the composition of the reflected light. Since this is the part of the light that was *not absorbed* by the pigment, a reflectance spectrum is a sort of negative of the absorption spectrum. Figure 18 shows the absorption and reflectance spectra for the pigment Monastral Blue.

Figure 18 Absorption and reflectance spectra of Monastral Blue

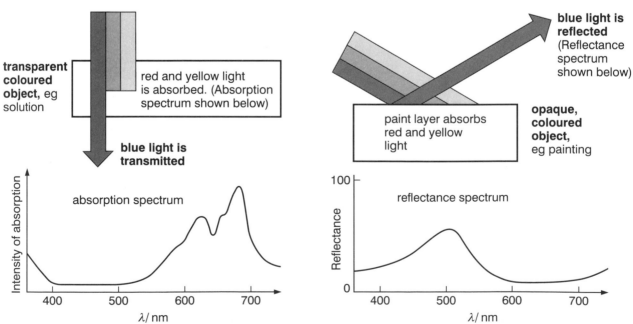

What happens to the radiation absorbed by coloured substances?

If you look back at **Section 6.2**, you will recall that the absorption of radiation in the ultra-violet and visible regions is associated with changes in the electronic energy of molecules. For this reason, ultra-violet and visible spectra are sometimes called **electronic spectra**.

When an electron is promoted to a higher energy level, we say the molecule is in an **excited electronic state**. It does not stay in the excited state for very long. The energy absorbed is re-emitted – but not necessarily all at once as visible radiation. The molecule may emit a smaller quantum of radiation and fall back to an intermediate energy level. The remaining energy can be converted to kinetic energy of the molecules. This means that the molecules move around more vigorously and the sample in the spectrometer becomes warmer.

PROBLEMS FOR 6.5

1   Look at the two absorption spectra in Figure 19. One is the spectrum of a synthetic dye. The other is the spectrum of a colourless compound.
   a   Which is the spectrum of the dye?
   b   What colour is the dye? (It will help you to refer to the visible spectrum in the Data Sheets, to find which colours correspond to different wavelengths.)

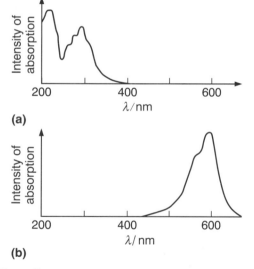

(a)

(b)

Figure 19

2   a   Draw a sketch of the absorption spectrum of
      i   a yellow dye
      ii a black dye.
   b   What would the reflectance spectrum of a black pigment look like?

3   Figure 20 shows the absorption spectra of two porphyrin pigments, haemoglobin and chlorophyll.
   a   Give the λ_{max} of the main absorption in each spectrum.
   b   Which spectrum corresponds to haemoglobin?

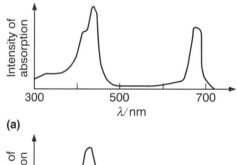

(a)

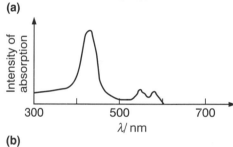

(b)

Figure 20

4   Chemists at the National Gallery recorded the reflectance spectrum shown in Figure 21 from a pigment on the surface of a painting.

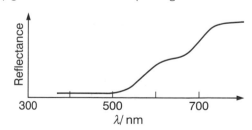

Figure 21

   a   Which of the following pigments do you think the spectrum most likely corresponds to?
      i   Red Ochre
      ii Malachite Green
      iii Prussian Blue.
   b   Give reasons for your choice in part **a**.

6.6 *Chemistry of colour*

Coloured or colourless?

Coloured substances absorb radiation in the visible region of the spectrum. The energy absorbed causes changes in *electronic energy* and electrons are promoted from the ground state to an *excited state*.

The electrons excited are the outermost ones. These are the electrons involved in bonding or present as lone pairs. The inner electrons are held tightly by the positive nucleus of the atom. The energy needed to excite these electrons is very large.

Not all electronic transitions are brought about by visible light. Many require greater energy corresponding to ultra-violet radiation. In this case, the compound would absorb ultra-light, but would appear colourless (**Section 6.5**). Figure 22 shows the energy needed to excite an electron in a coloured compound and in a colourless compound. This energy is called the **excitation energy**.

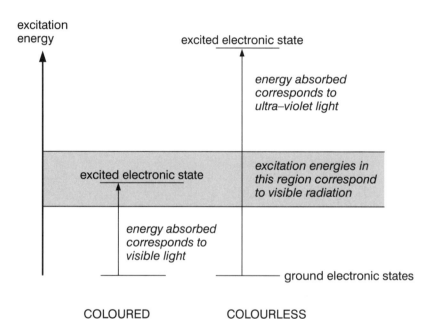

Figure 22 The energy needed to excite an electron in a coloured compound and in a colourless compound

Coloured organic compounds

These compounds often contain unsaturated groups such as C=C, C=O or —N=N—. These groups are usually part of an extended delocalised electron system called the **chromophore**. (In the past, the individual unsaturated groups such as —N=N— were called chromophores, and you will see the term used in this way in some textbooks.) Electrons in double bonds are more spread out and require less energy to excite than those in single bonds, particularly if the double bond is part of a conjugated system. The lower excitation energy means these compounds absorb in the visible region.

Functional groups such as —OH, —NH$_2$ or —NR$_2$ are often attached to chromophores to enhance or modify the colour of the molecules. These groups all contain lone pairs of electrons which become involved in the delocalised electron system. Small changes to the delocalised system can change the energy of the light absorbed by the molecule and so change the colour of the compound.

Often a dye molecule has a different colour in acidic and alkaline solutions. Compounds like this can be used as **acid-base indicators**. For example, Methyl Orange is red in acidic solutions below pH 3.5. Above this pH its solutions are yellow. In the red form, one of the N atoms has H^+ bonded to it, and this changes the energy absorbed by the delocalised electron system:

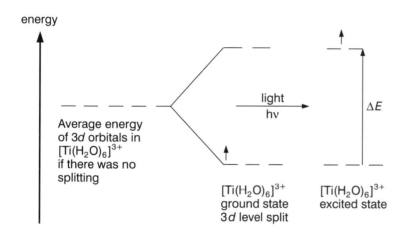

yellow   Methyl Orange   red

Coloured inorganic compounds

These compounds often contain transition metals. **Section 11.7** describes how the d sub-shell of electrons in a transition metal is affected by the presence of ligands. The five d orbitals no longer have the same energy but are split into two levels. Figure 23 shows the relative energy levels of the five $3d$ orbitals in the hydrated Ti^{3+} ion, $Ti(H_2O)_6^{3+}$.

energy

light
hv

ΔE

Average energy
of 3d orbitals in
$[Ti(H_2O)_6]^{3+}$
if there was no
splitting

$[Ti(H_2O)_6]^{3+}$
ground state
3d level split

$[Ti(H_2O)_6]^{3+}$
excited state

Figure 23 Relative energy levels for the five 3d orbitals of the hydrated Ti^{3+} ion

The energy needed to excite a d electron to the higher energy level also depends on the oxidation state of the metal. This is why redox reactions of transition metal compounds can be accompanied by spectacular colour changes. For example, vanadium shows a different colour in each of its oxidation states in aqueous solution.

V(+5) $\rightarrow$ V(+4) $\rightarrow$ V(+3) $\rightarrow$ V(+2)
yellow   blue   green   violet

Some ligands have a more powerful effect than others on the splitting of the d sub-shell. So changing the ligand complexed with a metal ion often results in a colour change.

$[Ni(H_2O)_6]^{2+}(aq) + 6NH_3(aq)$ $\rightleftharpoons$ $[Ni(NH_3)_6]^{2+}(aq) + 6H_2O$
light green   lilac/blue

Sometimes absorption of visible light can cause the transfer of an electron from the ground state of one atom to an excited state of *another*, adjacent atom. This is called **electron transfer**. It is a special sort of electronic transition and is responsible for some very bright pigment colours, such as Chrome Yellow and Prussian Blue.

PROBLEMS FOR 6.6

1 The blue colour of cornflowers and the red colour of poppies both derive from the same molecule, which has a purple colour. The structure of this molecule, called cyanidin, is shown below.

cyanidin

The sap of poppies is acidic and the cyanidin gains H$^+$, turning red. The sap of cornflowers is alkaline and the molecule loses H$^+$, turning blue.

red form

blue form

a Explain why molecules of this type absorb energy in the visible region producing coloured compounds.

b Why does the addition of H$^+$ to cyanidin change its colour from blue to red?

2 Different ligands in transition element complex ions can produce different stereochemical shapes for the ions. The change in shape can lead to dramatic changes in colour.

A particularly striking example is the colour change seen with cobalt chloride paper. When dry this paper is blue, but the colour changes to pink in the presence of water.

The blue colour is due to the tetrahedral complex ion $[CoCl_4]^{2-}$ and the pink colour is due to the octahedral complex ion $[Co(H_2O)_6]^{2+}$.

a Which of the two complex ions requires the greater quantity of energy for the electronic excitation of a *d* electron?

b Compare the magnitude of the splitting of the *d* orbitals in tetrahedral and octahedral environments.

c Can we be certain that the colour change is entirely due to the change in shape? Explain your answer.

6.7 *Infra-red spectroscopy*

The energy possessed by molecules is quantised, in other words they can only have a small number of definite energy values rather than any old energy. Analysis of the energy (or frequency of radiation) needed to produce a change from one energy level to another is the basis of most forms of spectroscopy.

In infra-red spectroscopy, substances are exposed to radiation in the frequency range 10^{14} Hz–10^{13} Hz, ie wavelengths 2.5 µm–15 µm. This makes vibrational energy changes occur in the molecules, which absorb infra-red radiation of specific frequencies.

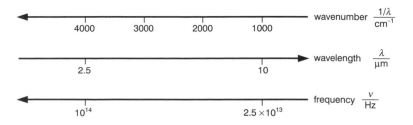

Figure 24 The relationship between wavenumber, wavelength and frequency

Frequency and wavelength are related by the equation: $c = \lambda\nu$ (where c is the speed of light, a constant). This means that the reciprocal of the wavelength $(1/\lambda)$ is a direct measure of frequency. It is this reciprocal, called the **wavenumber** of the radiation and usually measured in cm^{-1} units, which is recorded on an infra-red spectrum. Figure 24 shows the relationship between wavenumber, wavelength and frequency.

Bond deformation

Simple diatomic molecules such as HCl, HBr and HI, can only vibrate in one way, that is by *stretching*, where the atoms pull apart and then push together again (Figure 25).

For these molecules there is only one vibrational infra-red absorption. This corresponds to the molecules changing from their lowest vibrational energy state to the next higher level, in which the vibration is more vigorous.

The frequencies of the absorptions are different for each molecule. This is because the energy needed to excite a vibration depends on the strength of the bond holding the atoms together; weaker bonds require less energy. It is as if the bonds behave like springs of different strength holding the atoms together. This is illustrated in Table 2.

In more complex molecules, more bond deformations are possible. Most of these involve more than two atoms. For example, carbon dioxide can vibrate as shown in Figure 26.

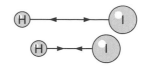

Figure 25 Stretching in the HI molecule

Compound	Bond enthalpy /kJ mol^{-1}	Infra-red absorption /cm^{-1}
HCl	432	2886
HBr	366	2559
HI	298	2230

Table 2 Bond enthalpies and infra-red absorptions for the hydrogen halides

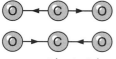 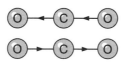 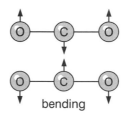

symmetric stretch   asymmetric stretch   bending

Figure 26 Vibrations in the CO$_2$ molecule

More complex molecules can have very many vibrational modes, with descriptions such as rocking, scissoring, twisting and wagging. Add these to the fact that the molecules also contain bonds with different bond energies, and you may have a very complicated spectrum. The important point to remember about infra-red spectroscopy is that you do not try to explain the whole spectrum; you look for one or two signals which are characteristic of particular bonds. Figure 27 shows the infra-red spectrum of ethanol, and Figure 28 shows some of the vibrations which give rise to the signals.

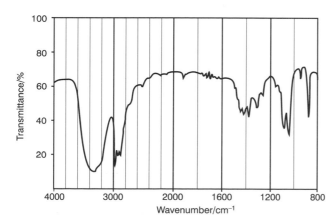

Figure 27 The infra-red spectrum of ethanol

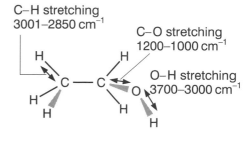

C–H stretching
3001–2850 cm^{-1}

C–O stretching
1200–1000 cm^{-1}

O–H stretching
3700–3000 cm^{-1}

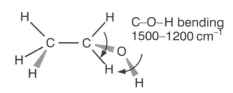

C–O–H bending
1500–1200 cm^{-1}

Figure 28 The ethanol molecule, showing the vibrations which give rise to some characteristic absorptions

The infra-red spectrometer

Infra-red radiation from a heated filament is split into two parallel beams, one of which passes through the sample, the other through a reference chamber. This ensures that unwanted absorptions from water and carbon dioxide in the air or from a solvent are cancelled out. The beams are then directed by mirrors so that they follow parallel paths (Figure 29).

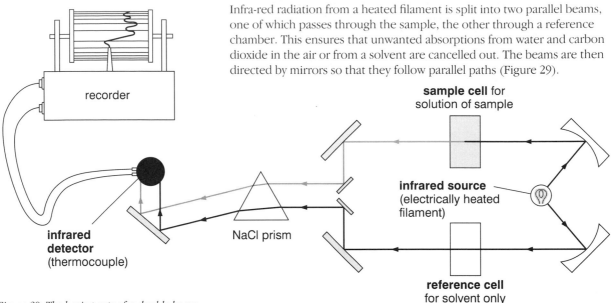

Figure 29 The basic parts of a double beam infra-red spectrometer

The beams are analysed by passing them through a prism of sodium chloride, which is transparent to infra-red radiation, or through a diffraction grating. Only light of one particular frequency will now be focussed onto the detector. The spectrum is produced by rotating the prism so that the detector scans the frequencies and records their intensities. When the sample is not absorbing there will be no difference between the two beams reaching the detector so no signal is recorded. When a vibration is being excited the sample beam intensity will be reduced and a signal generated.

The record of an infra-red spectrum seems to be upside down since the baseline is at the top and the signals (or bands) are recorded as downward movements on the chart recorder. As you will see from spectra like that in Figure 27, however, it is *transmittance* which is being recorded and this is at a maximum when no light is being absorbed.

Interpreting the spectra

The infra-red spectrum of oct-1-ene (Figure 30) shows most of the characteristic absorptions of a hydrocarbon. Some of these have been marked on the spectrum to show the bonds which are responsible. In general, we can match a particular bond to a particular absorption region. Table 3 on page 117 gives some examples. Note that the precise position of an absorption

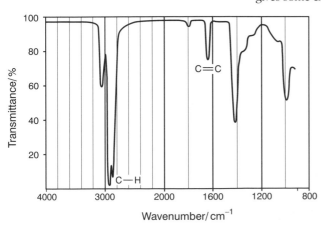

Figure 30 Infra-red spectrum of oct-1-ene

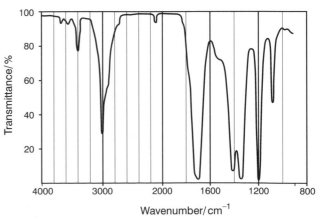

Figure 31 Infra-red spectrum of propanone

depends on the environment of the bond in the molecule, so we can only quote wavenumber *regions* in which we can expect absorptions to arise.

You have seen in Figure 27 (the infra-red spectrum of ethanol) that the C—H absorption which arises at around 3000 cm^{-1} is just on the shoulder of a much stronger and broader absorption from the O—H bond. This is more intense even though there are five times as many C—H bonds as O—H bonds in the ethanol molecule.

Similarly in Figure 31 (the infra-red spectrum of propanone), the C=O absorption at around 1720 cm^{-1} is very intense compared with the C=C absorption you saw in a similar region for octene.

Why are some absorptions intense while others are weaker? The strongest infra-red absorptions arise when there is a large change in bond polarity associated with the vibration. Hence O—H, C—O and C=O bonds, which are very polar, give more intense absorptions than the non-polar C—H, C—C and C=C bonds.

Using the combination of wavenumber and intensity it should be possible for you to interpret simple infra-red spectra for yourself. You will not be expected to remember the characteristic wavenumber regions! A reference table like Table 3 will always be available.

Bond	Location	Wavenumber/cm^{-1}	Intensity
C—H	alkanes	2850–2950	M–S
	alkenes, arenes	3000–3100	M–S
	alkynes	3300	S
C=C	alkenes	1610 to 1680	M
⬡	arenes	several peaks in range 1450 to 1650	variable
C≡C	alkynes	2070 to 2250	M
C=O	aldehydes, ketones, acids, esters	1680 to 1750	S
C—O	alcohols, ethers, esters	1000 to 1300	S
	aromatic esters and ethers	1300 to 1400	S
C≡N	nitriles	2200 to 2280	M
C—Cl		700 to 800	S
O—H	'free'	3580 to 3670	S
	hydrogen-bonded in alcohols, phenols	3230 to 3550	S (broad)
	hydrogen-bonded in carboxylic acids	2500 to 3300	M (broad)
N—H	primary amines	3100 to 3500	S

Table 3 Characteristic infra-red absorptions in organic molecules

M = medium,
S = strong

Before you try some problems, work through an interpretation of the spectrum of a sample of benzoic acid, shown in Figure 32.

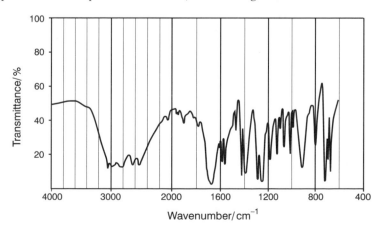

Figure 32 Infra-red spectrum of benzoic acid

Analysis of the spectrum confirms the following features:

- Strong, broad absorption in the region 2500–3300 cm⁻¹ suggests that an O—H bond is present, possibly that of an acid. Absorptions in this region also occur due to the carbon–hydrogen bonds.
- Absorption in the region 1680–1750 cm⁻¹ suggests that the C=O bond could be present.
- Absorption around the region 1300 cm⁻¹ again suggests that a C—O bond could be present.
- Absorption in the region 900–1100 cm⁻¹ could be due to the benzene ring or to the carbon–oxygen bond of the acid grouping.

The two absorptions in the region 700–800 cm⁻¹ are possible sources of confusion. These might have indicated a carbon–halogen bond but are probably due to the benzene ring.

The identity of the sample could now be confirmed by comparing its spectrum with that of an authentic sample of benzoic acid.

PROBLEMS FOR 6.7

1 Figure 33 shows the infra-red spectrum of butan-1-ol. Identify the key peaks in the spectrum, and the bond to which each corresponds.

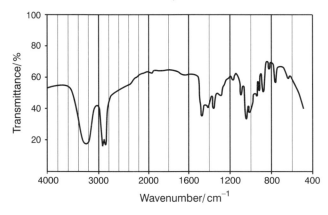

Figure 33 Infra-red spectrum of butan-1-ol

2 The infra-red spectra in Figure 34 represent three compounds A, B and C. The compounds are an ether, an ester and an alkyne, though not necessarily in that order. Identify the bands which are marked with an asterisk on each spectrum and so decide which spectrum represents which type of compound.

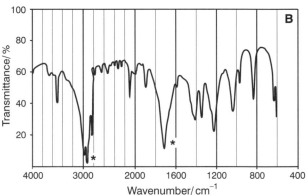

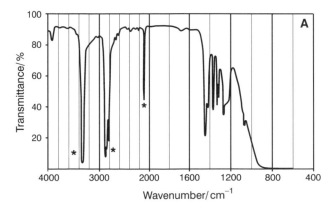

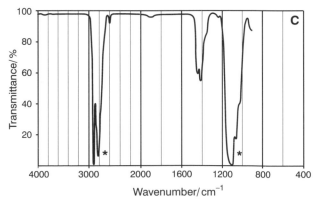

Figure 34 Infra-red spectra for compounds A, B and C

6.8 *Nuclear magnetic resonance spectroscopy*

What is n.m.r.?

Nuclear magnetic resonance spectroscopy (n.m.r.) is one of the most widely used analytical technique available to chemists. This might seem surprising. After all, our thinking in chemistry normally stops at the outermost electrons in a substance; the nucleus is just something we accept is there, small and heavy, at the centre of each atom.

However, n.m.r. provides us with very detailed information about the nuclei of certain atoms. For example, hydrogen nuclei behave differently in different molecular environments; n.m.r. tells us what the environments are and how many ^{1}H nuclei (usually just called protons) there are in each.

N.m.r can also be used to find out about certain other nuclei, particularly ^{13}C, but also ^{19}F and ^{31}P. However, in this section we will be concentrating on ^{1}H nuclei.

Such nuclei behave like tiny magnets and when they are placed in a strong magnetic field they align themselves with or against the magnetic field. If they are aligned *against* the magnetic field, they have a higher energy than if they are aligned *with* it. Slightly more than 50% of nuclei are in the lower energy level. If the correct frequency of radiation is applied, some of these nuclei move up to the higher level, absorbing energy ΔE as they do so (Figure 35). The energy corresponds to radio frequencies.

The energy needed for the nuclear 'magnet' to move to a higher energy level depends on the strength of the magnetic field. This is not quite the same as the field being applied by the instrument because the electrons associated with the neighbouring atoms and groups in the molecule give rise to tiny magnetic fields of their own.

These *local* fields are usually opposed to the *external* field. The *overall* field experienced by a proton is therefore slightly smaller than the external field, depending on the local field from the surrounding part of the molecule.

So for every type of molecular arrangement (*molecular environment*), there is a very slightly different magnetic field. The ^{1}H nuclei in these different environments have different energy gaps (ΔE) between their high and low energy levels, and so absorb different frequencies of radiation. They therefore give different n.m.r. peaks and we can find out how many hydrogen atoms of different types there are in a molecule.

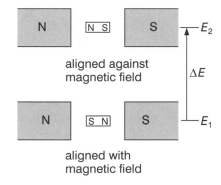

Figure 35 The principle of n.m.r.: a small magnet in a strong magnetic field can have two different energies

The n.m.r. instrument

The main features of a modern n.m.r. spectrometer are illustrated in Figure 36. There is a magnet which produces a strong magnetic field, a radio-frequency (RF) source, a detector and a recorder. The magnetic field is held constant and a band of radio frequencies is applied as a pulse to the sample.

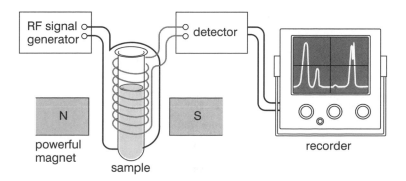

Figure 36 A simplified diagram of an n.m.r. spectrometer

Substances to be investigated are used as pure liquids, or are dissolved in solvents like CD_2Cl_2 or CD_3COCD_3, which contain no ^1H atoms. (If the solvent contained ^1H atoms, these would give a spectrum of their own.)

Immediately after the radio-frequency pulse there will be a greater than normal number of protons in the higher level. Some of these will therefore emit radiation corresponding to the energy difference ΔE, and return to the lower level. This emitted radiation is detected.

The radiation is weak and the process is over very quickly, so it is repeated many times in rapid succession to build up an accurate record. This is stored and analysed electronically.

Interpreting n.m.r. spectra

Ethanal has two types of protons; one in the CHO group and the other in the CH_3 group. The energy levels for the two types of protons in the same magnetic field are shown in Figure 37.

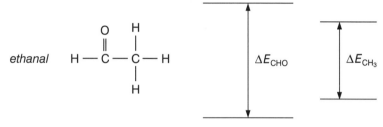

Figure 37 *Energy levels of the two types of protons in an ethanal molecule*

This molecule will emit two frequencies in an n.m.r. machine. The resulting spectrum is shown in Figure 38.

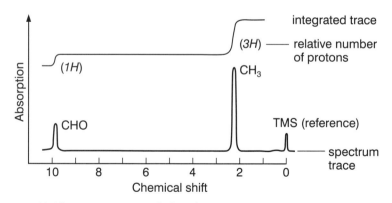

Figure 38 *The n.m.r. spectrum of ethanal*

Type of proton	Chemical shift in region of
$R-CH_3$	0.9
$R-CH_2-R$	1.3
R‑CH‑R with R above	2.0
$-C(=O)-CH_3$	2.3
$-O-CH_3$	3.8
$-O-CH_2R$	4.0
$-C=CH_2$	4.7
$-O-H$	5.0
$>C=CH-$	5.3
$-CH=CH-C(=O)-$	6.1
benzene ring $-H$	7.5
$-C(H)=O$	9.5 (aliphatic) 10 (aromatic)
$-C(O-H)=O$	11.0

Table 4 *Chemical shifts for some types of protons*

As well as the spectrum trace in Figure 38, the n.m.r. spectrometer has drawn an integrated spectrum trace. This goes upwards in steps which are proportional to the areas of the absorption signals, and therefore tells us how many protons are absorbing each time. Three times as many protons (the CH_3 group) absorb in the right-hand signal as in the left-hand signal (from the CHO group). In other spectra in this book we will omit the integrated trace, but we will label each peak with the relative numbers of protons.

Also shown on the spectrum is the absorption of tetramethylsilane, TMS (Figure 39). This has been chosen as a standard reference because it gives a sharp signal well away from most of the ones of interest to chemists. The extent to which a signal differs from TMS is called its **chemical shift**. TMS is set at a chemical shift at 0 and the shifts of other types of protons can be found from reference tables such as the one in Table 4.

$$CH_3 - Si - CH_3$$ with CH_3 above and CH_3 below

Figure 39 *Tetramethylsilane (TMS)*

It isn't quite as simple as that …

A sensitive n.m.r. machine actually produces a much more detailed spectrum then the one shown in Figure 38. Figure 40 shows what the detailed n.m.r. spectrum of ethanal actually looks like.

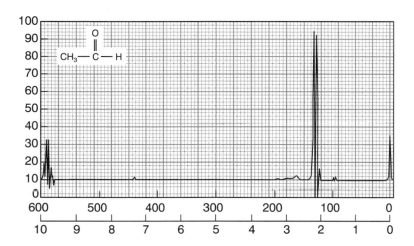

Figure 40 The detailed n.m.r. spectrum of ethanal

Figure 41 shows how the less detailed spectrum in Figure 38 is related to the detailed spectrum. How does the extra detail arise? The ^{1}H nuclei behave like tiny magnets, and they can be in one of two orientations depending on whether they are in the low or high energy level. In the CH_3 group, there are *four* combinations in which the *three* 'tiny proton magnets' can be arranged:

- all aligned with the external field
- two with and one against the external field
- one with and two against the external field
- all aligned against the external field

all aligned with the external field	S-N S-N S-N
two with and one against the external field	S-N S-N N-S
one with and two against the external field	S-N N-S N-S
all aligned against the external field	N-S N-S N-S

The magnetic effect of these arrangements in the CH_3 groups is transmitted to the neighbouring CHO protons so that they sense one of *four* magnetic fields. They can absorb four different frequencies and give *four* signals.

Because the CHO protons can only be arranged in one of two ways

S-N   or   N-S

the protons of the CH_3 groups sense only two different fields and they give only *two* signals.

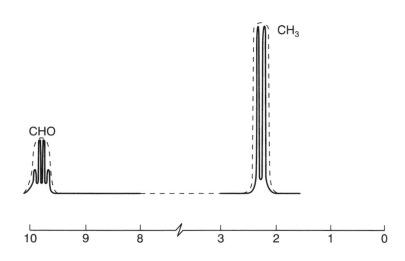

Figure 41 The dotted lines show how the less detailed spectrum in Figure 38 is related to the detailed spectrum in Figure 40

These more detailed spectra provide chemists with extra information about the compound being investigated. However, analysing high-resolution spectra is more tricky, and for this course we will just use the less detailed, low-resolution spectra.

PROBLEMS FOR 6.8

1   The n.m.r. spectrum of an alcohol, **E**, is shown in Figure 42.

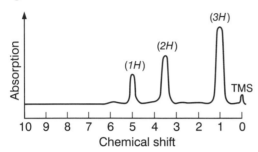

Figure 42 N.m.r. spectrum for alcohol **E**

a   Which types of protons give rise to each of the signals in the spectrum?
b   What are the relative numbers of each of the types of proton?
c   The relative molecular mass of the alcohol is 46. Identify the alcohol.

2   The n.m.r. spectrum of another alcohol, **F**, is shown in Figure 43.

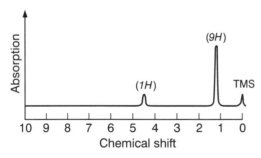

Figure 43 N.m.r. spectrum for alcohol **F**

a   Which types of protons give rise to each of the signals in the spectrum?
b   What are the relative numbers of each of the types of proton?
c   Identify the alcohol.

3   Compound **G** is a hydrocarbon with molecular formula C_8H_{10}. Use the n.m.r. spectrum in Figure 44 to identify **G**.

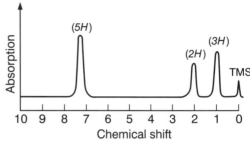

Figure 44 N.m.r. spectrum for compound **G**

4   The n.m.r. spectrum of an ester, **H**, is shown in Figure 45. Work out the structure of the ester.

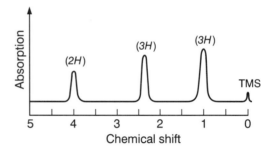

Figure 45 N.m.r. spectrum for compound **H**

5   Vanillin is a widely used flavouring which occurs naturally in the pods of the vanilla orchid. The structure of vanillin is shown in Figure 46.

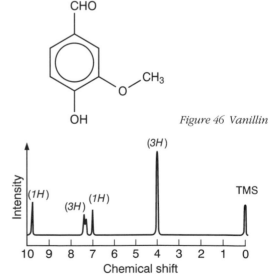

Figure 46 Vanillin

Figure 47 N.m.r. spectrum of vanillin

Figure 47 shows the n.m.r. spectrum of vanillin. Match each of the peaks on the spectrum with protons in the vanillin structure.

6.9 *Mass spectrometry*

The instrument

The mass spectrometer is an important instrument for chemists. It enables them to compare the mass of different particles. For example, we know that chlorine has two isotopes, ^{37}Cl and ^{35}Cl, because chemists have used the mass spectrometer to find the masses of the chlorine isotopes.

In this section we are particularly concerned with the way the mass spectrometer can be used to investigate molecular structure.

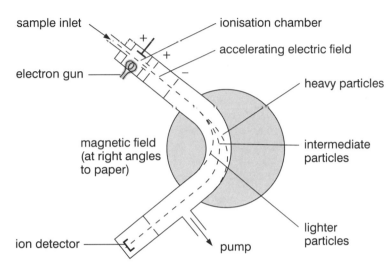

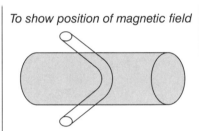

Figure 47 The mass spectrometer

There are five essential parts of the instrument:
- a sample inlet;
- an ionisation chamber;
- an electric field to accelerate a beam of positively charged ions;
- a magnetic field for deflecting the ion beam;
- an ion detector.

A very low pressure must be maintained in the mass spectrometer so that the ions can pass through the instrument without colliding with air molecules.

Sample inlet Small quantities of gases and liquids are injected. Most liquids vaporise at the low pressure in the machine but solids may need to be heated on special probes.

Ionisation chamber Here the sample is bombarded by a stream of high energy electrons from a heated filament. The bombardment knocks off electrons from the outside of the sample molecules producing positive ions. Occasionally two electrons may be dislodged, but this is unlikely and we usually work on the assumption that all the ions in the experiment carry a single positive charge. If the sample molecule is represented by X, the ionisation process can be summarised:

$$X(g) \quad + \quad e^- \quad \rightarrow \quad X^+(g) + 2e^-$$
$$\text{bombarding}$$
$$\text{electron}$$

Electrically charged plates These attract the positive ions from the mixture in the ionisation chamber and accelerate them. By carefully arranging the plates and choosing their voltages, a fine beam of ions with only a narrow range of kinetic energies is passed into the magnetic field.

Magnetic field The beam of charged ions travelling along the tube behaves like a current in a wire and the magnetic field forces it to one side so it follows a curved path. Lighter ions are pushed further off course than heavy ions. With a low magnetic field, these light ions can be targeted at the detector, but the field will have to be increased before a beam of heavy ions is deflected enough to take the same course.

Detector The detector produces an electric current when hit by ions. The detector signal is recorded as the magnetic field is gradually changed. This produces a **mass spectrum**, which is usually viewed on a computer screen, before being printed out.

Fragmentation

When we look at a typical mass spectrum from a compound, such as the ones for 2-ethoxybutane shown in Figure 48, we see that it is quite complex. The heaviest ion (M = 102) is the one corresponding to the ethoxybutane molecule with just one electron removed. This is called the **molecular ion**, X^+. However, this is not the only ion reaching the detector. The other ions are the result of **fragmentation** – breakdown of X^+ into lighter positively charged ions. The mass spectrometer analyses these too. The most abundant ion gives the strongest detector signal which is set to 100% in the spectrum. This is referred to as the **base peak** and the intensities of all the other peaks are expressed as a percentage fraction of its value. It is not uncommon for the molecular ion peak to be so weak as to be unnoticeable – the spectrum obtained then consists entirely of fragments.

Figure 48 Simplified mass spectrum for 2-ethoxybutane, showing the six largest peaks

These positive ions can undergo some strange chemistry, quite unlike the reactions you are used to with laboratory chemicals, so mass spectra can be quite complicated to interpret. However, simple ideas can often help us to make enough sense of what is going on to allow us to identify the substance we put in as the sample.

If we imagine the molecules of 2-ethoxybutane to be built like 'Lego' models that can be pulled apart into their constituent 'building blocks', we can identify nearly all the peaks in its mass spectrum. This is done for you in Figure 49.

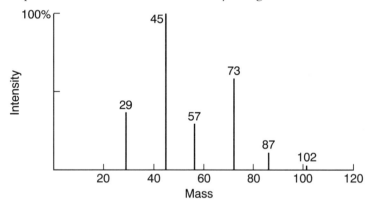

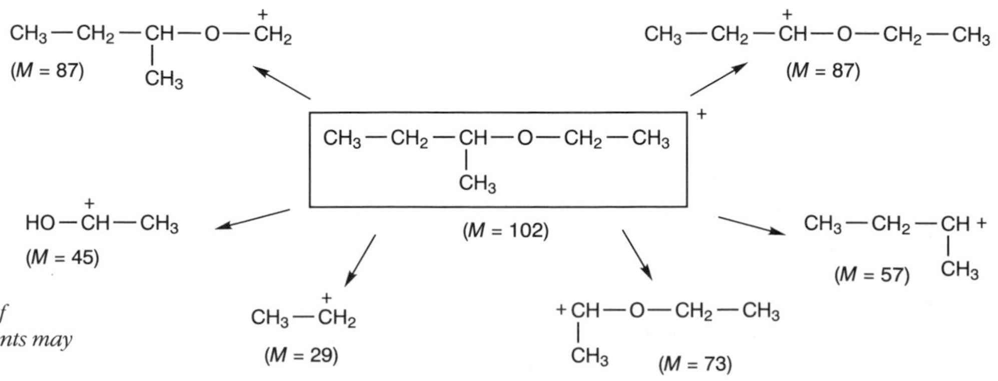

Figure 49 Fragmentation of 2-ethoxybutane (the fragments may be formed in several steps)

The 'Lego' approach enables you to make use of the mass differences between peaks (the masses of the bits which have fallen off) to help make sense of the spectrum. For example, in Figure 48, peaks 102 and 87 differ by 15, corresponding to loss of CH_3. This is a common process in mass spectrometry, so be on the lookout for gaps of 15 in your spectra – they mean you have a methyl group in your substance.

Table 5 gives some other common differences and the groups they suggest.

(Try using these ideas to do problems 1 to 3 at the end of this section.)

Isotope peaks

Some elements have more than one stable, naturally-occurring isotope. Chlorine is an example; natural chlorine is made up of the isotopes ^{35}Cl and ^{37}Cl, in the ratio 75% : 25%. Ions containing isotopic atoms have different masses, and so they will separate in the mass spectrometer. In the case of a molecule containing a chlorine atom there will be two sorts of ions; one containing ^{35}Cl atoms and giving a signal at mass M, the other containing ^{37}Cl atoms giving a signal at mass M + 2. The heights of the two peaks will be in the ration 75% : 25%, i.e. 3 : 1. A pair of peaks like this is a tell-tale sign that Cl is present. For example, look at Figure 50, which shows the mass spectrum for chloroethane, C_2H_5Cl. It shows two pairs of peaks with the tell-tale 3 : 1 ratio.

64 and 66 due to $[C_2H_5{}^{35}Cl]^+$ and $[C_2H_5{}^{37}Cl]^+$ (the molecular ion)

49 and 51 due to $[CH_2{}^{35}Cl]^+$ and $[CH_2{}^{37}Cl]^+$ (the molecular ion minus CH_3)

You can see that each type of fragment gives two peaks: one containing ^{35}Cl and one containing ^{37}Cl.

Mass difference	Group that is suggested
15	CH_3
17	OH
28	$C{=}O$ or C_2H_4
29	C_2H_5
45	COOH
77	C_6H_5

Table 5 Some common mass differences, and the groups they suggest

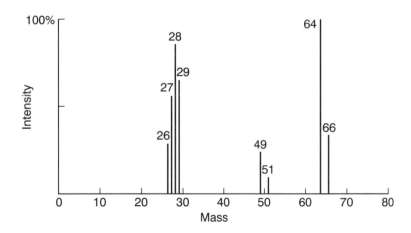

Figure 50 The mass spectrum of chloroethane

One isotope which can be very useful in helping to work out mass spectra of more complicated molecules is ^{13}C. This accounts for only 1.1% of a sample of carbon atoms. However, when there are several carbon atoms in a compound the abundance of ions containing a ^{13}C becomes significant enough for their signal to be clearly detectable. The more carbon atoms in a molecule, the greater is the chance of one of them being ^{13}C.

For example, consider the mass spectrum of decan-1-ol, $C_{10}H_{21}OH$. You would expect a molecular ion peak at 158. However, with a molecule containing 10 carbon atoms there is approximately a $10 \times 1.1\% = 11\%$ chance of one atom being ^{13}C. This makes the molecular ion heavier by 1 unit. So, out of every 100 molecular ions, 11 will have mass 159, and the other 89 will have mass 158. There will, therefore, be two molecular ion peaks, at 158 and 159, with intensities in the ratio 89 : 11.

PROBLEMS FOR 6.9

1   The mass spectrum of a hydrocarbon with molecular formula C_4H_{10} is shown below.

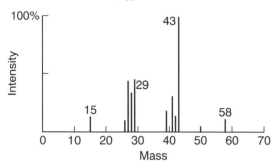

Figure 51 The mass spectrum of C_4H_{10}

  a   What ions give rise to the peaks at mass 58, 43, 29, and 15?

  b   Using this pattern of peaks, draw the full structural formula for this compound.

2   Compounds **A** and **B** both have the same molecular formula, C_3H_6O, but their molecular structures are different. They are isomers. The mass spectra of **A** and **B** are shown in Figure 52.

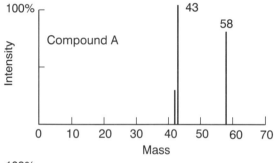

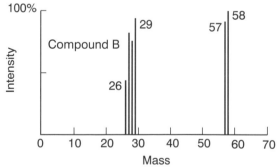

Figure 52 The mass spectra of compounds A and B

  a   For compound **A**, which group of atoms could be lost when the ion of mass 43 forms from the ion of mass 58?

  b   For compound **B**, what atom or groups of atoms could be lost when

    i   the molecular ion changes into the ion of mass 57?

    ii   the ion of mass 57 changes into the ion of mass 29?

  c   Suggest formulae for the ions of mass

    i   43 in compound **A**

    ii   28, 29 and 57 in compound **B**.

  d   Identify **A** and **B**.

3   Compound **C** contains only carbon, hydrogen and oxygen. Its accurate molecular mass was found to be 72.0573. A data base gave four compounds with masses in this region: $C_2H_4N_2O$, $C_3H_8N_2$, $C_3H_4O_2$ and C_4H_8O. Use accurate atomic masses to work out the formula of **C** and then use the mass spectrum in Figure 53 to work out its structure. Identify as many of the peaks in the spectrum as possible.

  Accurate atomic masses: H = 1.0078
                                   O = 15.9949
                                   N = 14.0031
                                   C = 12.0000

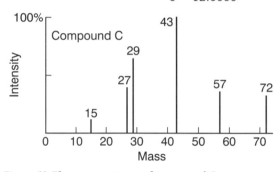

Figure 53 The mass spectrum of compound C

4   Compound **D** is an ester (see **Section 13.5**) with empirical formula C_2H_4O, and a relative molecular mass of 88.

  a   Draw four structures for esters which could correspond to **D**.

  b   Work out which of these structures would give a mass spectrum like the one in Figure 54.

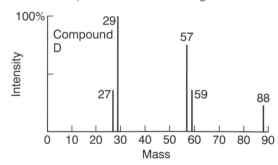

Figure 54 The mass spectrum of compound D

  c   Which fragments do you think give rise to the peaks at masses 29, 57, 59?

5 Bromine consists of the isotopes ^{79}Br and ^{81}Br in approximately 50% : 50% ratio. Use this information, together with the mass spectrum for chloroethane in Figure 50, to draw a sketch of what you think the mass spectrum of bromoethane looks like.

6 The mass spectrum of compound **X** is shown in Figure 55. The table shows data for each peak listed by a computer linked to the mass spectrometer.

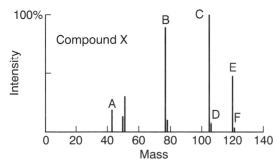

Figure 55 *The mass spectrum of compound X*

Peak	Intensity	Mass	Peak	Intensity	Mass
A	17.37%	43.0	**C**	100.00%	105.0
	0.43%	44.0	**D**	7.75%	106.0
	11.09%	50.0	**E**	44.13%	120.0
	29.29%	51.0	**F**	3.96%	121.0
B	84.19%	77.0			
	9.06%	78.0			

a Calculate the intensity ratios for peaks **F** and **E**, and **D** and **C**.

b The natural abundance of ^{13}C is about 1.1%. Explain the occurrence of peaks **C**, **D**, **E** and **F**, and estimate the number of carbon atoms in ion **C** and ion **E**.

c Using the peaks at masses 43 and 44, estimate the number of carbon atoms in ion **A**.

d What is the relative molecular mass of compound **X**?

e The principal feature of the infra-red spectrum of compound **X** is a strong absorption around 1700 cm^{-1}. What functional group could give rise to this absorption band?

f Use the information you have gathered to deduce the structural formula of compound **X**, and suggest formulae for the ions responsible for peaks **A**, **D**, **C** and **E**.

7

EQUILIBRIUM IN CHEMISTRY

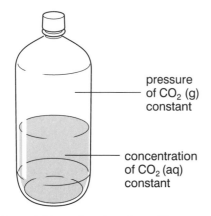

Figure 1 *The carbon dioxide equilibrium on a macroscopic scale*

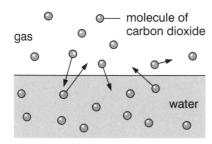

Figure 2 *The carbon dioxide equilibrium on a microscopic scale*

7.1 *Chemical equilibrium*

Dynamic equilibrium

The general meaning of the term *equilibrium* is a state of balance in which nothing changes. For example, a see-saw with two people of equal mass, sitting one on each side, is in a state of equilibrium.

In chemistry, a state of equilibrium is also a state of balance, but it has a special feature: chemical equilibrium is **dynamic equilibrium**.

Consider the sealed bottle of soda water shown in Figure 1. The bottle and its contents make up a closed system. Nothing can enter or leave the bottle. In this system, carbon dioxide is present in two states: as a gas above the liquid, $CO_2(g)$, and as an aqueous solution in water, $CO_2(aq)$.

Suppose the bottle has been standing at a steady temperature for some time. If you measure the pressure of carbon dioxide gas above the water you find it is constant. If you measure the concentration of carbon dioxide dissolved in the water, you find that is constant too. The system is at equilibrium, and nothing appears to be changing, at least on the **macroscopic scale** – the scale which we humans are used to.

If you were able to see how the individual molecules are behaving – that is on a **microscopic scale** – the picture would be rather different.

In any gas, the molecules are constantly moving. They move rapidly in random directions, and inevitably they collide with the molecules in the surface of the water (see Figure 2). Some bounce back into the gas phase, but some dissolve and enter the aqueous phase.

At the same time, the molecules of carbon dioxide in the water are also constantly moving around, colliding with water molecules and with each other. Near the surface of the water, some of these molecules escape into the gaseous phase (see Figure 2).

So there are molecules entering the aqueous phase, and molecules leaving it. *When the system is at equilibrium, the molecules enter and leave at the same rate.* On the macroscopic scale it *seems* as though nothing is changing, but, on the molecular scale, molecules are constantly moving from one phase to the other. That is why the situation is described as *dynamic* equilibrium. It can be represented by the equation

$$CO_2(g) \rightleftharpoons CO_2(aq)$$

The $\rightleftharpoons$ sign represents dynamic equilibrium.

Other examples of dynamic equilibrium

Dynamic equilibrium is a basic principle of chemistry. You can find examples everywhere – and you can find two more without going any further than the bottle of soda water shown in Figure 1.

Water in equilibrium

The water itself is present in two phases: liquid and gas. Water molecules from the liquid are constantly escaping into the gaseous phase. At the same time, molecules from the gas are constantly colliding with the surface of the liquid and condensing. In a closed system (the bottle with the top on), molecules reach the situation after a while where they are escaping and condensing at the same rate. Then the system is in dynamic equilibrium, and it is represented by:

$$H_2O(l) \rightleftharpoons H_2O(g)$$

Carbon dioxide reacts with water

Neither of the two examples just mentioned involve chemical reactions. They are physical processes involving the liquid/gas phase change. But when carbon dioxide dissolves in water, a small proportion of it actually *reacts* with the water instead of just dissolving. Hydrogencarbonate ions and hydrogen ions are formed,

$$CO_2(aq) + H_2O(l) \rightleftharpoons HCO_3^-(aq) + H^+(aq)$$

hydrogencarbonate ion

Only a small proportion of the carbon dioxide reacts in this way. The majority of it remains as $CO_2(aq)$, with only a small amount converted to $HCO_3^-(aq)$.

On the molecular scale, CO_2 molecules are constantly reacting with H_2O molecules to form HCO_3^- and H^+. At the same time, HCO_3^- and H^+ are constantly reacting together to reform CO_2 and H_2O. It is a **reversible reaction**.

After the system in the soda water bottle has had a chance to settle down, the forward and reverse reactions go at the same rate. Nothing *seems* to change on the overall, macroscopic level. If you measure the pH of the soda water, you find it is constant; the concentration of $H^+(aq)$ ions is constant, and so are the concentrations of all the other substances involved. It is another example of dynamic equilibrium.

Starting to put numbers to equilibrium

Figure 3 shows a device for making soda water. It forces carbon dioxide under pressure into water. The same principle is used to make most fizzy drinks.

Normally, carbon dioxide gas is injected into the water until it reaches a certain pressure, then escapes from under the seal. But if you control the valve you can *vary* the pressure of carbon dioxide in the space above the water. Some of this carbon dioxide gas will then dissolve in the water, and after a while it will reach equilibrium. If you then measure the pH of the water, you can find out the concentration of H^+ ions in the aqueous phase and use this to work out the concentration of carbon dioxide in the aqueous phase. If you introduce a pressure gauge, you can measure the pressure of carbon dioxide in the gas phase.

Table 1 shows the figures that were obtained when measurements of this kind were carried out at 292 K.

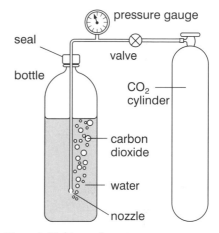

Figure 3 Making soda water

Experiment	Pressure of $CO_2(g)$ /atm	Concentration of $CO_2(aq)$ /mol dm^{-3}
1	1.0	0.035
2	1.2	0.043
3	1.5	0.053
4	1.9	0.065
5	2.3	0.080

Table 1 Measurements of the pressure of carbon dioxide in the gas phase and the concentration of carbon dioxide in the aqueous phase when the system is in equilibrium at 292K

If, for each of these experiments, you work out the ratio

$$\frac{\text{concentration of } CO_2(aq)\ (/\text{mol dm}^{-3})}{\text{pressure of } CO_2(g)\ (/\text{atm})}$$

you will find the value is always the same.

The ratio $\dfrac{\text{concentration of } CO_2(aq)}{\text{pressure of } CO_2(g)}$ or $\dfrac{[CO_2(aq)]}{p_{CO_2(g)}}$

is found always to be constant at a particular temperature. This makes sense if you think about it: if you have a greater pressure of carbon dioxide gas, you would expect proportionately more of it to dissolve in the water. (Note the use of square brackets to represent the *concentration* of $CO_2(aq)$ in mol dm^{-3}).

The above ratio is an example of an **equilibrium constant**, and it is given the symbol K. If you take an average of the five ratios you worked out from the data in Table 1, you will have a value for the equilibrium constant for the reaction

$$CO_2(g) \rightleftharpoons CO_2(aq)$$

Similarly, there is an equilibrium constant for the reaction

$$CO_2(aq) + H_2O(l) \rightleftharpoons HCO_3^-(aq) + H^+(aq)$$

It has a value of $4.3 \times 10^{-7}\,mol\,dm^{-3}$ at 292 K.

Now think what happens if we try to force more carbon dioxide to dissolve in water by using a higher pressure for the gas. Table 1 shows you that the concentration of dissolved gas will indeed rise, so that the equilibrium constant remains at its fixed value. We say that the **position of equilibrium** shifts to the right. This will have an effect on the reaction of carbon dioxide with water to form hydrogencarbonate ion. If it didn't, the increased amount of dissolved carbon dioxide would mean that you had raised the concentration of reactants but left the products alone, and then the reactants and products would no longer be in equilibrium.

To keep K at $4.3 \times 10^{-7}\,mol\,dm^{-3}$, the reaction must create some more products – make more H^+ ions – and so the pH will fall. Hence we can use the pH of the solution to measure the pressure of carbon dioxide in the gas phase.

In **Activities A8.2 and A8.3** you can use these ideas to investigate the concentrations of CO_2 in both the aqueous phase and in the gas phase.

Position of equilibrium

Once a reaction system is at equilibrium, it is impossible to tell whether the equilibrium mixture was arrived at by starting with reactants or with products.

There are many equilibrium mixtures possible for a given reaction system, depending on the concentrations of the substances you mix and the conditions, such as temperature and pressure. The important thing is that the *equilibrium constant has a constant value at a particular temperature*. The concentrations of all the substances will change until this is achieved.

We often use the term **position of equilibrium** to describe one particular set of equilibrium concentrations for a reaction. If one of the concentrations is changed, the system is no longer in equilibrium, and the concentrations of all the substances will change until a new position of equilibrium is reached.

Chemical equilibria and steady state systems

Strictly speaking a chemical equilibrium can only be established in a **closed system** which is sealed off from its surroundings. For example, if you heat calcium carbonate in a closed container, it decomposes and an equilibrium mixture is established.

$$CaCO_3(s) \rightleftharpoons CaO(s) + CO_2(g)$$

If you heat calcium carbonate in an *open* tube, carbon dioxide is lost into the air. The equilibrium position shifts to the right, until eventually all the calcium carbonate is converted into calcium oxide. It is an open system, so equilibrium is never established.

Sometimes in an open system, a series of reactions can reach a **steady state**, where the concentrations of reactants and products remain constant. A blue Bunsen burner flame is in a steady state. It doesn't seem to change, but reactants are being used up as fast as they arrive, so it certainly isn't in equilibrium.

Another example of a steady state is the series of reactions which produce and destroy ozone in the stratosphere.

$O + O_2 \rightarrow O_3$ ⟶ ozone production

$\left.\begin{array}{l} O_3 + h\nu \rightarrow O_2 + O \\ O + O_3 \rightarrow O_2 + O_2 \end{array}\right\}$ ozone destruction

None of these reactions comes to equilibrium, but left to themselves they will reach a point at which ozone is being produced as fast as it is being used up. We say the series of reactions has reached a steady state.

PROBLEMS FOR 7.1

1 a Use the values in Table 1 to plot a graph of concentration of carbon dioxide in the aqueous phase, $[CO_2(aq)]$, against its pressure in the gas phase, $P_{CO_2(g)}$.
   b Use your graph to work out an accurate value for the equilibrium constant, at 292 K, of the reaction
      $CO_2(g) \rightleftharpoons CO_2(aq)$
   c Why is this better than taking an average of the five ratios you worked out from the data in Table 1?

2 You will be very familiar with the equilibrium
   $H_2O(l) \rightleftharpoons H_2O(g)$
   a Why is this described as a *dynamic* equilibrium?
   b One place where this equilibrium is established is in a plastic bag full of wet socks. Use the idea of dynamic equilibrium to explain why the socks in the bag never get dry.
   c You can dry the socks by taking them out of the bag. Once they are out of the bag the system is no longer in equilibrium. Explain why.
   d In which direction does this equilibrium move
      i when dew forms in the evening?
      ii when you demist a car window?

3 Iodine is only sparingly soluble in pure water. When an aqueous solution of iodine is required, it is dissolved in aqueous potassium iodide. In this solution the iodine molecules are in equilibrium with the red-brown triiodide ion, $I_3^-(aq)$.
   $I_2(aq) + I^-(aq) \rightleftharpoons I_3^-(aq)$
   Addition of dilute aqueous sodium hydroxide to this solution causes the red-brown colour to become yellow and, with excess alkali, almost colourless. The alkali reacts with the iodine molecules to produce iodate(I) ions, $IO^-(aq)$;
   a Explain the disappearance of the red-brown colour in terms of the equilibrium involving the triiodide ions.
   b Explain why the triiodide equilibrium enables chemists to use the solution as though it is aqueous iodine.
   c Predict what you might see on adding excess dilute hydrochloric acid drop by drop to the colourless solution obtained on adding alkali to aqueous iodine.

4 If solid bismuth trichloride, $BiCl_3$, is added to water, a reaction occurs producing bismuth chlorate(I), BiClO, as a white precipitate.
   $BiCl_3(aq) + H_2O(l) \rightleftharpoons BiClO(s) + 2HCl(aq)$
   A colourless aqueous solution of $BiCl_3$ may be prepared by dissolving the solid in concentrated hydrochloric acid.
   a Explain, in terms of the equilibrium, why concentrated hydrochloric acid enables an aqueous solution of bismuth trichloride, $BiCl_3(aq)$, to be formed.
   b Predict what you would see if BiClO(s) was treated with concentrated hydrochloric acid. Explain your prediction in terms of the equilibrium.
   c Predict what you would see if $BiCl_3(aq)$ was diluted with a large volume of water. Explain your prediction in terms of the equilibrium.

7.2 *Equilibrium constants*

To investigate chemical equilibrium further, we need to ask questions about the *quantities* of substances involved. For a reaction that is going on in solution, the quantities that matter are *concentrations*.

In **Section 7.1**, you looked at the ratio of concentrations of CO_2 in gaseous and aqueous phases in the process

$CO_2(g) \rightleftharpoons CO_2(aq)$

You found that the ratio of the concentrations in the two phases is constant. This is called the **equilibrium constant**.

You can write equilibrium constants for *all* equilibrium processes. Table 2 shows data obtained for the hydrolysis of an ester, ethyl ethanoate,

$$CH_3COOC_2H_5(l) + H_2O(l) \rightleftharpoons CH_3COOH(l) + C_2H_5OH(l)$$

The table shows the equilibrium concentrations of different reaction mixtures. 'Equilibrium concentration' means the concentration when the reaction has reached equilibrium. It is indicated by the subscript 'eq', though this is often omitted.

Equilibrium concentrations/mol dm^{-3}	$[CH_3COOC_2H_5(l)]_{eq}$	$[H_2O(l)]_{eq}$	$[CH_3COOH(l)]_{eq}$	$[C_2H_5OH(l)]_{eq}$
Experiment 1	0.090	0.531	0.114	0.114
Experiment 2	0.204	0.118	0.082	0.082
Experiment 3	0.151	0.261	0.105	0.105

Table 2 Equilibrium concentrations for the hydrolysis of ethyl ethanoate at 293 K

If you look at the data in the table, you will find that the expression

$$K_c = \frac{[CH_3COOH(l)]\ [C_2H_5OH(l)]}{[CH_3COOC_2H_5(l)]\ [H_2O(l)]}$$

is constant for the three experiments. The constant **K_c** is the **equilibrium constant** for this reaction.

The concentrations in the expression for K_c *must* be those at equilibrium. From the data given in Table 2 you can check that the value of K_c is constant at about 0.28 for the hydrolysis of ethyl ethanoate at 293 K. This value, which is less than 1, tells you that, at equilibrium, a substantial proportion of the reactants is left unreacted – the reaction is incomplete.

Table 3 shows results obtained for another reaction, between hydrogen and iodine to form hydrogen iodide

$$H_2(g) + I_2(g) \rightleftharpoons 2HI(g)$$

In the first three experiments, mixtures of hydrogen and iodine were put into sealed reaction vessels. In the final two experiments hydrogen iodide alone was sealed into the vessel (Figure 4). The mixtures were held at a constant temperature of 731 K until equilibrium was reached. The concentrations of all three substances were then recorded.

For this reaction the expression which is constant at equilibrium is

Figure 4 The equilibrium $H_2(g) + I_2(g) \rightleftharpoons 2HI(g)$ can be approached from two opposite directions (Table 3)

$$K_c = \frac{[HI(g)]^2}{[H_2(g)]\ [I_2(g)]}$$

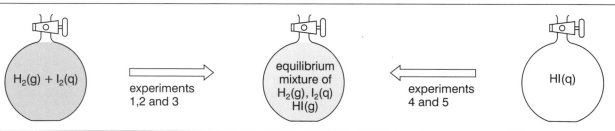

Experiment	Initial concentrations/mol dm^{-3}			Equilibrium concentrations/mol dm^{-3}			K_c
	$[H_2(g)]$	$[I_2(g)]$	$[HI(g)]$	$[H_2(g)]$	$[I_2(g)]$	$[HI(g)]$	
1	2.40×10^{-2}	1.38×10^{-2}	0	1.14×10^{-2}	0.12×10^{-2}	2.52×10^{-2}	46.4
2	2.44×10^{-2}	1.98×10^{-2}	0	0.77×10^{-2}	0.31×10^{-2}	3.34×10^{-2}	46.7
3	2.46×10^{-2}	1.76×10^{-2}	0	0.92×10^{-2}	0.22×10^{-2}	3.08×10^{-2}	46.9
4	0	0	3.04×10^{-2}	0.345×10^{-2}	0.345×10^{-2}	2.35×10^{-2}	46.9
5	0	0	7.58×10^{-2}	0.86×10^{-2}	0.86×10^{-2}	5.86×10^{-2}	46.4

Table 3 Initial and equilibrium concentrations for the reaction $H_2(g) + I_2(g) \rightleftharpoons 2HI(g)$

The values of K_c are shown in Table 3. You can see that the equilibrium constant is the same whether we start from a mixture which is all $H_2 + I_2$ (Experiments 1 to 3) or one which is all HI (Experiments 4 and 5). The graphs in Figure 5 show how the concentrations of reactants and products change as the reaction approaches equilibrium.

The mean value of K_c for this reaction at 731 K is about 49.7. This means that, at equilibrium, most of the H_2 and I_2 has been converted to HI, with a bit left unreacted.

Writing the expression for K_c

The rules for writing K_c expressions have been discovered by using the results of many experiments. This makes it possible to write an expression for K_c for any reaction, without having to examine data.

Start by writing a balanced equation for the reaction. For example, take the reaction in which insulin dimers, In_2, break down to form insulin monomers, In,

$$3In_2(aq) \rightleftharpoons 6In(aq)$$

In the expression for the K_c, the products of the forward reaction appear on the top line and the reagents on the bottom line. The power to which you raise the concentration of a substance is the same as the number which appears in front of it in the balanced equation.

$$K_c = \frac{[In(aq)]^6}{[In_2(aq)]^3}$$

In general, if an equilibrium mixture contains substances A, B, C, and D which react according to the equation.

$$aA + bB \rightleftharpoons cC + dD$$

then $K_c = \dfrac{[C]^c\,[D]^d}{[A]^a\,[B]^b}$

Values of K_c vary enormously, as you can see from the values in Table 4 for three reactions at the same temperature.

All reactions are equilibrium reactions. Even reactions that seem to go to completion actually have a little bit of reactant left in equilibrium with the product. In some cases the equilibrium position is so far towards products that it is hard to detect the tiny concentration of reactants left in the equilibrium mixture. Such reactions have very large equilibrium constants.

What are the units of K_c?

The units of K_c vary. It depends on the expression for K_c for the particular reaction you are studying.

Example 1

$$H_2(g) + Br_2(g) \rightleftharpoons 2HBr(g)$$

$$K_c = \frac{[HBr(g)]^2}{[H_2(g)]\,[I_2(g)]}$$

The units of K_c are $\dfrac{(mol\,dm^{-3})^2}{(mol\,dm^{-3})\,(mol\,dm^{-3})}$

So K_c for this reaction has no units, since they cancel out on the top and bottom of the expression.

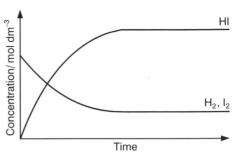

(a) A mixture of pure hydrogen and iodine reaches equilibrium.

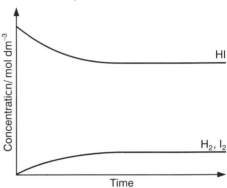

(b) Hydrogen iodide decomposes to hydrogen and iodine at the same temperature.

Figure 5 Equilibrium can be reached from either a mixture of hydrogen and iodine, or from pure hydrogen iodide

Reaction	K_c
$H_2(g) + I_2(g) \rightleftharpoons 2HI(g)$	~10
$H_2(g) + Br_2(g) \rightleftharpoons 2HBr(g)$	~10^9
$H_2(g) + Cl_2(g) \rightleftharpoons 2HCl(g)$	~10^{17}

Table 4 Some values of K_c

Example 2

$$3In_2(aq) \rightleftharpoons 6In(aq)$$

$$K_c = \frac{[In(aq)]^6}{[In_2(aq)]^3}$$

The units of K_c are $\dfrac{(mol\,dm^{-3})^6}{(mol\,dm^{-3})^3} = \mathbf{mol^3\,dm^{-9}}$

You can see that the units of K_c vary from reaction to reaction, and need to be worked out from the K_c expression. When you quote a value for K_c you must always show a balanced equation for the reaction.

What happens to the position of equilibrium if concentrations are changed?

Suppose a system is at equilibrium and you suddenly disturb it, say by adding more of a reagent. The composition of the system will change until equilibrium is reached again. The composition of the mixture will always adjust to keep the value of K_c constant, provided the temperature stays constant.

Lets look at an example. In an experiment involving the formation of ethyl ethanoate

$$CH_3COOH(l) + C_2H_5OH(l) \rightleftharpoons CH_3COOC_2H_5(l) + H_2O(l)$$

the system was allowed to reach equilibrium (Experiment 1). The equilibrium concentrations for Experiment 1 are shown in Figure 6.

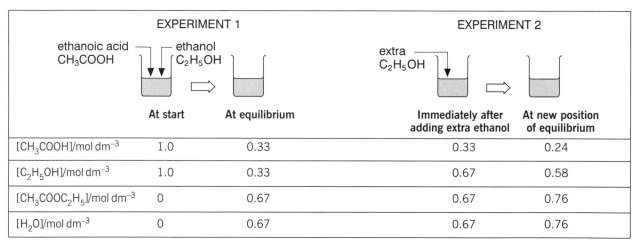

	EXPERIMENT 1		EXPERIMENT 2	
	At start	At equilibrium	Immediately after adding extra ethanol	At new position of equilibrium
$[CH_3COOH]$/mol dm^{-3}	1.0	0.33	0.33	0.24
$[C_2H_5OH]$/mol dm^{-3}	1.0	0.33	0.67	0.58
$[CH_3COOC_2H_5]$/mol dm^{-3}	0	0.67	0.67	0.76
$[H_2O]$/mol dm^{-3}	0	0.67	0.67	0.76

Figure 6 In Experiment 1, equilibrium is set up, starting with equal concentrations of ethanoic acid and ethanol. In Experiment 2, the equilibrium is disturbed by adding extra ethanol

Using the equilibrium concentrations from Experiment 1,

$$K_c = \frac{[CH_3COOC_2H_5]\,[H_2O]}{[CH_3COOH]\,[C_2H_5OH]}$$

$$= \frac{(0.67\,mol\,dm^{-3})\,(0.67\,mol\,dm^{-3})}{(0.33\,mol\,dm^{-3})\,(0.33\,mol\,dm^{-3})}$$

$$= 4.0$$

In Experiment 2, one of the concentrations was deliberately changed by adding more C_2H_5OH to give a new concentration of 0.67 mol dm^{-3}.

Immediately after adding the extra C_2H_5OH, before any changes occur, the new concentration ratio is

$$\frac{(0.67\,mol\,dm^{-3})\,(0.67\,mol\,dm^{-3})}{(0.33\,mol\,dm^{-3})\,(0.67\,mol\,dm^{-3})} = 2, \text{ much smaller than } K_c.$$

In order to restore the value of K_c to 4.0, some C_2H_5OH and CH_3COOH must react (making the bottom line smaller) to produce $CH_3COOC_2H_5$ and H_2O (making the top line bigger).

The system was now left to reach equilibrium again, and the new equilibrium concentrations were measured. The values are shown in Figure 6.

Substituting these concentrations in the expression for K_c does give a value close to 4.0, as expected.

$$\frac{(0.76 \text{ mol dm}^{-3})\ (0.76 \text{ mol dm}^{-3})}{(0.24 \text{ mol dm}^{-3})\ (0.58 \text{ mol dm}^{-3})} = 4.1$$

So although the equilibrium was disturbed, it moved in such a way that K_c remained constant.

Le Chatelier's principle

The position of equilibrium is affected not only by concentration changes, but also by changes in other factors, such as temperature and pressure.

By studying data from many reactions Henri Le Chatelier was able (in 1888) to propose a simple general rule:

If a system is at equilibrium, and a change is made in any of the conditions, then the system responds to counteract the change as much as possible.

This rule allows you to make qualitative predictions about the effect of changes on equilibria. For example, adding more reactant will shift an equilibrium towards the product side. Removing product will have a similar effect: the system will respond to the change by reacting to make more product.

Adding product will shift an equilibrium towards the reactant side because the system responds by reacting to use up some of the added product.

You can find out about the effect of other factors, such as temperature and pressure, in **Section 7.4**.

PROBLEMS FOR 7.2

1 Write expressions for K_c for the following reactions. In each case, give the units of K_c.
   a $2NO(g) + O_2(g) \rightleftharpoons 2NO_2(g)$
   b $C_2H_6(g) \rightleftharpoons C_2H_4(g) + H_2(g)$
   c $2HI(g) \rightleftharpoons H_2(g) + I_2(g)$
   d $2NO_2(g) \rightleftharpoons N_2O_4(g)$
   e $2CO(g) + O_2(g) \rightleftharpoons 2CO_2(g)$
   f $C_5H_{10}(l) + CH_3COOH(l) \rightleftharpoons CH_3COOC_5H_{11}(l)$

2 A mixture of nitrogen and hydrogen was sealed into a steel vessel and held at 1000 K until equilibrium was reached. The contents were then analysed. The results are given in the following table.

Substance	Equilibrium concentration/mol dm^{-3}
$N_2(g)$	0.142
$H_2(g)$	1.84
$NH_3(g)$	1.36

   a Write an expression for K_c for the reaction
      $N_2(g) + 3H_2(g) \rightleftharpoons 2NH_3(g)$
   b Calculate a value for K_c. Remember to give units.

3 For the reaction
   $2H_2(g) + S_2(g) \rightleftharpoons 2H_2S(g)$
   K_c was found to be $9.4 \times 10^5 \text{ mol}^{-1} \text{dm}^3$ at 1020 K. Equilibrium concentrations were measured as
      $[H_2(g)] = 0.234 \text{ mol dm}^{-3}$
      $[H_2S(g)] = 0.442 \text{ mol dm}^{-3}$
   a Write an expression for K_c for the reaction.
   b What is the equilibrium concentration of $S_2(g)$ in the above mixture?

4 The reaction of iodine with propanone slowly reaches an equilibrium with the formation of iodopropanone.

   $CH_3COCH_3(aq) + I_2(aq) \rightleftharpoons CH_3COCH_2I(aq) + I^-(aq) + H^+(aq)$
   propanone   *iodopropanone*   (reaction 1)

   The reaction is catalysed by acid.
   As the iodine is present in the mixture mainly as the triiodide ion, $I_3^-(aq)$, the equation is better written as

   $CH_3COCH_3(aq) + I_3^-(aq) \rightleftharpoons CH_3COCH_2I(aq) + 2I^-(aq) + H^+(aq)$
                                                          (reaction 2)

   The equilibrium between iodine and triiodide is
   $I_2(aq) + I^-(aq) \rightleftharpoons I_3^-(aq)$ (reaction 3)

A mixture of aqueous iodine (in an excess of potassium iodide), excess propanone and excess acid was prepared and allowed to reach equilibrium at 298 K. The solution was then analysed, and the equilibrium concentrations were found to be

$[I_3^-(aq)]$ $= 4.25 \times 10^{-3} \, \text{mol dm}^{-3}$
$[CH_3COCH_2I(aq)]$ $= 5.75 \times 10^{-3} \, \text{mol dm}^{-3}$
$[CH_3COCH_3(aq)]$ $= 0.828 \, \text{mol dm}^{-3}$
$[H^+(aq)]$ $= 0.60 \, \text{mol dm}^{-3}$
$[I^-(aq)]$ $= 0.20 \, \text{mol dm}^{-3}$

a   Write expressions, K_1, K_2, K_3 for the equilibrium constants for reactions 1, 2 and 3.

b   Show that $K_1 = K_2 \times K_3$

c   Use your expression for K_2 and the equilibrium concentrations given to calculate a value for K_2. Remember to give its units.

d   The value for K_3 is $714 \, \text{mol}^{-1} \, \text{dm}^3$. Calculate a value for K_1.

e   A second reaction mixture was prepared using more iodine but keeping the other concentrations the same. The new equilibrium concentration of $I_3^-(aq)$ was $9.50 \times 10^{-3} \, \text{mol dm}^{-3}$. Use your value of K_2 to calculate the new concentration of iodopropanone in the mixture. (As the propanone, acid and iodide were in excess it is a good approximation to assume that their equilibrium concentrations are effectively the same as their original concentrations.)

f   Explain the change in iodopropanone concentration in terms of the equilibria involved.

7.3 *Partition equilibrium*

Most of the equilibrium systems you have met so far have involved reactants and products in the same phase, either all gases or all miscible liquids.

However you also get equilibrium between substances in *different* phases, such as two liquids which do not mix. For example, ammonia gas is extremely soluble in water, and also dissolves in trichloromethane. The two liquids do not mix.

Suppose you dissolve some ammonia in water, add an approximately equal volume of trichloromethane, and shake the two liquids together in a separating funnel (Figure 7). Ammonia then passes between the two liquids until equilibrium is established.

$$NH_3(aq) \rightleftharpoons NH_3(CHCl_3)$$

You could run the two layers into separate containers, and find the equilibrium concentrations of ammonia by titration with standard acid solution. Table 5 shows equilibrium concentrations for different initial concentrations of ammonia at 298 K.

You can write an equilibrium constant as before.

$$K_c = \frac{[NH_3(CHCl_3)]}{[NH_3(aq)]}$$

You can use the data in the table to show that K_c is indeed a constant, with a value of about 0.04. This particular kind of equilibrium constant measures how a substance distributes, or partitions, itself between two phases. It is called a **partition coefficient.**

What determines how a solute distributes itself between two solvents? Substances which are ionic or polar are more soluble in water than in non-polar organic solvents. Some values of partition coefficients are given in Table 6 to illustrate this behaviour.

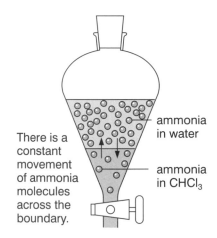

There is a constant movement of ammonia molecules across the boundary.

Figure 7 An equilibrium is set up between ammonia in water and ammonia in CHCl₃

Equilibrium concentrations of ammonia	
in water/mol dm⁻³	in trichloromethane /mol dm⁻³
0.60	0.025
0.93	0.040
1.40	0.060
1.86	0.080

Table 5 Equilibrium concentrations of ammonia partitioned between water and trichloromethane at 298 K

Solute	Solvents	$K = \dfrac{[\text{solute(organic)}]}{[\text{solute(aq)}]}$
Cl_2	CCl_4 / H_2O	10
I_2	CCl_4 / H_2O	83
CH_3COOH	C_6H_6 / H_2O	0.063
$CH_2ClCOOH$	C_6H_6 / H_2O	0.036
DDT	octanol/H_2O	9.5×10^5

Table 6 Partition coefficients at 298 K for some solutes between organic solvents and water

Ion-exchange is another example of partition equilibrium. Ions establish equilibrium between aqueous solution and the surface of the ion-exchange resin:

$$resin–H^+(s) + Na^+(aq) \rightleftharpoons resin–Na^+(s) + H^+(aq)$$

PROBLEMS 7.3

1 One gram of butanedioic acid was placed in a separating funnel with $25\,cm^3$ water and $25\,cm^3$ ethoxyethane. The funnel was shaken until all the acid had dissolved and equilibrium was reached. The mixture was then left to stand while the two liquids separated. Portions of each layer were titrated with a standard solution of sodium hydroxide. A series of experiments was carried out at the same temperature with different initial quantities of butanedioic acid. Some results are given in the table below.

Experiment number	Equilibrium concentration of butanedioic acid/mol dm^{-3}	
	in ethoxyethane layer	in water layer
1	0.023	0.152
2	0.036	0.242
3	0.044	0.300
4	0.055	0.381

a Calculate an average value for the partition coefficient of butanedioic acid between ethoxyethane and water at this temperature.

b If the equilibrium concentration of butanedioic acid in the water layer is $0.036\,mol\,dm^{-3}$, then what will be the equilibrium concentration in the ethoxyethane layer?

2 The insecticide permethrin is sprayed onto crops at a concentration of $200\,g\,ha^{-1}$ (1 hectare, ha, $= 1 \times 10^4\,m^2$). One cubic decimetre of spray is needed for one square metre.

After two months the permethrin concentration in the water in the soil has dropped to 2% of the spray concentration.

a Calculate
  i the mass of permethrin in one cubic decimetre of spray;
  ii the concentration in $mol\,dm^{-3}$ of permethrin in the spray
    (M_r of permethrin $= 391$)
  iii the concentration of permethrin in the water in the soil after two months.

b The partition coefficient, K_{ow}, for permethrin between water and octanol is 1.00×10^5.
  i Suppose that after two months, the permethrin in the soil water is allowed to come to partition equilibrium between the soil water and octanol. Calculate the equilibrium concentration of permethrin in the octanol. Assume that you have a very large excess of water relative to octanol. It is then a good approximation to assume that the concentration of permethrin in the soil moisture is unchanged. (A similar situation would exist between an insect and the soil water.)
  ii Given that the toxic level of permethrin in insects is 5×10^{-7} mol per kg of body mass, comment on the likely effect of the concentration of permethrin in the soil water after two months.

7.4 *What decides the position of equilibrium?*

In **Section 7.2** you met Le Chatelier's principle, a simple rule which helps you predict the effect of changes on the position of equilibrium:

If a system is at equilibrium, and a change is made in any of the conditions, then the system responds to counteract the change as much as possible.

This rule allows you to make qualitative predictions about the effect of changes on equilibria.

What happens if the pressure is changed?

If the *total* pressure of a gas mixture at equilibrium is increased by compressing it, then Le Chatelier's principle tells us the system will respond by reducing the pressure again.

The pressure exerted by a gas depends on the number of gas molecules present. So the equilibrium shifts to the side with the smaller number of gas molecules.

In the ammonia system

$$N_2(g) + 3H_2(g) \rightleftharpoons 2NH_3(g)$$

there are four moles of gas on the reagent side, and only two moles of gas on the product side. What will happen if the total pressure is increased by compressing the gases? The system reacts to reduce the pressure by reducing the total number of moles of gas. Nitrogen and hydrogen combine to produce more ammonia. Table 7 shows some experimental results. As the pressure is increased, the proportion of ammonia increases.

The effect of increasing total pressure can be summarised by saying

> *increasing the pressure shifts the equilibrium position towards the side with the smaller number of moles of gas.*

If there is the same number of moles of gas on each side, then increasing the total pressure has no effect on the equilibrium position.

Total pressure/atm	Equilibrium % of ammonia
1.0	0.24
5.0	9.5
100	16.2
200	25.3

Table 7 Percentage of ammonia in equilibrium mixtures at different total pressures. The temperature is 723 K

What happens if the temperature is changed?

According to Le Chatelier's principle any increase in temperature of a mixture at equilibrium will cause change which counteracts the increase.

For the ammonia system, the forward reaction is exothermic.

$$N_2(g) + 3H_2(g) \rightleftharpoons 2NH_3(g); \quad \Delta H^\circ = -92\,kJ\,mol^{-1}$$

The back reaction is endothermic. An increase in temperature will favour the back reaction. The reason for this is that when the back reaction occurs the reaction mixture cools down because the reaction absorbs energy. This counteracts the increase in temperature.

So in general, increasing the temperature makes the equilibrium position move in the direction of the *endothermic* reaction.

How is the equilibrium constant affected?

Changes in concentration and pressure may alter the composition of equilibrium mixtures, but *they do not alter the value of the equilibrium constant itself*, provided the temperature does not change. The proportions of reactants and products alter in such a way as to keep the ratio in the K_c expression unchanged.

However, *changes in temperature actually alter the value of the equilibrium constant itself.*

Table 8 The effect of temperature on the equilibrium constant of two reacting systems; the value of ΔH° is for the forward reaction in each case

$N_2(g) + 3H_2(g) \rightleftharpoons 2NH_3(g)$; $\Delta H^\circ = -92\,kJ\,mol^{-1}$		$N_2O_4 \rightleftharpoons 2NO_2(g)$; $\Delta H^\circ = +57\,kJ\,mol^{-1}$	
T/K	K_c/mol^{-2} dm^6	*T*/K	K_c/mol dm^{-3}
400	4.39×10^4	200	5.51×10^{-8}
600	4.03	400	1.46
800	3.00×10^{-2}	600	3.62×10^2
$K_c = \dfrac{[NH_3(g)]^2}{[N_2(g)]\,[H_2(g)]^3}$		$K_c = \dfrac{[NO_2(g)]^2}{[N_2O_4(g)]}$	

Table 8 shows some experimental measurements of how temperature changes affect K_c for two reactions.

You can see from the data that for endothermic reactions, a rise in temperature favours the products and increases K_c. For exothermic reactions a rise in temperature favours the reactants and decreases K_c.

The effect of changing conditions on equilibrium mixtures is summarised in Table 9.

Does a catalyst affect the equilibrium position?

No, catalysts do not affect the position of equilibrium. They alter the *rate* at which equilibrium is attained but not the composition of the equilibrium mixture.

Change in:	Composition	K_c
concentration	changed	unchanged
total pressure	may change	unchanged
temperature	changed	changed

Table 9 The effect of changing conditions on equilibrium mixtures

---

PROBLEMS FOR 7.4

1 State the direction in which the position of equilibrium of each system would move (if at all) if the pressure were increased by compressing the reaction mixture. Give your answer as 'left → right', 'right → left', or 'no change'.

  a $2NO(g) + O_2(g) \rightleftharpoons 2NO_2(g)$
  b $C_2H_6(g) \rightleftharpoons C_2H_4(g) + H_2(g)$
  c $2HI(g) \rightleftharpoons H_2(g) + I_2(g)$
  d $2NO_2(g) \rightleftharpoons N_2O_4(g)$
  e $2CO(g) + O_2(g) \rightleftharpoons 2CO_2(g)$

2 In the manufacture of ammonia, hydrogen can be obtained from natural gas by an endothermic reaction with steam,

$$CH_4(g) + H_2O(g) \rightleftharpoons CO(g) + 3H_2(g);$$
$$\Delta H^\circ = +206.1 \text{ kJ mol}^{-1}$$

  a Write an expression for K_c for this reaction.
  b For each of the changes below, say what will happen to the value of K_c and explain why.
   i The temperature is increased.
   ii The pressure is increased by compressing the gas mixture.
   iii A catalyst is used.
  c How will the composition of the equilibrium mixture be affected by
   i increasing the pressure?
   ii increasing the temperature?
   iii using a catalyst?

3 Consider the reaction between hydrogen and oxygen to produce steam.
  a Write an equation for the reaction.
  b Write an expression for K_c.
  c State how the equilibrium position is affected by
   i an increase in temperature
   ii an increase in pressure.

4 The following data indicate the effect of temperature and pressure on the equilibrium concentration of the product, Z, in an equilibrium involving gases.

Temperature/K	Percentage of Z in the equilibrium mixture at pressure of		
	1 atm	100 atm	200 atm
820	0.077	6.70	11.9
920	0.032	3.02	5.71
1020	0.016	1.54	2.99
1120	0.009	0.87	1.68

  a Use the data to deduce whether the production of Z is accompanied by an increase or a decrease in pressure (assuming the volume stays constant). Explain your reasoning.
  b Use the data to deduce whether the production of Z is an exothermic or endothermic process. Explain your reasoning.
  c What conditions of pressure and temperature would give the highest equilibrium yield of Z? (Give a qualitative answer – no numbers are needed.)

7.5 *Ion exchange*

In ion-exchange reactions, ions from solution change places with ions held by a solid, called the ion exchanger. The solid may be a synthetic resin or a natural material. Natural materials include alumino-silicate minerals, especially zeolites. Zeolites do not have the sheet structure of the clay minerals. Instead the silicate tetrahedra are joined together to give open frameworks with channels and cavities into which cations can fit (Figure 8). There is more about zeolites in the **Developing Fuels** storyline.

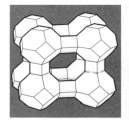 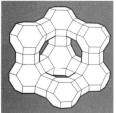

Type A   Type Y

Figure 8 Two types of zeolite structure

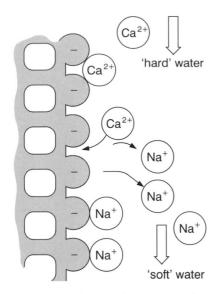

Figure 9 *Ion exchange columns are used to soften water. The ion exchange resin removes calcium ions from the water, and replaces them with sodium ions*

Zeolites have high exchange capacities compared to clays. Artificial zeolites, called 'Permutits' were developed at the beginning of this century for water softening. Since then many different materials have been developed for both cation and anion exchange.

When the ion exchanger comes into contact with ions in solution an equilibrium is set up, for example:

$$resin–H^+(s) + Na^+(aq) \rightleftharpoons resin–Na^+(s) + H^+(aq)$$

resin–H^+ stands for the resin in its hydrogen form.

You may already know about the use of ion exchange columns for purifying water. Water which contains more than 0.2 mol dm^{-3} of dissolved calcium and magnesium salts is called 'hard water'. It can cause problems in domestic appliances, and is a serious problem in industry where pipes can be quickly clogged up with insoluble calcium and magnesium carbonates. To remove the 'hardness' the water is passed down a column of resin in its sodium form (Figure 9).

Calcium and magnesium ions in the water change places with sodium ions on the resin. Because the calcium and magnesium ions have a double positive charge they are held more firmly than sodium ions.

The reaction is reversible, but because water is flowing continuously through the column the net effect is to remove calcium and magnesium ions from the water and replace them with sodium ions. We can show the reaction as

$$2\ resin–Na^+(s) + M^{2+}(aq) \rightleftharpoons (resin)_2–M^{2+}(s) + 2Na^+(aq)$$

where M^{2+} represents a calcium or magnesium ion.

When the ion exchanger becomes exhausted (all the Na^+ ions replaced by Ca^{2+} and Mg^{2+} ions) it can be regenerated by passing a concentrated solution of sodium chloride through the column.

What decides how strongly an ion binds to a resin?

Resins bind different cations more or less firmly. The strength with which an ion is held to the surface depends on both the nature of the ion-exchanger, and the ion being held. The *size* and the *charge* of the ion are important: there is more about ionic size in **Section 3.2**. Some typical trends are described below.

Singly charged ions

strongly held $\quad$ $Cs^+ > Rb^+ > K^+ > NH_4^+ > Na^+ > Li^+$ $\quad$ weakly held

The smaller Li^+ ions are strongly hydrated in solution and the layers of water molecules around the lithium ion reduce the force between the resin and the lithium ions. Caesium ions do not hold as many water molecules around them as Li^+ ions do. They are therefore more firmly held by the ion exchanger than Li^+ ions. Table 10 shows the radii of some hydrated ions.

Doubly charged ions tend to be held more strongly than similarly sized singly charged ions.

Ion	Radius of hydrated ion/nm
Li$^+$	1.00
Na$^+$	0.79
NH$_4^+$	0.54
K$^+$	0.53
Cs$^+$	0.51

Table 10 *The radii of some hydrated ions*

Doubly charged ions

strongly held $\quad$ $Ba^{2+} > Sr^2 > Ca^2 > Cu^{2+} > Zn^{2+} > Mg^{2+}$ $\quad$ weakly held

There are variations from resin to resin as to which ions are most strongly bound to the resin.

Ion exchange finds a wide range of applications, for example in waste treatment, the recovery of pharmaceutical products from fermentation tanks, and in purification processes including the treatment of water containing radioactive ions like $^{137}Cs^+$.

Ion exchange processes are also very important in the soil. In some ways, soil behaves like a huge ion-exchange column.

PROBLEMS FOR 7.5

1 The equilibrium constant for the ion-exchange process
$$clay–Mg^{2+} + Ca^{2+}(aq) \rightleftharpoons clay–Ca^{2+} + Mg^{2+}(aq)$$
is close to 1. What would be the effect on the position of equilibrium of increasing the concentration of calcium ions, $Ca^{2+}(aq)$, in the soil solution?

2 An ion exchange resin can be regenerated to its hydrogen form by washing it in a column with a large volume of concentrated acid solution. Explain why
   a   this regenerates the resin,
   b   a high concentration of acid is used,
   c   a large volume of acid is used.

3 After a nuclear accident, the levels of certain radioisotopes may persist for some time. For example, despite the fact that caesium compounds are very soluble in water, radioactive caesium may continue to find its way into livestock, making meat and milk unfit for human consumption. In the UK there was a problem following the nuclear accident at Chernobyl. Sheep in upland areas of Wales and Cumbria with high rainfall were particularly affected.
   Explain why the uptake of radioactive caesium by the sheep continued for some time after the pollution had occurred, despite the fact that caesium compounds are very soluble and quickly washed into the soil.

7.6 *Chromatography*

Chromatography – the general principle

Chromatography is an important analytical technique used by chemists to separate and identify the components of a mixture.

There are a number of different types of chromatography. They all depend on the equilibrium set up when a compound distributes itself between two phases. One phase stands still (the **stationary phase**) while the other moves over it (the **mobile phase**). Different compounds distribute themselves between the two phases to different extents, and so move along with the mobile phase at different speeds.

You will probably be familiar with paper chromatography, and you may also have used **thin-layer chromatography** (t.l.c.) shown in Figure 10. Here a small sample of a mixture is spotted onto a solid support material (the *stationary phase*). This might be silica gel spread in a thin uniform layer on a plastic plate. A suitable solvent now rises up the plate. This is the *mobile phase*.

R_f values

The distance a particular substance travels in paper or thin-layer chromatography depends on
● the nature of the substance
● the total distance travelled by the solvent front
● the conditions under which chromatography is carried out – such as temperature, type of paper or thin-layer plate, and the type of solvent used.

The R_f **value** for a substance is the distance that substance travels relative to the solvent front (Figure 11). R_f values are constant for a particular set of conditions, so we can use them to identify the different spots on a chromatogram.

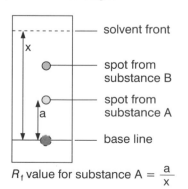

R_f value for substance A $= \dfrac{a}{x}$

Figure 11 R_f values

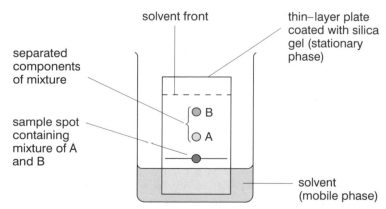

Figure 10 Thin-layer chromatography

What you see when you examine the developed plate is a series of spots, one for each compound in the mixture. Figure 12 shows in a simplified way how two components get separated in thin-layer chromatography.

Figure 12 A simplified explanation of thin-layer chromatography. This magnified view shows the behaviour of components A and B as the solvent passes between the particles of silica gel in the thin layer

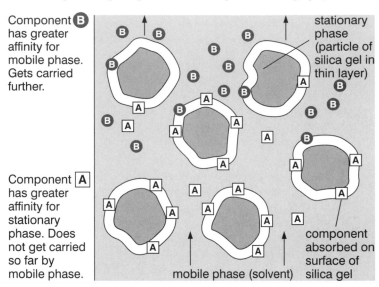

Component **B** has greater affinity for mobile phase. Gets carried further.

stationary phase (particle of silica gel in thin layer)

Component **A** has greater affinity for stationary phase. Does not get carried so far by mobile phase.

mobile phase (solvent)   silica gel

component absorbed on surface of silica gel

Gas-liquid chromatography

In **gas-liquid chromatography** (g.l.c.) the principle is the same as in thin-layer chromatography. However, in this case the mobile phase is an unreactive gas such as nitrogen, called the **carrier gas**. The stationary phase is a small amount of a high boiling point liquid held on a finely-divided inert solid support. This material is packed into a long thin tube called a **column** (Figure 13). The column is coiled inside an oven.

The main parts of a simple gas-liquid chromatograph are shown in Figure 14. The sample to be analysed is injected into the gas stream just before it enters the column. The components of the mixture are carried through the column in a stream of gas.

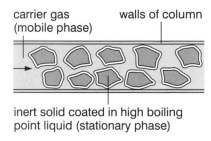

carrier gas (mobile phase)   walls of column

inert solid coated in high boiling point liquid (stationary phase)

Figure 13 Inside a g.l.c. column. (In reality there are far more particles of solid than this)

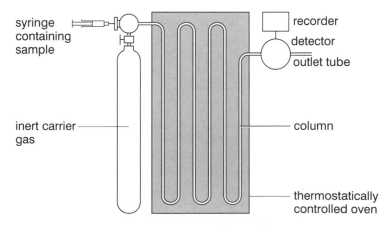

syringe containing sample

recorder
detector
outlet tube

inert carrier gas

column

thermostatically controlled oven

Figure 14 A gas-liquid chromatograph

Each component has a different affinity for the stationary phase compared with the mobile phase. Each distributes itself to different extents between the two phases and so emerges from the column at a different time. Those compounds that favour the mobile phase (the carrier gas) are carried along more quickly. The most volatile compounds usually emerge first. The compounds that favour the stationary phase (the liquid) get held up in the column and come out last.

A **detector** on the outlet tube monitors compounds coming off the column. Signals from the detector are plotted out by a **recorder** as a **chromatogram**. This shows the recorder response against the time which has elapsed since the sample was injected onto the column. Each component of the mixture gives rise to a peak. Figure 15 shows a typical gas chromatogram of a premium grade petrol (see **Developing Fuels** storyline).

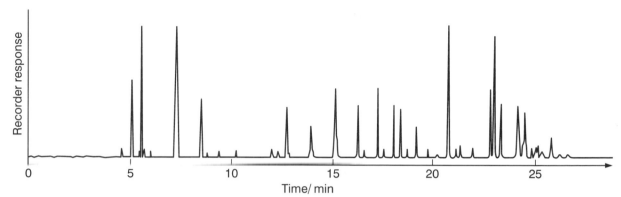

Figure 15 Gas chromatogram of a premium grade petrol

The time a compound is held on a column under given conditions is characteristic of the compound and is called its **retention time**.

Lots of factors can affect the retention time – such as the length and packing of the column, the nature and flow rate of the carrier gas, the temperature of the column. So you have to calibrate the instrument with known compounds and keep the conditions constant throughout the analysis.

The area under each peak depends on the amount of compound present, so you can use a gas chromatogram to work out the *relative amount* of each component in the mixture. (If the peaks are very sharp, their relative *heights* can be used.)

The technique is very sensitive and very small quantities can be detected, such as traces of explosives or drugs in forensic tests. With larger instruments, a pure sample of each compound can be collected as it emerges from the outlet tube. In more sophisticated instruments, the outlet tube is connected to a mass spectrometer, so that each compound can be identified directly.

PROBLEMS FOR 7.6

1   Explain why it is important to enclose the column of a gas-liquid chromatograph inside an oven kept at a constant temperature.

2   What types of substances do you think would not be suitable for analysis using a gas-liquid chromatograph? Explain your answer.

3   The chromatogram in Figure 16 was obtained from a mixture containing a pair of *cis* and *trans* isomers.

Pure *cis*-isomer had a retention time of 9.0 min under identical conditions. The pure *trans*-isomer had a retention time of 12.5 min. What were the proportions of *cis* and *trans* isomers in the mixture?

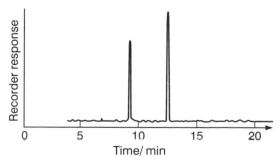

Figure 16 Gas chromatogram for a mixture containing a pair of cis and trans isomers

7.7 *Solubility equilibria*

We often describe substances as being either 'soluble' or 'insoluble', but this is an oversimplification. As you read through **The Oceans** storyline, you will realise that no substance is totally insoluble; there will always be some quantity in solution, however small. Thus rainwater dissolves silicon(IV) oxide from rocks and transports it to the sea, even though the sand on the beach appears insoluble. Calcium carbonate dissolves and reprecipitates in various ocean processes.

A sparingly soluble ionic solid like calcium carbonate, in contact with a saturated solution of its ions is an example of a chemical equilibrium. We can write

$$CaCO_3(s) \rightleftharpoons Ca^{2+}(aq) + CO_3^{2-}(aq)$$

$$K_c = \frac{[Ca^{2+}(aq)]\,[CO_3^{2-}(aq)]}{[CaCO_3(s)]}$$

Adding more solid will not cause the equilibrium to shift further to the product side because the solution is saturated at that temperature. In other words, the equilibrium is not affected by the amount of solid present. We can omit the solid from the expression for the equilibrium constant, which becomes

$$K_{sp} = [Ca^{2+}(aq)]\,[CO_3^{2-}(aq)]$$

Solubility products

K_{sp} stands for **solubility product**. In this case it is the solubility product of $CaCO_3(s)$. Its value is $5.0 \times 10^{-9}\,mol^2\,dm^{-6}$ at 298 K.

K_{sp} represents the conditions for equilibrium between a sparingly soluble solid and its saturated solution. At 298 K, whenever we have solid calcium carbonate in equilibrium with its solution, we will always find that $[Ca^{2+}(aq)]\,[CO_3^{2-}(aq)] = 5 \times 10^{-9}\,mol^2\,dm^{-6}$.

We can use the K_{sp} value to *predict whether a precipitate will form from a solution*. Suppose we have a solution in which $[Ca^{2+}(aq)] = 10^{-5}\,mol\,dm^{-3}$ and $[CO_3^{2-}(aq)] = 10^{-5}\,mol\,dm^{-3}$ (Figure 17a). In this case, $[Ca^{2+}(aq)]\,[CO_3^{2-}(aq)] = 10^{-10}\,mol^2\,dm^{-6}$. This is *less* than $5.0 \times 10^{-9}\,mol^2\,dm^{-6}$, so no precipitate of $CaCO_3$ will form.

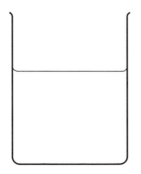

(a) $[Ca^{2+}(aq)] = 10^{-5}\,mol\,dm^{-3}$
$[CO_3^{2-}(aq)] = 10^{-5}\,mol\,dm^{-3}$
$[Ca^{2+}(aq)][CO_3^{2-}(aq)] = 10^{-10}\,mol^2\,dm^{-6}$
This is less than $5.0 \times 10^{-9}\,mol^2\,dm^{-6}$
⟶ NO PRECIPITATE

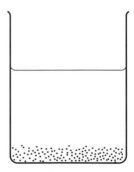

(b) $[Ca^{2+}(aq)] = 10^{-4}\,mol\,dm^{-3}$
$[CO_3^{2-}(aq)] = 10^{-4}\,mol\,dm^{-3}$
$[Ca^{2+}(aq)][CO_3^{2-}(aq)] = 10^{-8}\,mol^2\,dm^{-6}$
This is greater than $5.0 \times 10^{-9}\,mol^2\,dm^{-6}$
⟶ PRECIPITATE FORMS

Figure 17 We can use K_{sp} to decide whether a precipitate will form from a solution

However, if we have a solution in which $[Ca^{2+}(aq)] = 10^{-4}\,mol\,dm^{-3}$ and $[CO_3^{2-}(aq)] = 10^{-4}\,mol\,dm^{-3}$ (Figure 17b) then $[Ca^{2+}(aq)]\,[CO_3^{2-}(aq)] = 10^{-8}\,mol^2\,dm^{-6}$. This is *greater* than $5.0 \times 10^{-9}\,mol^2\,dm^{-6}$, so $CaCO_3(s)$ precipitates out of solution, until the concentrations of $Ca^{2+}(aq)$ and $CO_3^{2-}(aq)$ have decreased so that $[Ca^{2+}(aq)]\,[CO_3^{2-}(aq)] = 5.0 \times 10^{-9}\,mol^2\,dm^{-6}$. Then no more precipitate forms, and the solution is in equilibrium with the solid. So K_{sp} gives us the *maximum* concentrations of ions that can be in equilibrium with the solid.

K_{sp} can be used like any other equilibrium constant, and there are solubility products for all sparingly soluble ionic solids, such as $AgCl(s)$, $CaF_2(s)$ and $BaSO_4(s)$. For $CaF_2(s)$, for example, the equilibrium law allows us to write K_{sp} as

$$K_{sp}\,(CaF_2) = [Ca^{2+}(aq)]\,[F^-(aq)]^2$$

Some K_{sp} values are collected together in Table 11. Note that K_{sp} changes with temperature, so we must always quote the temperature to which a K_{sp} value relates.

We can use K_{sp} for silver chloride to calculate the silver ion concentration in a saturated solution of silver chloride:

$AgCl(s) \rightleftharpoons Ag^+(aq) + Cl^-(aq)$
$K_{sp}(AgCl) = [Ag^+(aq)]\,[Cl^-(aq)] = 2.0 \times 10^{-10}\,mol^2\,dm^{-6}$
Since equal amounts of Ag^+ and Cl^- go into solution,
$[Ag^+(aq)] = [Cl^-(aq)]$
Therefore, $2.0 \times 10^{-10}\,mol^2\,dm^{-6} = [Ag^+(aq)]^2$
and $[Ag^+(aq)] = 1.4 \times 10^{-5}\,mol\,dm^{-3}$

Sometimes, other compounds which are in solution can affect the solubility of a sparingly soluble compound. The solubility of silver chloride, for example, will be affected by the presence of any other compound which dissolves to produce Ag^+ or Cl^- ions in the solution. Le Chatelier's principle tells us that silver ions or chloride ions will cause the solubility equilibrium to shift towards the $AgCl(s)$ side, making silver chloride less soluble.

Imagine that we put some solid silver chloride into a $0.1\,mol\,dm^{-3}$ solution of sodium chloride. What will be the dissolved silver ion concentration at equilibrium? This time we can not assume that $[Ag^+(aq)] = [Cl^-(aq)]$. In fact, we can neglect the small amount of $Cl^-(aq)$ which comes from silver chloride, and assume that the concentration of Cl^- ions in the solution is effectively equal to the concentration of the sodium chloride solution. In which case:

$[Cl^-(aq)] = 0.1\,mol\,dm^{-3}$.
Therefore, $2.0 \times 10^{-10}\,mol^2\,dm^{-6} = [Ag^+(aq)] \times 0.1\,mol\,dm^{-3}$

$$so\ [Ag^+(aq)] = \frac{2.0 \times 10^{-10}\,mol^2\,dm^{-6}}{0.1\,mol\,dm^{-3}}$$

or $[Ag^+(aq)] = 2.0 \times 10^{-9}\,mol\,dm^{-3}$ (at 298 K)

The silver ion concentration is a lot lower than in a solution of silver chloride alone, and so, of course, is the actual *solubility* of the silver chloride. But the product of the silver ion concentration and the chloride ion concentration is the same in both solutions. It is constant: $2.0 \times 10^{-10}\,mol^2\,dm^{-6}$ at 298 K.

Compound	K_{sp}	Units
$CaCO_3$	5.0×10^{-9}	$mol^2\,dm^{-6}$
$CaSO_4$	2.0×10^{-5}	$mol^2\,dm^{-6}$
$BaSO_4$	1.0×10^{-10}	$mol^2\,dm^{-6}$
PbI_2	7.1×10^{-9}	$mol^3\,dm^{-9}$
$AgCl$	2.0×10^{-10}	$mol^2\,dm^{-6}$
AgI	8.0×10^{-17}	$mol^2\,dm^{-6}$
PbS	1.3×10^{-28}	$mol^2\,dm^{-6}$
$Fe(OH)_3$	8.0×10^{-40}	$mol^4\,dm^{-12}$

Table 11 Values for some solubility products at 298 K

PROBLEMS FOR 7.7

1 Write down the expressions for K_{sp} for the following compounds and give the units for K_{sp} in each case.
   a $BaSO_4(s)$ c $PbI_2(s)$
   b $Ag_2CO_3(s)$ d $Fe(OH)_3(s)$

2 K_{sp} for thallium chloride, TlCl, is $1.75 \times 10^{-4}\,mol^2\,dm^{-6}$. Calculate the concentration of aqueous sodium chloride needed to just cause a precipitate of thallium chloride to form when mixed with an equal volume of $7.0 \times 10^{-3}\,mol\,dm^{-3}$ thallium nitrate(V), $TlNO_3$.

3 K_{sp} for silver bromate(V), $AgBrO_3$, is $6.0 \times 10^{-5}\,mol^2\,dm^{-6}$ at 298 K. A student added $100\,cm^3$ of $0.01\,mol^{-3}$ silver nitrate(V) solution to $100\,cm^3$ of $0.01\,mol\,dm^{-3}$ potassium bromate(V) solution at 298 K. Use a calculation to predict whether or not a precipitate of silver bromate(V) would be formed. (Remember that two solutions are being mixed together in this example and that this will affect the concentrations of the solutes.)

4 The Italian Dolomites are much favoured by climbers for their hard limestone which produces spectacular mountains as well as long, demanding climbing routes. This limestone differs from most English limestone in being composed largely of magnesium carbonate. The solubility products of calcium and magnesium carbonates are $5.0 \times 10^{-9}\,mol^2\,dm^{-6}$ and $1.0 \times 10^{-5}\,mol^2\,dm^{-6}$, respectively.
   a Calculate the concentration of magnesium ions in pure water which has reached equilibrium with magnesium carbonate.

   b Which carbonate, $CaCO_3$ or $MgCO_3$, is more soluble in pure water? Explain your answer.
   c A sample of natural water contains $9.1 \times 10^{-4}\,mol\,dm^{-3}$ of magnesium ions and $8.3 \times 10^{-4}\,mol\,dm^{-3}$ of calcium ions. On adding aqueous $1.00\,mol\,dm^{-3}$ sodium carbonate drop by drop to a $100\,cm^3$ sample of this water, a white precipitate forms. What is the composition of the white precipitate, initially and on adding excess of the aqueous sodium carbonate? Explain your answer in terms of solubility products. Assume 1 drop of the aqueous sodium carbonate has a volume of $0.10\,cm^3$.

5 a Use the value of K_{sp} for $BaSO_4(s)$ at 298 K given in Table 11 to calculate the barium ion concentration in a saturated solution of barium sulphate at this temperature.
   b One way of analysing for the barium content of a solution is to add a solution of sulphate ions (for example, using sodium sulphate), and to weigh the barium sulphate which is precipitated. An excess of sulphate ions is used so that as much barium sulphate as possible is precipitated. Calculate the barium ion concentration which would be left in a solution at 298 K in which the remaining concentration was $0.1\,mol\,dm^{-3}$ with respect to sulphate ions.

ACIDS AND BASES

Properties of acidic solutions

- turn litmus red
- neutralised by bases
- pH < 7
- liberate CO_2 from carbonates

8.1 *Acid-base reactions*

What do we mean by acid and base?

We say that hydrochloric acid is an 'acid' because of the things it does. It turns universal indicator red, reacts with carbonates to give CO_2 and is neutralised by bases – all properties that we expect of acids.

Chemists try to *explain* properties in terms of what goes on at the level of atoms, molecules and ions. They explain the characteristic properties of acids by saying that acids have *the ability to transfer H^+ ions to something else*. The substance which accepts the H^+ ion is called a **base**.

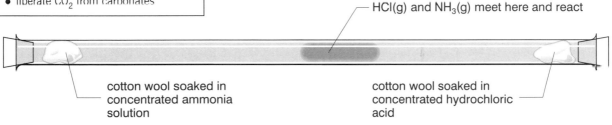

Figure 1 *The reaction of hydrogen chloride with ammonia – an acid-base reaction*

For example, take the reaction of hydrogen chloride with ammonia. You may have seen this reaction demonstrated in apparatus like that shown in Figure 1. The reaction forms a salt, ammonium chloride:

$$NH_3(g) + HCl(g) \rightarrow NH_4^+Cl^-(s)$$

In this reaction, the HCl transfers H^+ to NH_3 (Figure 2). The HCl is behaving as an acid, and the NH_3 is behaving as a base.

Notice that this is not a redox reaction. The oxidation states of H, Cl and N are unchanged during the reaction – check for yourself.

The general definition of an acid is that it is a substance which donates H^+ in a chemical reaction. The substance that accepts the H^+ is a base. The reaction in which this happens is called an **acid-base reaction**. This theory of H^+ transfer is known as the *Brønsted-Lowry theory* of acids and bases.

Since a hydrogen atom consists of only a proton and an electron, an H^+ ion corresponds to just one proton. Sometimes we refer to acids as *proton donors* and bases as *proton acceptors*.

An **alkali** is a special kind of base: a base that dissolves in water to produce hydroxide ions, ^-OH. This is illustrated in Figure 3. Some alkalis, such as sodium hydroxide, already contain ^-OH ions. Others, such as sodium carbonate, Na_2CO_3, form hydroxide ions when they react with water.

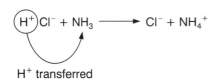

Figure 2 *HCl transfers an H^+ ion to NH_3*

The Brønsted-Lowry theory

An acid is an H^+ donor
A base is an H^+ acceptor

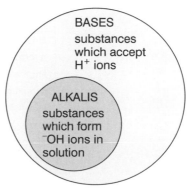

Figure 3 *The relationship between alkalis and bases*

Water and oxonium ions

Hydrogen chloride is a gas and contains HCl molecules. Water is almost totally made up from H_2O molecules. Yet hydrochloric acid, a solution of hydrogen chloride in water, conducts electricity, so it must contain ions. There must be a reaction between the HCl molecules and the H_2O molecules which produces these ions. The reaction is

$$HCl\,(aq)\ +\ H_2O(l)\ \rightarrow\ H_3O^+(aq)\ +\ Cl^-(aq)$$
  acid   base

In this reaction, H_2O is behaving as a base.

You may not previously have come across the ion with the formula H_3O^+. It is called the **oxonium ion**. It is a very common ion – it is present in every solution of an acid in water. In other words it occurs in every *acidic* solution. The acid donates H^+ to H_2O to form H_3O^+. Figure 4 shows how the H_2O bonds to H^+.

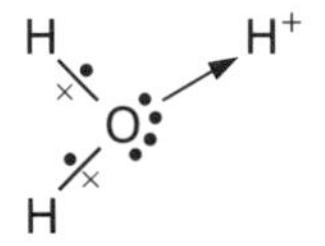

The H_3O^+ ion can itself act as an acid – it can donate H^+ and turn into an H_2O molecule. The familiar properties of acidic solutions are all properties of the H_3O^+ ion. You will often see the formula $H_3O^+(aq)$ shortened to $H^+(aq)$.

*Figure 4 The bonding in the H_3O^+ ion. A lone pair on the O atom forms a **dative** bond to H^+*

Acid-base pairs

Once an acid has donated an H^+ ion, there is always the possibility that it will take it back again. For example, consider ethanoic acid (acetic acid), the acid in vinegar. When ethanoic acid behaves as an acid, this is the change that happens:

$$CH_3COOH \rightarrow CH_3COO^- + H^+$$

But if you add a strong acid to a solution containing ethanoate ions, CH_3COO^-, the ethanoate ions accept H^+ from the stronger acid and go back to ethanoic acid:

$$CH_3COO^- + H^+ \rightarrow CH_3COOH$$

In this reaction, the ethanoate ion is behaving as a *base*. It is called the **conjugate base** of ethanoic acid.

Every acid has a conjugate base, and every base has a conjugate acid. They are called a **conjugate acid-base pair.** If we represent a general acid as HA, then we have

$$HA\ \ \rightarrow\ \ H^+\ \ +\ \ A^-$$
  conjugate   conjugate
    acid   base

For example, the conjugate base of HCl is Cl^-. The conjugate acid of NH_3 is NH_4^+. Table 1 shows more examples.

Acid		Base
chloric(VII) acid	$HClO_4 \longrightarrow$	$H^+ + ClO_4^-$
hydrochloric acid	$HCl \longrightarrow$	$H^+ + Cl^-$
sulphuric acid	$H_2SO_4 \longrightarrow$	$H^+ + HSO_4^-$
oxonium ion	$H_3O^+ \longrightarrow$	$H^+ + H_2O$
ethanoic acid	$CH_3COOH \longrightarrow$	$H^+ + CH_3COO^-$
hydrogen sulphide	$H_2S \longrightarrow$	$H^+ + HS^-$
ammonium ion	$NH_4^+ \longrightarrow$	$H^+ + NH_3$
water	$H_2O \longrightarrow$	$H^+ + HO^-$
ethanol	$C_2H_5OH \longrightarrow$	$H^+ + C_2H_5O^-$

Table 1 Conjugate acid-base pairs

Notice in Table 1 that water, H_2O, can be both an acid and a base. It all depends on what the water is reacting with. If it is with a strong acid such as HCl, water acts as a base, accepting H^+ and forming H_3O^+. If it is with a strong base such as CaO, water acts as an acid, donating H^+ to form ^-OH ions.

Many other substances can behave as both acid and base: they are described as **amphoteric**.

Strength of acids and bases

Not all acids have the same strength. Some are powerful H^+ donors and are described as *strong acids*. Others are *weak acids*; they are moderate or weak H^+ donors. Table 2 shows the range of strengths found in acids.

	Acid		Base
STRONGEST ACID	$HClO_4 \longrightarrow$	$H^+ + ClO_4^-$	WEAKEST BASE
	$HCl \longrightarrow$	$H^+ + Cl^-$	
increasing	$H_2SO_4 \longrightarrow$	$H^+ + HSO_4^-$	increasing
acid	$H_3O^+ \longrightarrow$	$H^+ + H_2O$	base
strength	$CH_3COOH \longrightarrow$	$H^+ + CH_3COO^-$	strength
	$H_2S \longrightarrow$	$H^+ + HS^-$	
	$NH_4^+ \longrightarrow$	$H^+ + NH_3$	
	$H_2O \longrightarrow$	$H^+ + HO^-$	
WEAKEST ACID	$C_2H_5OH \longrightarrow$	$H^+ + C_2H_5O^-$	STRONGEST BASE

Table 2 The strength of acids and their conjugate bases.

Notice that a strong acid has a weak conjugate base, and vice-versa. This makes sense: if an acid has a strong tendency to donate H^+ ions, its conjugate base will have a weak tendency to accept them back.

Table 2 includes some substances, such as ethanol, that we do not normally think of as acids. Ethanol has a very small tendency to donate protons. It *can* behave as an acid, but only in the presence of a very strong base.

Indicators

Acid-base indicators such as litmus are coloured organic substances which are themselves weak acids. The special thing about indicators is that the conjugate acid and base forms have *different colours*. For example, the acid form of litmus is red, and its conjugate base is blue. If we represent litmus as HIn, we have

$$HIn(aq) \rightleftharpoons H^+(aq) + In^-(aq)$$
acid   base
red   *blue*

When you add H_3O^+ from an acid to blue litmus, this reaction occurs:

$$In^-(aq) + H_3O^+(aq) \rightarrow HIn(aq) + H_2O(l)$$
blue   *red*

The blue litmus turns red.

When you add an alkali containing ^-OH to red litmus, this reaction happens:

$$HIn(aq) + {}^-OH(aq) \rightarrow In^-(aq) + H_2O(l)$$
red   *blue*

The red litmus turns blue.

PROBLEMS FOR 8.1

1 Write out each of the equations which follow and indicate which of the reactants is the acid and which is the base.

a $HNO_3 + H_2O \rightarrow H_3O^+ + NO_3^-$
b $NH_4^+ + OH^- \rightarrow NH_3 + H_2O$
c $SO_4^{2-} + H_3O^+ \rightarrow HSO_4^- + H_2O$
d $H_2O + H^- \rightarrow H_2 + OH^-$
e $H_3O^+ + OH^- \rightarrow 2H_2O$
f $NH_3 + HCl \rightarrow NH_4^+ + Cl^-$

2 Classify each of the equations which follow as either acid-base or redox.

a $NH_4^+ + CO_3^{2-} \rightarrow NH_3 + HCO_3^-$
b $H_2S + 2OH^- \rightarrow S^{2-} + 2H_2O$
c $I_2 + 2OH^- \rightarrow I^- + IO^- + H_2O$
d $Mg + 2H^+ \rightarrow Mg^{2+} + H_2$

3 Write balanced equations for each of the following reactions between acids and bases. In each case identify the two conjugate acid/base pairs.

For example, for hydrochloric acid and hydroxide ion the answer would be

Equation: $HCl + {}^-OH \rightarrow Cl^- + H_2O$
Conjugate pairs:

	acid	base
	HCl	Cl$^-$
	H$_2$O	$^-$OH

a ethanoic acid and hydroxide ion
b hydrogencarbonate ion, HCO_3^-, and hydrochloric acid producing carbonic acid, H_2CO_3,

4 The blue of cornflowers and the red of poppies is based on the same purple molecule, cyanidin. Cornflower sap is alkaline and removes an H^+ ion from the purple molecule to form a blue compound. Poppy sap is acidic and donates an H^+ ion to the purple molecule to form a scarlet compound. (You can try this for yourself by putting cornflowers to stand for a few days in vinegar, or poppies in aqueous washing soda.) The purple form of cyanidin can be represented as **(cyanidin)H**; the red form can be represented as **(cyanidin)H$_2$$^+$** and the blue form as **(cyanidin)$^-$**.

Write equations and identify the conjugate acid/base pairs for the reaction of cyanidin with:
a the oxonium ion, H_3O^+;
b the hydroxide ion, ^-OH.

8.2 *Weak acids and pH*

Strong and weak acids

The Brønsted-Lowry theory of acids and bases was introduced in **Section 8.1**. This theory describes an **acid** as an H^+ **donor**. An important consequence of this is that, in aqueous solution, acids donate H^+ to water molecules to produce **oxonium ions**, H_3O^+. The presence of oxonium ions gives rise to the familiar acidic properties of all aqueous solutions of acids.

Acids vary in **strength**. Different acids donate H^+ to differing extents. **Strong acids** have a strong tendency to donate H^+: the donation of H^+ is essentially complete. The reaction with water can be regarded as going to completion and can be described by the following equation, where HA represents the strong acid. (We can assume that *no* unreacted HA remains in solution.)

$$HA(aq) + H_2O(l) \rightarrow H_3O^+(aq) + A^-(aq)$$

We can simplify this to

$$HA(aq) \rightarrow H^+(aq) + A^-(aq)$$

by leaving out the water, which is present in excess.

HCl and H_2SO_4 are examples of strong acids.

If a substance is a **weak acid**, its tendency to donate H^+ is weaker and the reaction with water is incomplete. Some $H^+(aq)$ ions are formed but there is still some unreacted acid in solution. If we represent the weak acid as HA, its reaction with water can be represented as

$$HA(aq) \rightleftharpoons H^+(aq) + A^-(aq)$$

The equilibrium sign shows that a significant concentration of HA is present along with the A^- and $H^+(aq)$ ions formed from it.

Ethanoic acid (CH_3COOH) and chloric(I) acid (HClO) are examples of weak acids.

The pH scale

We can find out if a compound is a weak acid or a strong acid by measuring the extent to which it forms ions in solution. We could measure the electrical conductivity of the solution, which would tell us about *any* ions present, or we could be more selective and measure the concentration of $H^+(aq)$ ions – the only ions which are common to *all* acidic solutions.

Values for the concentration of $H^+(aq)$ in aqueous solutions vary widely from about 1×10^{-15} mol dm^{-3} to about 10 mol dm^{-3}. The numbers are often inconveniently small, and it is common to use the **pH scale**, where pH is defined as

pH = –lg [H$^+$(aq)]

If you are unfamiliar with the lg function, you can think of it as a technique for bringing numbers of very different magnitudes onto the same scale. Try entering the following numbers into your calculator and finding their lg values:

1×10^{-14}; 1×10^{-7}; 1×10^{-3}; 1

Notice that the pH scale runs in the opposite direction to the scale of $[H^+(aq)]$ values: a *low* pH corresponds to a *high* concentration of $H^+(aq)$. This is because pH is equal to *minus* $\lg[H^+(aq)]$.

You can use indicator paper to measure pH by comparing the colour of the paper with a colour chart. You will get a more accurate answer, and you can record a numerical value directly, if you use a special **glass electrode** connected to a specially calibrated voltmeter called a **pH meter.**

Calculating pH values

Strong acids

You can find the pH of a solution of a strong acid by a straightforward calculation. Since the reaction with water goes effectively to completion, the amount in moles of $H^+(aq)$ ions is equal to the amount in moles of acid (HA) put into solution. Therefore a 0.01 mol dm^{-3} solution of HA has $[H^+(aq)]$ equal to 0.01 mol dm^{-3} and has a pH of 2; for a 1 mol dm^{-3} solution $[H^+(aq)] = 1$ mol dm^{-3} and its pH is zero.

Weak acids

For a weak acid,
$$HA(aq) \rightleftharpoons H^+(aq) + A^-(aq)$$
We can write an equilibrium constant in the usual way as

$$K_a = \frac{[H^+(aq)]\ [A^-(aq)]}{[HA(aq)]}$$

This equilibrium constant is called the **acidity constant** (or often the *acid dissociation constant*) and so is given the symbol K_a. Table 3 gives values for some weak acids.

Acid		K_a/mol dm^{-3}
methanoic	HCOOH	1.6×10^{-4}
benzoic	C_6H_5COOH	6.3×10^{-5}
ethanoic	CH_3COOH	1.7×10^{-5}
chloric(I)	HClO	3.7×10^{-8}
hydrocyanic	HCN	4.9×10^{-10}
nitrous (nitric(III))	HNO_2	4.7×10^{-4}

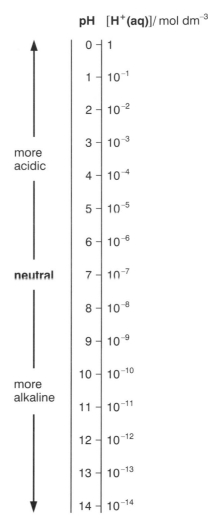

pH	[H$^+$(aq)]/ mol dm^{-3}
0	1
1	10^{-1}
2	10^{-2}
3	10^{-3}
4	10^{-4}
5	10^{-5}
6	10^{-6}
7	10^{-7}
8	10^{-8}
9	10^{-9}
10	10^{-10}
11	10^{-11}
12	10^{-12}
13	10^{-13}
14	10^{-14}

more acidic

neutral

more alkaline

Figure 5 The pH scale

Table 3 Acidity constants for some weak acids at 298 K

To work out the pH, we need first to find $[H^+(aq)]$. In doing so, we can make two assumptions which simplify the calculation considerably:

1 *$[H^+(aq)] = [A^-(aq)]$* At first sight this might seem obvious because equal amounts of $H^+(aq)$ and A^- are formed from HA. But there is another source of $H^+(aq)$ – from water itself. This is through the equilibrium

$$2H_2O(l) \rightleftharpoons H^+(aq) + {}^-OH(aq)$$

Water produces far fewer $H^+(aq)$ ions than most weak acids, so we do not introduce a significant inaccuracy into the calculation by neglecting the ionisation of water.

2 *the amount of HA at equilibrium is equal to the amount of HA put into the solution.* In other words when we calculate the concentration of HA at equilibrium, we can neglect the fraction of HA which has lost H^+. We can do this because we are dealing with weak acids and this fraction is very small. For example, in a solution of HA of concentration $0.1 \, mol \, dm^{-3}$ we can assume that when equilibrium has been established, $[HA(aq)] = 0.1 \, mol \, dm^{-3}$.

Let us use these ideas to calculate the value of $[H^+(aq)]$ and the pH for ethanoic acid solutions at 298 K with concentrations of **a** $1 \, mol \, dm^{-3}$, **b** $0.01 \, mol \, dm^{-3}$. (K_a for ethanoic acid is $1.7 \times 10^{-5} \, mol \, dm^{-3}$ at 298 K.)

We can represent the reaction between ethanoic acid and water by the general equation

$$HA(aq) \rightleftharpoons H^+(aq) + A^-(aq)$$

and the acidity constant is therefore given by

$$K_a = \frac{[H^+(aq)] \, [A^-(aq)]}{[HA(aq)]} = 1.7 \times 10^{-5} \, mol \, dm^{-3} \text{ at 298 K}$$

Assumptions **1** and **2** then allow us to write

a
$$1.7 \times 10^{-5} \, mol \, dm^{-3} = \frac{[H^+(aq)]^2}{1 \, mol \, dm^{-3}}$$

Therefore $[H^+(aq)]^2 = 1.7 \times 10^{-5} \, mol^2 \, dm^{-6}$

So $\quad [H^+(aq)] = 4.12 \times 10^{-3} \, mol \, dm^{-3}$ and pH = 2.39 (at 298 K)

b
$$1.7 \times 10^{-5} \, mol \, dm^{-3} = \frac{[H^+(aq)]^2}{0.01 \, mol \, dm^{-3}}$$

Therefore $[H^+(aq)]^2 = 1.7 \times 10^{-7} \, mol^2 \, dm^{-6}$

So $\quad [H^+(aq)] = 4.12 \times 10^{-4} \, mol \, dm^{-3}$ and pH = 3.39 (at 298 K)

Notice that when $[HA(aq)]$ is decreased by a factor of 100, $[H^+(aq)]$ falls by a factor of 10 and pH increases by 1 unit. The calculations on strong and weak acids are compared in Table 4.

Table 4 A comparison of strong and weak acids

	Strong acid	**Typical weak acid**
acid is diluted by a factor of 100	$[H^+(aq)]$ becomes $100 \times$ smaller	$[H^+(aq)]$ becomes $10 \times$ smaller
	pH increases by 2	pH increases by 1

Strong or concentrated?

Measurement of $[H^+(aq)]$ or pH alone does not allow us to distinguish strong and weak acids. For example, solutions of both could have $[H^+(aq)] = 1 \times 10^{-4} \, mol \, dm^{-3}$, but the solutions would have different concentrations of acid. The strong acid would have a concentration of $1 \times 10^{-4} \, mol \, dm^{-3}$; the weak acid would have to be much more concentrated to have a pH as low as 4.

It is important to distinguish carefully between *concentration* and *strength*. Concentration is a measure of the amount of substance in a given volume of solution – typically measured in $mol\,dm^{-3}$. Strength is a measure of the extent to which an acid can donate H^+. The two terms have very different meanings in chemistry but are often interchangeable in everyday language.

A table of K_a values, like Table 3, provides an easy way of telling the strength of an acid: the greater the value of K_a, the stronger the acid. Like $[H^+(aq)]$ values, K_a values are spread over a wide range, and it is often convenient to define a term pK_a.

$$pK_a = -\lg K_a$$

(like $pH = -\lg [H^+(aq)]$) as a measure of acid strength. Just as a lower pH means a greater $[H^+(aq)]$, so a smaller pK_a corresponds to a greater acid strength. Table 5 gives the pK_a values for the acids in Table 3.

Acid		$K_a/mol\,dm^{-3}$	pK_a
methanoic	HCOOH	1.6×10^{-4}	3.8
benzoic	C_6H_5COOH	6.3×10^{-5}	4.2
ethanoic	CH_3COOH	1.7×10^{-5}	4.8
chloric(I)	HClO	3.7×10^{-8}	7.4
hydrocyanic	HCN	4.9×10^{-10}	9.3
nitrous (nitric(III))	HNO_2	4.7×10^{-4}	3.3

Table 5　K_a and pK_a values for some weak acids

Ionisation of water

We do not think of water as an ionic substance, but water does in fact ionise slightly. When water ionises, it behaves as both an acid and a base. In the equilibrium

$$H_2O(l) \rightleftharpoons H^+(aq) + {}^-OH(aq)$$

water is acting like other weak acids because H–OH can be thought of as HA, and ^-OH as A^-. We can write an expression for K_a for water as

$$K_a = \frac{[H^+(aq)]\,[^-OH(aq)]}{[H_2O(l)]}$$

However, we can leave out the $[H_2O(l)]$ term, which is effectively constant because water is present in excess. The expression becomes

$$K_w = [H^+(aq)]\,[^-OH(aq)]$$

Notice the special symbol, K_w, to denote the **ionic product** of water.

At 298 K,
$K_w = 10^{-14}\,mol^2\,dm^{-6}$. In pure water $[H^+(aq)] = [^-OH(aq)]$, so
$K_w = 10^{-14}\,mol^2\,dm^{-6} = [H^+(aq)]^2$
Therefore　$[H^+(aq)] = 10^{-7}\,mol\,dm^{-3}$
and pH = 7

The idea that pure water, or a neutral solution, has pH = 7 should be familiar to you. But it is only true at 298 K. Ionisation of water is endothermic. Le Chatelier's principle predicts that K_w will increase with increasing temperature. $[H^+(aq)]$ will therefore increase in a similar way and the pH of water will fall as it gets hotter.

K_w also allows us to calculate the pH of a solution of a strong base – a solution in which the production of hydroxide ions is complete. For example, sodium hydroxide is completely ionised in solution, so in

0.1 mol dm^{-3} NaOH(aq), [OH$^-$(aq)] = 0.1 mol dm^{-3}. We can neglect the small amount of OH$^-$ formed from water, so

$$K_w = 10^{-14} \text{mol}^2 \text{dm}^{-6} = [\text{H}^+(\text{aq})] \times 0.1 \text{mol dm}^{-3}$$

Therefore, $[\text{H}^+(\text{aq})] = \dfrac{10^{-14} \text{mol}^2 \text{dm}^{-6}}{0.1 \text{mol dm}^{-3}}$

or $\qquad [\text{H}^+(\text{aq})] = 10^{-13} \text{mol dm}^{-3}$

and $\qquad$ pH = 13

PROBLEMS FOR 8.2

1 Calculate the pH values at 298 K of the following solutions. Parts **a** and **b** deal with strong acids; for the weak acids in parts **c** to **e** you will need to look up K_a values in Table 3.
   a 0.1 mol dm^{-3} hydrochloric acid
   b 0.1 mol HClO$_4$(l) in 250 cm^3 of aqueous solution
   c 1 × 10^{-3} mol dm^{-3} benzoic acid
   d 0.01 mol dm^{-3} nitric(III) acid
   e 0.1 mol methanoic acid in 100 cm^3 of solution

2 Calculate the pH of the following solutions of strong bases at 298 K.
   a 1 mol dm^{-3} KOH(aq)
   b 0.01 mol dm^{-3} NaOH(aq)
   c 0.1 mol dm^{-3} Ba(OH)$_2$(aq)

3 Indicators are weak acids. The acidic form, HIn, of the indicator is one colour, with its conjugate base, In$^-$, being a different colour. At the end point for a titration, [HIn(aq)] = [In$^-$(aq)].

For the indicator phenolphthalein, HIn is colourless and In$^-$ is pink.

HIn(aq) ⇌ H$^+$(aq) + In$^-$(aq)

   a What is the colour of phenolphthalein in alkaline solution? Explain your answer.
   b Write an expression for the acidity constant, K_a, for phenolphthalein.
   c Use this expression to calculate [H$^+$(aq)] and hence the pH at the end point of a titration when phenolphthalein is used as an indicator. The pK_a for phenolphthalein is 9.3.

4 Use the pH of the following acid solutions to calculate their acidity constants, K_a.
   a 0.10 mol dm^{-3} HCN, pH = 5.15
   b 0.005 mol dm^{-3} phenol, pH = 6.10
   c 1.00 mol dm^{-3} HF, pH = 1.66

8.3 *Buffer solutions*

Where do we find buffer solutions?

Many processes in living systems must take place under fairly precise pH conditions. If the pH changes to a value outside a narrow range, the process will not occur at the correct rate, or it may not take place at all, and the organism will die. The pH ranges for some fluids in our bodies are shown in Table 6.

Small organisms must also be surrounded by liquid at the correct pH. This is true, for example, for the bacteria and moulds used for fermentation in biotechnology processes, as well as for many life-forms in the sea.

Controlling pH may not be a matter of life or death in a chemical manufacturing process, but it is often very important. For example, using reactive dyes to colour fabrics can be ineffective at the wrong pH.

So, during evolution and in the practice of chemistry, solutions have been developed which can *resist changes in pH despite the addition of acid or alkali*. Such solutions are called **buffer solutions**. Their pH stays approximately constant even if small amounts of acid or alkali are added.

Buffer solutions are usually made from either a weak acid and one of its salts, or a weak base and one of its salts. An example of the first type is ethanoic acid plus sodium ethanoate. The salt has an ionic structure and acts as a source of ethanoate ions. The presence of ethanoic acid leads to an acidic pH.

Fluid	pH range
stomach juices	1.6–1.8
saliva	6.4–6.8
blood	7.35–7.45

Table 6 pH ranges of some body fluids

Ammonia solution plus ammonium chloride is an example of a buffer made from a weak base. The salt provides NH_4^+ ions and the ammonia leads to an alkaline pH. The action of both types of buffer depends on the presence of an H^+ donor (CH_3COOH or NH_4^+) and and H^+ acceptor (CH_3COO^- or NH_3).

By choosing suitable pairs of conjugate acids and bases, we can make buffer solutions with almost any desired pH. Living systems often use H_2CO_3 and HCO_3^-, or $H_2PO_4^-$ and HPO_4^{2-}, in combination with proteins.

How do buffers work?

The action of a buffer solution depends on the weak acid equilibrium which was discussed in **Section 8.2**, and which we can represent as

$$HA(aq) \rightleftharpoons H^+(aq) + A^-(aq)$$

We can make two very good approximations about the species present in this equilibrium in the case of a buffer solution.

1 *All the A^- ions come from the salt*. The weak acid, HA, supplies very few A^- ions in comparison with the fully ionised salt.

2 *Almost all the HA molecules put into the buffer remain unchanged*. We make the same assumption when calculating the pH of a weak acid.

What happens if an acidic substance is added to a buffer? Any rise in $H^+(aq)$] would disturb the equilibrium. Some A^- ions from the salt react with the extra $H^+(aq)$ ions to form HA and water. A significant fall in pH is prevented.

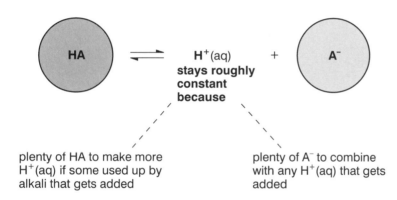

plenty of HA to make more $H^+(aq)$ if some used up by alkali that gets added

plenty of A^- to combine with any $H^+(aq)$ that gets added

Figure 6 How a buffer solution keeps the pH constant

If alkali is added, $H^+(aq)$ ions are removed from the solution. The buffer solution counteracts this because $H^+(aq)$ can be regenerated from the acid HA. A significant pH rise is avoided.

The presence of *both* a weak acid *and* its salt are necessary for a buffer to work. There must be plenty of HA to act as a *source* of extra $H^+(aq)$ ions when they are needed, and plenty of A^- to act as a *sink* for any extra $H^+(aq)$ ions which have been added (Figure 6).

Calculations with buffers

All we need to enable us to do calculations on buffer solutions is the K_a expression for the relevant weak acid.

$$K_a = \frac{[H^+(aq)]\ [A^-(aq)]}{[HA(aq)]} = \frac{[H^+(aq)] \times [A^-(aq)]}{[HA(aq)]}$$

If we make use of assumptions **1** and **2**, on page 152 we get

$$K_a = H^+(aq)] \times \frac{[salt]}{[acid]}$$

The value of $[H^+(aq)]$, and therefore the pH, depends on *two* factors.

1 K_a This provides a 'coarse tuning' of the buffer's pH. K_a values normally lie in the range $10^{-4}\,mol\,dm^{-3}$ to $10^{-10}\,mol\,dm^{-3}$. Choice of a particular weak acid determines which *region* of the pH range the buffer is in, from about pH = 4 to pH = 10.

2 *Ratio of [salt] : [acid]* This provides a 'fine tuning' of the buffer pH. Changing the ratio from about 3 : 1 to about 1 : 3 changes $[H^+(aq)]$ by a factor of 9, and alters pH by approximately 1 unit. The ratio should not be too far outside this range, otherwise there will be insufficient HA or A$^-$ for the buffer to be effective.

The expression

$$K_a = H^+(aq)] \times \frac{[salt]}{[acid]}$$

shows that the pH of a buffer is not affected by dilution. When you add water, the concentrations of both salt and acid are reduced equally. Therefore the ratio of their concentrations remains the same, and the pH is unchanged.

Finally, let's use this expression to calculate the pH of a buffer solution which contains $0.1\,mol\,dm^{-3}$ ethanoic acid and $0.2\,mol\,dm^{-3}$ sodium ethanoate. K_a for ethanoic acid is $1.7 \times 10^{-5}\,mol\,dm^{-3}$ at 298 K.

$$K_a = [H^+(aq)] \times \frac{[CH_3OO^-(aq)]}{[CH_3COOH(aq)]}$$

Using assumptions **1** and **2** we can write

$$1.7 \times 10^{-5}\,mol\,dm^{-3} = [H^+(aq)] \times \frac{0.2\,mol\,dm^{-3}}{0.1\,mol\,dm^{-3}}$$

Therefore $[H^+(aq)] = 1.7 \times 10^{-5}\,mol\,dm^{-3} \times \dfrac{0.1\,mol\,dm^{-3}}{0.2\,mol\,dm^{-3}}$

i.e. $[H^+(aq)] = 8.5 \times 10^{-6}\,mol\,dm^{-3}$

and pH = 5.07 at 298 K

PROBLEMS FOR 8.3

1 Calculate the pH values of the following buffer solutions at 298 K. (You will find K_a values in Table 3, page 151.)

a A solution in which the concentrations of methanoic acid and potassium methanoate are both $0.1\,mol\,dm^{-3}$.

b A solution made by dissolving 0.01 mol benzoic acid and 0.03 mol sodium benzoate in 1 dm^3 of solution.

c A solution made by mixing equal volumes of $0.1\,mol\,dm^{-3}$ methanoic acid and $0.1\,mol\,dm^{-3}$ potassium methanoate.

2 Which of the acids listed in Table 3 on page 151 would be the most suitable choice for the preparation of a buffer with pH 5.2? Give a reason for your answer.

3 Explain why a mixture of hydrochloric acid (a strong acid) and sodium chloride would not be an effective buffer solution.

4 Many biological systems are buffered by a mixture of dihydrogen phosphate and hydrogen phosphate ions:

$H_2PO_4^-(aq) \rightleftharpoons H^+(aq) + HPO_4^{2-}(aq)$

Write the acidity constant expression for the above equilibrium.

a Predict qualitatively, giving your reasoning, how the pH of a solution containing equal concentrations of these two ions will behave on adding a few drops of:

  i dilute hydrochloric acid;

  ii dilute sodium hydroxide.

b Suppose you want to make a buffer solution of pH 6.9. What concentration of $HPO_4^{2-}(aq)$ would be needed in a solution that is already $0.100\,mol\,dm^{-3}$ in $H_2PO_4^-(aq)$? The acidity constant, K_a of $H_2PO_4^-$ is $6.2 \times 10^{-8}\,mol\,dm^{-3}$.

REDOX

9.1 Oxidation and reduction

When we set fire to a piece of magnesium ribbon in air, the magnesium is oxidised – it gains oxygen. The product, magnesium oxide, is an ionic compound.

$$Mg(s) + \tfrac{1}{2}O_2(g) \rightarrow Mg^{2+}O^{2-}(s)$$

If you think carefully about what is happening, you will see that the reaction is composed of two *half-reactions* which can be described by two *half-equations*:

$$Mg \rightarrow Mg^{2+} + 2e^- \quad \text{and} \quad \tfrac{1}{2}O_2 + 2e^- \rightarrow O^{2-}$$

The magnesium is losing electrons in one half-reaction, and the oxygen is gaining them in the other. Another way of looking at oxidation is to say that *oxidation is the loss of electrons.*

Reduction is the opposite of oxidation, so *reduction* must occur when *electrons are gained*. The oxygen is being reduced in this example.

Reduction and oxidation occur together when magnesium burns in air: it is a reduction-oxidation or **redox** reaction (Figure 1).

If you had burned magnesium in chlorine, you would have performed another redox reaction. The half-equations show this:

$$Mg \rightarrow Mg^{2+} + 2e^- \quad \text{and} \quad Cl_2 + 2e^- \rightarrow 2Cl^-$$

Magnesium loses electrons and is oxidised; chlorine gains electrons and is reduced.

Redox is a very important type of chemical reaction. Many reactions can be classified as redox reactions using the ideas in the box on the left.

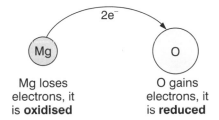

Figure 1 *A redox reaction occurs when electrons are transferred from magnesium to oxygen*

Oxidation is loss of electrons.	Reduction is gain of electrons.
An oxidising agent removes electrons from something else.	A reducing agent gives electrons to something else.

Oxidation state

How would you describe the reaction which takes place when hydrogen burns in oxygen?

$$H_2(g) + \tfrac{1}{2}O_2(g) \rightarrow H_2O(l)$$

The equation is so similar to the one for the magnesium–oxygen reaction that it must be a redox reaction; but the product, water, is molecular not ionic. This means that we cannot break the overall equation down into half-equations involving electron transfer, as we did for magnesium and oxygen.

We need to extend the idea of redox to include reactions like this – and there are many of them. We can do this by using the idea of **oxidation state** (also known as **oxidation number**).

Oxidation states are assigned to atoms in molecules or ions to show how much they are oxidised or reduced. Two very useful rules about oxidation states are
- atoms in *elements* are in oxidation state *zero*
- in *simple ions* the oxidation state is the same as the *charge* on the ion. Thus bromine has oxidation state 0 in Br_2, but oxidation state –1 in Br^-. Magnesium has oxidation state 0 in Mg, but oxidation state +2 in Mg^{2+}.

Since *compounds* have no overall charge, the oxidation states of all the constituent elements must add up to zero.

Some elements have oxidation states that rarely vary, whatever the compound they are in. For example, fluorine's oxidation state is always –1. Table 1 shows some more elements whose oxidation states are invariable, with a few exceptions.

An element is oxidised when its oxidation state increases.	A decrease in oxidation state corresponds to reduction.

Element	Oxidation state
F	–1
O	–2 (except in O_2^{2-} and OF_2)
H	+1 (except in H^-)
Cl	–1 (except when combined with O or F)

Table 1 *Oxidation states of elements in compounds*

Compound	Oxidation states of elements	
H_2O	H +1	O −2
CH_4	H +1	C −4
BrF_3	F −1	Br +3
SO_2	O −2	S +4
PCl_3	Cl −1	P +3

Table 3

Ion	Oxidation states	
NH_4^+	H +1	N −3
ClO_3^-	O −2	Cl +5
VO^{2+}	O −2	V +4

Table 4

Here is an example of how to use these rules to work out the oxidation states of the elements in compounds:

In CO_2, the oxidation state of O must be −2 (it is one of the rules).

There are two oxygens, so the total contribution of oxygen to the oxidation states in the compound comes to $-2 \times 2 = -4$.

To balance this, the oxidation state of the carbon must be +4.

This makes the oxidation states add up to zero.

Table 3 shows some further examples.

The same rules can be used for the elements which make up *ions*, such as PO_4^{3-}; but this time the oxidation states of the constituent elements add up to the overall charge on the ion.

For example, in PO_4^{3-} the oxidation state of O must be −2. There are four oxygens, so the contribution of oxygen to the oxidation states in the ion is −8. The oxidation state of P in this ion must be +5 if a charge of −3 is to be left over for the total charge on the ion.

Table 4 gives some other examples.

Summary

Chemists use the following ideas for oxidation and reduction:

Something is oxidised if	Something is reduced if
• it gains oxygen	• it loses oxygen
• it loses electrons	• it gains electrons
• its oxidation state increases	• its oxidation state decreases

Summary of rules for assigning oxidation states
- The oxidation state of atoms in elements is 0.
- In compounds, the sum of all the oxidation states is 0.
- In ions, the sum of all the oxidation states is equal to the charge on the ion.
- In compounds and ions, oxidation states are assigned as follows:

$$F \quad -1$$
$$O \quad -2 \text{ (except in } O_2^{2-} \text{ and } OF_2)$$
$$H \quad +1 \text{ (except in } H^-)$$
$$Cl \quad -1 \text{ (except when combined with O or F).}$$

For example, look at the reaction

$$Cl_2(aq) + 2I^-(aq) \rightarrow 2Cl^-(aq) + I_2(aq)$$

We can use oxidation states to find what has been oxidised and what has been reduced.

Oxidation state of chlorine:

in Cl_2 0
in Cl^- −1

The oxidation state of chlorine *decreases*, so chlorine has been *reduced*.

Oxidation state of iodine:

in I^- −1
in I_2 0

The oxidation state of iodine *increases*, so iodine has been *oxidised*.

Try some more of these in problem 3 at the end of this section.

Oxidation states in names

Oxidation states can also help us to give systematic names to compounds and ions which contain elements capable of existing in more than one oxidation state. For example, the oxides of iron, FeO and Fe_2O_3, are called

FeO   iron(II) oxide
Fe_2O_3 iron(III) oxide

Notice that
- Roman numerals are used
- the number shows us the oxidation state of the element
- the number is placed close up to the element it refers to – there is no space between the name and the number.

Here are some other examples:

$CuCl_2$ copper(II) chloride
Cu_2O copper(I) oxide
MnO_2 manganese(IV) oxide

With ions, oxidation states are used to help clarify the names of **oxyanions** – negative ions which contain oxygen. You can tell an oxyanion because its name ends in -*ate*; for example, chlorate. The first part of the name tells you which element is combined with the oxygen – chlorine in this example.

There are several different chlorate ions, with formulae ClO^-, ClO_2^-, ClO_3^- and ClO_4^-. The oxidation states of Cl are given in Table 5, together with the names of the oxyanions.

Figure 2 shows another example.

Ion	Oxidation state		Name of ion
ClO^-	Cl	+1	chlorate(I)
ClO_2^-	Cl	+3	chlorate(III)
ClO_3^-	Cl	+5	chlorate(V)
ClO_4^-	Cl	+7	chlorate(VII)

Table 5

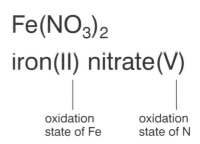

Figure 2 Oxidation states in the naming of $Fe(NO_3)_2$

PROBLEMS FOR 9.I

1 a Insert electrons (e^-) on the appropriate side of the following half-equations in order to balance and complete them, so that the electrical charges on both sides are equal.

  i $K \rightarrow K^+$
  ii $H_2 \rightarrow 2H^+$
  iii $O \rightarrow O^{2-}$
  iv $Cu^+ \rightarrow Cu^{2+}$
  v $Cr^{3+} \rightarrow Cr^{2+}$

  b For each completed half-equation describe the process as oxidation or reduction.

2 Write down the oxidation states of the elements in the following examples.

  a Ag^+ **f** CO_2
  b Br_2 **g** MgF_2
  c P_4 **h** ClO^-
  d N^{3-} **i** SO_4^{2-}
  e O_3

3 Some of the reactions of the halogens are shown below. They are all examples of redox reactions. In each case state which element is oxidised and which is reduced, and give the oxidation states before and after the reaction.

For example:

$$Cl_2 + 2Br^- \rightarrow 2Cl^- + Br_2$$
Oxidation states Cl(0)   Cl(−1)   *Cl is reduced*
                    Br(−1)   Br(0)
                                      Br is oxidised

  a $2AgCl$ $\rightarrow 2Ag + Cl_2$
  b $Cl_2 + 2OH^-$ $\rightarrow Cl^- + ClO^- + H_2O$
  c $2Fe + 3Cl_2$ $\rightarrow 2FeCl_3$
  d $3Br_2 + 6OH^-$ $\rightarrow BrO_3^- + 5Br^- + 3H_2O$
  e $2Br^- + 2H^+ + H_2SO_4$ $\rightarrow Br_2 + SO_2 + 2H_2O$
  f $8I^- + 8H^+ + H_2SO_4$ $\rightarrow 4I_2 + H_2S + 4H_2O$
  g $4IO_3^-$ $\rightarrow 3IO_4^- + I^-$
  h $H_2 + Cl_2$ $\rightarrow 2HCl$
  i $2F_2 + 2H_2O$ $\rightarrow 4HF + O_2$

4 Use oxidation states to name the compounds and ions with formulae

  a SnO **d** $CoCl_3$ **g** NO_3^- **j** CrO_4^{2-}
  b SnO_2 **e** FeS **h** SO_3^{2-}
  c $CoCl_2$ **f** NO_2^- **i** MnO_4^-

5 Write formulae for the following compounds:
  a potassium chlorate(III)
  b sodium vanadate(V)
  c manganese(III) hydroxide
  d iron(III) nitrate(V)
  e nickel(II) nitrate(III)

9.2 *Redox reactions and electrode potentials*

Redox reactions

Redox reactions involve electron transfer. Redox reactions can be split into two *half-reactions*, one producing electrons, and one accepting them. For example, when zinc is added to copper(II) sulphate solution a redox reaction takes place. The blue colour of the solution becomes paler, and copper metal deposits on the zinc. The temperature rises; it is an exothermic reaction. The overall equation is

$$Zn(s) + CuSO_4(aq) \rightarrow ZnSO_4(aq) + Cu(s)$$

but this isn't the most helpful way of showing what is happening.

Copper(II) sulphate solution contains a mixture of copper ions and sulphate ions moving around independently of each other. Zinc sulphate solution contains zinc ions and sulphate ions. The sulphate ions play no part in the reaction and can be left out of the equation. The important part of the reaction can be shown by the ionic equation

$$Zn(s) + Cu^{2+}(aq) \rightarrow Cu(s) + Zn^{2+}(aq)$$

What the reaction amounts to is Zn atoms transferring electrons to Cu^{2+} ions (Figure 3).

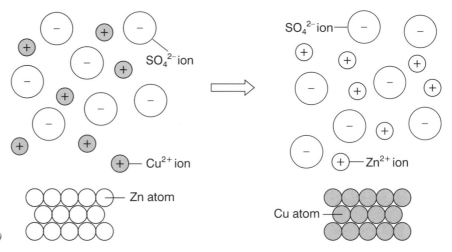

Figure 3 The reaction of zinc with copper(II) sulphate solution:
$Zn(s) + Cu^{2+}(aq) \rightarrow Cu(s) + Zn^{2+}(aq)$

The ionic equation can be written as two half-equations:

$Zn(s) \rightarrow Zn^{2+}(aq) + 2e^-$ oxidation, electrons produced in this half-reaction

$Cu^{2+}(aq) + 2e^- \rightarrow Cu(s)$ reduction, electrons accepted in this half-reaction

Zinc provides the electrons which reduce Cu^{2+} to Cu, so we say that zinc is a **reducing agent**. Similarly, Cu^{2+} is an **oxidising agent**.

If copper is added to zinc sulphate solution no change is observed. Reaction does not occur in the reverse direction. Zinc reacts with copper ions, but copper ions do not react with zinc. However, if copper is added to silver nitrate(V) solution the copper *does* react. A grey precipitate forms, and the solution turns from colourless to blue. The overall reaction is

$$Cu(s) + 2Ag^+(aq) \rightarrow Cu^{2+}(aq) + 2Ag(s)$$

and the half-reactions are

$$Cu(s) \rightarrow Cu^{2+}(aq) + 2e^-$$ oxidation
$$2Ag^+(aq) + 2e^- \rightarrow 2Ag(s)$$ reduction

No reaction is observed when silver is added to copper(II) sulphate solution.

Individual half-reactions are reversible. They can go either way.

$$Cu^{2+}(aq) + 2e^- \rightleftharpoons Cu(s)$$

The actual direction they take depends on what they are reacting with. For example, *zinc atoms* can supply electrons to *copper ions*, so that the copper half-reaction is

$$Cu^{2+}(aq) + 2e^- \rightarrow Cu(s)$$

But *copper atoms* supply electrons to *silver ions*, so in this case the copper half-reaction is

$$Cu(s) \rightarrow Cu^{2+}(aq) + 2e^-$$

Photosynthesis as a redox reaction

Photosynthesis is an important redox reaction. The overall equation for photosynthesis is

$$H_2O + CO_2 \rightarrow (CH_2O) + O_2$$

and the two half-reactions are

$$2H_2O \rightarrow O_2 + 4H^+ + 4e^-$$ oxidation
$$CO_2 + 4H^+ + 4e^- \rightarrow (CH_2O) + H_2O$$ reduction

Combining half-equations

Once we know the direction in which each half-reaction will go, we can add together the half-equations to get an equation for the overall reaction. For example, if you add zinc to silver ions, the zinc atoms supply electrons to the silver ions. The half-equations are

$$Zn(s) \rightarrow Zn^{2+}(aq) + 2e^-$$
$$Ag^+(aq) + e^- \rightarrow Ag(s)$$

To combine the two half-equations together, we need to make sure the number of electrons is the same in each half-equation – because every electron released by a zinc atom must be accepted by a silver ion.

This means we have to multiply the silver half-equation by two so there are $2e^-$ in each half-equation:

$$Zn(s) \rightarrow Zn^{2+}(aq) + 2e^-$$
$$2Ag^+(aq) + 2e^- \rightarrow 2Ag(s)$$

Now we can add the two half-equations together to give the overall equation:

$$Zn(s) + 2Ag^+(aq) \rightarrow 2Ag(s) + Zn^{2+}(aq)$$

The $2e^-$ disappear because they are on both sides of the equation.

Electrochemical cells

Something must control the direction of electron transfer in a redox reaction. To find out more about redox reactions, and what makes them go in a particular direction, we need to be able to study the half-reactions.

We can arrange for the two half-reactions to occur separately with electrons flowing through an external wire from one half-reaction to the other.

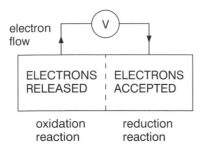

Figure 4 *The general arrangement for an electrochemical cell*

A system like this is used in all batteries and 'dry' cells. In one part of the cell an oxidation reaction occurs. Electrons are produced and transferred through an external circuit to the other part of the cell where a reduction reaction takes place, accepting the electrons. The two parts are called **half-cells**, which, when combined, make an **electrochemical cell**. Figure 4 shows the general arrangement, and Figure 5 shows a familiar example.

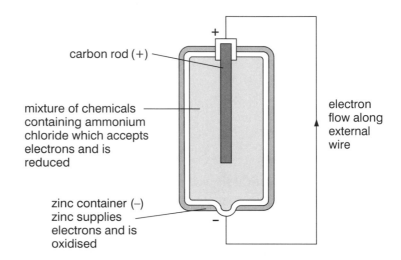

Figure 5 *An ordinary dry cell – the kind you use in a torch*

The energy given out, instead of heating the surroundings, becomes available as electrical energy which we use to do work for us.

Cells are labelled with positive and negative terminals and a voltage. The voltage measures the potential difference between the two terminals.

As current flows in a circuit, the voltage can drop. The higher the current drawn, the lower the voltage the cell may give.

If we want to compare cells, and half-cells, by measuring voltages, we do need to be careful to compare like with like. We can do this if we measure the potential difference between the terminals of the cell when *no* current flows. This potential difference is given the symbol E_{cell}. (It is sometimes called the electromotive force, or emf of the cell, although this is not a very good term because it is not a force).

To measure E_{cell} we use a high resistance voltmeter so that almost zero current flows. We record the maximum potential difference between the electrodes of the two half-cells. The potential difference is a measure of how much each electrode is tending to release or accept electrons.

Metal ion–metal half-cells

We can set up a simple half-cell by using a strip of metal dipping into a solution of metal ions. For example, the copper-zinc cell consists of two half-cells (Figure 6).

Electrical units

Electric charge is measured in *coulombs* (C)

Electric current is a flow of charge and is measured in *amps* (A).

One amp is a flow of charge of one coulomb per second.

The *potential difference* between the terminals of the cell is measured in *volts* (V). The voltage of the cell tells you the number of joules of energy transferred whenever one coulomb of charge flows round the circuit.

$1\,V = 1\,J\,C^{-1}$

For example if one coulomb of charge flows through a potential difference of 3 V, then 3 J are transferred.

Figure 6 *The copper and zinc half-cells*

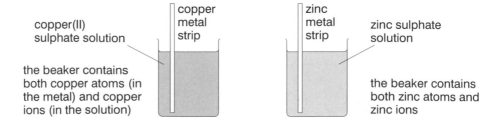

Each of these half-cells has its own electrode potential. Take the zinc half-cell, for example (Figure 7). The Zn atoms in the zinc strip form Zn^{2+} ions by releasing electrons:

$$Zn(s) \rightarrow Zn^2(aq) + 2e^-$$

The electrons released make the Zn strip negatively charged relative to the solution, so there is a potential difference between the zinc strip and the solution. The Zn^{2+} ions in the solution accept electrons, re-forming Zn atoms:

$$Zn^{2+}(aq) + 2e^- \rightarrow Zn(s)$$

When Zn^{2+} ions are turning back to Zn as fast as they are being formed, an equilibrium is set up:

$$Zn^{2+}(aq) + 2e^- \rightleftharpoons Zn(s)$$

For a general metal, M:

$$M^{2+}(aq) + 2e^- \rightleftharpoons M(s)$$

The position of this equilibrium determines the size of the potential difference (the electrode potential) between the metal strip and the solution of metal ions. The further to the right the equilibrium lies, the greater the tendency of the electrode to accept electrons, and the more positive the electrode potential.

When we put two half-cells together, the one with the more positive potential will become the positive terminal of the cell, and the other one will become the negative terminal.

Making a cell from two half-cells

A connection is needed between the two solutions, but the solutions should not mix together. A strip of filter paper soaked in saturated potassium nitrate(V) solution can be used as a junction, or **salt bridge**, between the half-cells. The potassium and nitrate(V) ions carry the current in the salt bridge so that there is electrical contact between the solutions, but no mixing. The complete set up is shown in Figure 8.

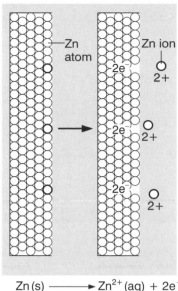

$$Zn(s) \longrightarrow Zn^{2+}(aq) + 2e^-$$

Figure 7 Zinc atoms form zinc ions, releasing electrons and setting up a potential difference

An **electrochemical cell** consists of two half-cells, connected by a salt bridge. We measure the maximum potential difference between the two electrodes of the half-cells with a high resistance voltmeter, so that negligible current flows.

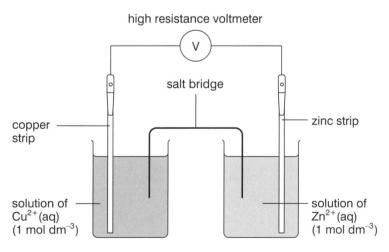

Figure 8 A copper-zinc cell

The circuit is completed by a metal wire connecting the copper and zinc strips. A high resistance voltmeter can be included in the circuit to measure the maximum voltage, E_{cell}, produced by the cell.

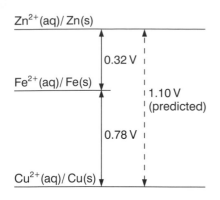

Figure 9 A chart showing the potential differences between three half-cells

Table 6 shows data obtained from a number of different cells.

Positive half-cell	Negative half-cell	E_{cell}/V
$Cu^{2+}(aq)/Cu(s)$	$Zn^{2+}(aq)/Zn(s)$	1.10
$Zn^{2+}(aq)/Zn(s)$	$Mg^{2+}(aq)/Mg(s)$	1.61
$Au^{3+}(aq)/Au(s)$	$Zn^{2+}(aq)/Zn(s)$	2.26
$Cu^{2+}(aq)/Cu(s)$	$Mg^{2+}(aq)/Mg(s)$	2.71
$Ag^+(aq)/Ag(s)$	$Zn^{2+}(aq)/Zn(s)$	1.57

Table 6 The potential differences, E_{cell}, generated by some cells. A shorthand notation is used to represent each half-cell. The oxidised form is always written first

Can you detect any patterns in the voltages in Table 6? It is difficult. This is because we are measuring *differences* between the electrode potentials of the various half-cells.

Figure 9 shows one way of sorting out some of the data. It shows the potential differences between three half-cells: copper, iron and zinc. The potential difference measured between the electrodes of the copper and iron half-cells is 0.78 V, with the copper positive. The potential difference measured between the iron and zinc half-cells is 0.32 V with the iron positive.

What potential difference would you expect between the electrodes of the copper and zinc half-cells? Using Figure 9, we can predict it will be 1.10 V. The value we actually get when we measure the voltage is indeed 1.10 V.

It would help to sort things out if we select one reference half-cell and measure all the others against it. We would then have a common reference point, and we could construct a list of electrode potentials relative to it. A **standard hydrogen half-cell** is chosen as the reference (Figure 10). The half-reaction in this half-cell is:

$$2H^+(aq) + 2e^- \rightarrow H_2(g)$$

Standard conditions for the hydrogen half-cell

$[H^+(aq)] = 1.00\ mol\ dm^{-3}$

pressure of H_2 gas = 1 atm

$T = 298\ K$

Standard conditions

Electrode potentials vary with temperature, and so for all cells a standard temperature is defined. This is 298 K. Altering the concentrations of any ions appearing in the half-reactions also affects the voltages and so a standard concentration of $1\ mol\ dm^{-3}$ is chosen. Standard pressure is 1 atmosphere (101.3 kPa)

The potential of the standard hydrogen half-cell is *defined* as 0.00 V, a value chosen for convenience.

The **standard electrode potential** of a half cell, $E^\ominus$, is defined as the potential difference between it and a standard hydrogen half-cell.

$E^\ominus$ values have a sign depending on whether the half-cell is at a more or less positive potential than the standard hydrogen half-cell. Measurements are made at 298 K, with the metal dipping into a $1.00\ mol\ dm^{-3}$ solution of a salt of the metal. Some values are shown in Table 7.

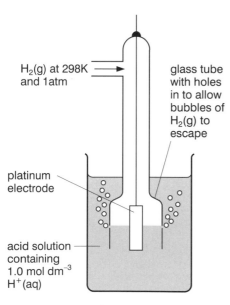

H$_2$(g) at 298K and 1atm

glass tube with holes in to allow bubbles of H$_2$(g) to escape

platinum electrode

acid solution containing 1.0 mol dm^{-3} H$^+$(aq)

Figure 10 The standard hydrogen half-cell (sometimes called a standard hydrogen electrode)

Half-cell	Half-reaction	$E^\ominus$/V
$Mg^{2+}(aq)/Mg(s)$	$Mg^{2+} + 2e^- \rightarrow Mg(s)$	−2.37
$Zn^{2+}(aq)/Zn(s)$	$Zn^{2+}(aq) + 2e^- \rightarrow Zn(s)$	−0.76
$2H^+(aq)/H_2(g)$	$2H^+(aq) + 2e^- \rightarrow H_2(g)$	0 (by definition)
$Cu^{2+}(aq)/Cu(s)$	$Cu^{2+}(aq) + 2e^- \rightarrow Cu(s)$	+0.34
$Ag^+(aq)/Ag(s)$	$Ag^+(aq) + e^- \rightarrow Ag(s)$	+0.81

Table 7 The standard electrode potentials of some half-cells

Listed like this, with the most positive potential at the bottom, the series is called the **electrochemical series**.

The half-cells at the bottom of the series have the greatest tendency to *accept* electrons. The half-cells at the top have the least tendency to accept electrons and the most negative $E^\ominus$ values. In fact, the half-cells at the top have the greatest tendency to go in the reverse direction and *release* electrons. For this reason, the most reactive metals are at the top of the series. You will notice that the order is very similar to the 'reactivity series' of metals you will have met in earlier studies.

Other half-cell reactions

Metal ion/metal reactions are only one case of redox reaction. There are many others. For example, between ions:

$$Fe^{3+}(aq) + e^- \rightarrow Fe^{2+}(aq)$$
$$Cr_2O_7^{2-}(aq) + 14H^+(aq) + 6e^- \rightarrow 2Cr^{3+}(aq) + 7H_2O(l)$$
$$MnO_4^-(aq) + 8H^+(aq) + 5e^- \rightarrow Mn^{2+}(aq) + 4H_2O(l)$$

and between molecules and ions:

$$Cl_2(aq) + 2e^- \rightarrow 2Cl^-(aq)$$

All these half-equations can be set up as half-cells. However, there is no metal in the half-reaction to make electrical contact, so an electrode made of an unreactive metal such as platinum is used. It dips into a solution containing all the ions and molecules involved in the half-reaction. For example, Figure 11 shows the set-up for a $Fe^{3+}(aq)/Fe^{2+}(aq)$ half-cell.

Table 8 shows values of standard electrode potential for a selection of half-cells. There is a fuller table on the Data Sheets.

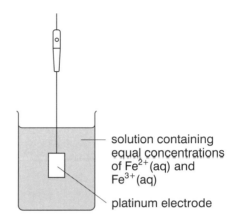

solution containing equal concentrations of $Fe^{2+}(aq)$ and $Fe^{3+}(aq)$

platinum electrode

Figure 11 A standard half-cell for the $Fe^{3+}(aq)/Fe^{2+}(aq)$ half-reaction

Half-reaction	$E^\ominus/V$
$I_2(aq) + 2e^- \rightarrow 2I^-(aq)$	+0.54
$Br_2(aq) + 2e^- \rightarrow 2Br^-(aq)$	+1.09
$Cl_2(g) + 2e^- \rightarrow 2Cl^-(aq)$	+1.36
$MnO_4^-(aq) + 8H^+(aq) + 5e^- \rightarrow Mn^{2+}(aq) + 4H_2O(l)$	+1.51

Table 8 Some standard electrode potentials

Working out E_{cell} from standard electrode potentials

An electrochemical cell consists of two half-cells. If you know the electrode potential of each half-cell, you can work out the potential, E_{cell}, of the cell as a whole.

For example, suppose we set up a cell using the two half-reactions:

$$Zn^{2+}(aq) + 2e^- \rightarrow Zn(s); \quad E^\ominus = -0.76\,V$$
$$Cu^{2+}(aq) + 2e^- \rightarrow Cu(s); \quad E^\ominus = +0.34\,V$$

(This is called a Daniell cell – it was invented in 1836).

First, we have to decide the direction of the cell reaction. It isn't possible for both reactions to accept electrons; one half-reaction must go in reverse so it releases electrons. The one that gets reversed is always the one with the more negative electrode potential – in this case the zinc, which will be the negative electrode in the cell. So the half-reactions that actually happen when you connect these two half-cells together are:

$$Zn(s) \rightarrow Zn^{2+}(aq) + 2e^-; \quad E^\ominus = +0.76\,V$$
$$Cu^{2+}(aq) + 2e^- \rightarrow Cu(s); \quad E^\ominus = +0.34\,V$$

Notice that the sign of the zinc half-cell reaction has reversed, because the direction of the half-reaction has reversed.

We now know the contribution that each half-cell makes to the overall cell potential. To get $E^\ominus{}_{cell}$ we simply add them together:

$$E^\ominus{}_{cell} = +0.76\,V + 0.34\,V = +1.10\,V$$

with zinc being the negative electrode.

To find the overall reaction occurring in the cell as a whole, we can add together the two half-equations.

$$Zn(s) \rightarrow Zn^{2+}(aq) + 2e^-$$

$$\underline{Cu^{2+}(aq) + 2e^- \rightarrow Cu(s)}$$

$$Zn(s) + Cu^{2+}(aq) \rightarrow Cu(s) + Zn^{2+}(aq)$$

This is the reaction we discussed at the very beginning of this section.

A table of standard electrode potentials allows us to calculate the maximum voltage obtainable from any cell under standard conditions. But electrode potentials are useful in other ways. We can use them to *predict* the reactions which can take place when two half-cells are connected so that electrons can flow. In fact, we can use them to make predictions about any redox reaction, whether or not it occurs in a cell. **Section 9.3** explains how.

PROBLEMS FOR 9.2

Note You may need to consult the Data Sheets in order to answer some of these questions. In some questions it may help to draw up an electrode potential chart.

1 a Copy the following half-equations and write an oxidation state under each element:

   i $Mg(s) \rightarrow Mg^{2+}(aq)$

   ii $2I^-(aq) \rightarrow I_2(aq)$

   iii $Ti^{3+}(aq) \rightarrow Ti^{4+}(aq)$

   iv $CO_2(g) + 2H^+(aq) \rightarrow CO(g) + H_2O(l)$

   v $MnO_4^-(aq) + 8H^+(aq) \rightarrow Mn^{2+}(aq) +$
                                 $4H_2O(l)$

b Insert electrons on either side of each half-equation in order to balance and complete it.

c For each complete half-equation describe the process as oxidation or reduction.

2 a Classify each of the following reactions as **redox, acid-base, complex formation** or **none of these**.

   i $Fe_2O_3(s) + 3CO(g) \rightarrow 2Fe(s) + 3CO_2(g)$

   ii $CuO(s) + H_2(g) \rightarrow Cu(s) + H_2O(l)$

   iii $NaOH(aq) + HCl(aq) \rightarrow NaCl(aq) + H_2O(l)$

   iv $2Na(s) + 2HCl(aq) \rightarrow 2NaCl(aq) + H_2(g)$

   v $NH_3(aq) + H^+(aq) \rightarrow NH_4^+(aq)$

   vi $Cu^{2+}(aq) + 4NH_3(aq) \rightarrow [Cu(NH_3)_4]^{2+}(aq)$

   vii $NaHS(s) + HCl(aq) \rightarrow NaCl(aq) + H_2S(g)$

   viii $2NH_3(g) \rightarrow N_2(g) + 3H_2(g)$

b For each of the reactions which you classified as redox, say which element is being oxidised and which reduced.

3 For each of the following redox reactions, write equations for the two half-reactions. Omit ions which are unchanged in the reaction.

  a $Zn(s) + Pb^{2+}(aq) \rightarrow Zn^{2+}(aq) + Pb(s)$

  b $2Al(s) + 6H^+(aq) \rightarrow 2Al^{3+}(aq) + 3H_2(g)$

  c $2Ag^+(aq) + Cu(s) \rightarrow Cu^{2+}(aq) + 2Ag(s)$

  d $Cl_2(g) + 2I^-(aq) \rightarrow I_2(aq) + 2Cl^-(aq)$

  e $Zn(s) + S(s) \rightarrow ZnS(s)$

4 When a cell is made from a standard cadmium half-cell, and a standard copper half-cell, the reading on a high resistance voltmeter is 0.74 V. The copper electrode forms the positive terminal of the cell. What is the electrode potential of a standard cadmium half-cell?

(You will need to look up the standard electrode potential of the copper half-cell.)

5 A cell is made from a $Co^{2+}(aq)/Co(s)$ half-cell and a $Cu^{2+}(aq)/Cu(s)$ half-cell. $E^\ominus{}_{cell} = +0.62$ V with the copper half-cell positive. Calculate the standard electrode potential of the cobalt half-cell.

6 If a metal is placed high in the electrochemical series its ability to release electrons and form hydrated positive ions is high. The metal is a strong reducing agent. Consult a table of electrode potentials and arrange the following metals in order of their strength as reducing agents:

   Ag Ce Sn Ni Cd K

7 a What would be the maximum potential difference of a cell made up from a standard $Mg^{2+}(aq)/Mg(s)$ half-cell and a standard $Pb^{2+}(aq)/Pb(s)$ half-cell?

b In the above cell which electrode will form the positive terminal, the Mg or the Pb electrode?

c Write an equation for the half-reaction that can occur at the positive electrode.

8 a Calculate $E^\ominus{}_{cell}$ values for cells made from the following standard half-cells:

   i $Fe^{2+}(aq)/Fe(s)$ and $Ni^{2+}(aq)/Ni(s)$

   ii $Zn^{2+}(aq)/Zn(s)$ and $S(s)/S^{2-}(aq)$

   iii $Sn^{4+}(aq)/Sn^{2+}(aq)$ and $MnO_4^-(aq)/Mn^{2+}(aq)$

   In each case identify which electrode will form the positive terminal of the cell.

b Write an equation for the overall reaction occurring in each cell.

9 A series of electrochemical cells was set up. The table shows the half-cells used and readings obtained on a high resistance voltmeter

Work out values for the standard electrode potentials of each half-cell. (Constructing a chart will help you to do this.)

Positive half-cell	Negative half-cell	$E^{\ominus}_{cell}$/V
$2H^+(aq), H_2(g)$/Pt	$Pb^{2+}(aq)$/Pb(s)	0.13
$Cd^{2+}(aq)$/Cd(s)	$Cr^{3+}(aq)$/Cr(s)	0.34
$Pb^{2+}(aq)$/Pb(s)	$Cd^{2+}(aq)$/Cd(s)	0.27
$Ag^+(aq)$/Ag(s)	$Cd^{2+}(aq)$/Cd(s)	1.20

9.3 *Predicting the direction of redox reactions*

Using electrode potentials to predict the direction reactions can take

Electrode potentials measure the tendency of a half-reaction to accept electrons. Figure 12 illustrates this for three half-reactions. If a half-reaction has a large tendency to *accept* electrons, it will have a small tendency to *supply* them – and vice-versa.

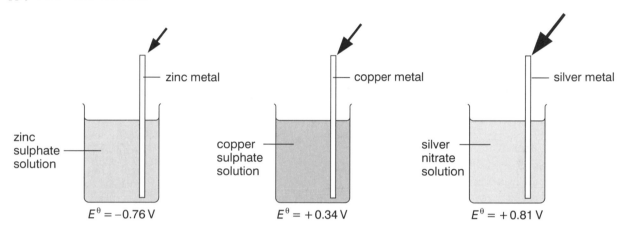

zinc metal

zinc sulphate solution

$E^{\theta} = -0.76$ V

copper metal

copper sulphate solution

$E^{\theta} = +0.34$ V

silver metal

silver nitrate solution

$E^{\theta} = +0.81$ V

Figure 12 The size of the arrows indicates the tendency of the half-cell to accept electrons

Copper ions will accept electrons supplied by zinc. Silver ions will accept electrons supplied by copper.

For zinc to supply electrons the reaction must be

$$Zn(s) \rightarrow Zn^{2+}(aq) + 2e^-$$

For the copper ions to accept electrons, the reaction must be

$$Cu^{2+}(aq) + 2e^- \rightarrow Cu(s)$$

The prediction for the overall reaction agrees with the observed changes

$$Zn(s) + Cu^{2+}(aq) \rightarrow Zn^{2+}(aq) + Cu(s)$$

When the copper and silver half-cells are connected, the predicted changes are:

$$2Ag^+(aq) + 2e^- \rightarrow 2Ag(s)$$
$$Cu(s) \rightarrow Cu^{2+}(aq) + 2e^-$$

The overall reaction predicted is

$$Cu(s) + 2Ag^+(aq) \rightarrow Cu^{2+}(aq) + 2Ag(s)$$

This again agrees with the observed changes.

Electrode potential charts provide a useful way of displaying and using the data. We can use them to make predictions about the direction a particular redox reaction will take. Figure 13 shows an electrode potential chart for the three half-reactions that we have just discussed.

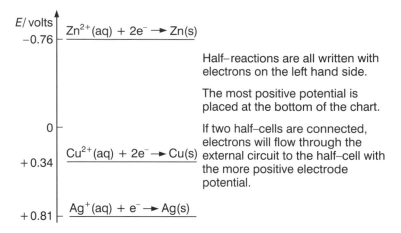

Figure 13 An electrode potential chart

Using electrode potential charts

Here is a worked example:

Problem Table 9 below gives electrode potentials of some half-reactions. What reactions can occur if we connect the $MnO_4^-(aq)/Mn^{2+}(aq)$ and $Fe^{3+}(aq)/Fe^{2+}(aq)$ half-cells so that electrons can flow? Write an equation for the overall reaction.

Half-cell	Half-reaction	$E^\ominus/V$
$I_2(aq)/2I^-(aq)$	$I_2(aq) + 2e^- \rightarrow 2I^-(aq)$	+0.54
$Fe^{3+}(aq)/Fe^{2+}(aq)$	$Fe^{3+}(aq) + e^- \rightarrow Fe^{2+}(aq)$	+0.77
$Br_2(aq)/2Br^-(aq)$	$Br_2(aq) + 2e^- \rightarrow 2Br^-(aq)$	+1.09
$Cl_2(g)/2Cl^-(aq)$	$Cl_2(aq) + 2e^- \rightarrow 2Cl^-(aq)$	+1.36
$MnO_4^-(aq)/Mn^{2+}(aq)$	$MnO_4^-(aq) + 8H^+(aq) + 5e^- \rightarrow Mn^{2+}(aq) + 4H_2O(l)$	+1.51

Table 9 Standard electrode potentials for a number of half-cells

Method 1 Construct an electrode potential chart
2 Use it to make predictions about half-reactions
3 Use the half-equations to give an overall equation

Solution 1 The electrode potential chart for the half-reactions in Table 9 is shown in Figure 14 below.

Figure 14 Electrode potential chart for the half-reactions in Table 9

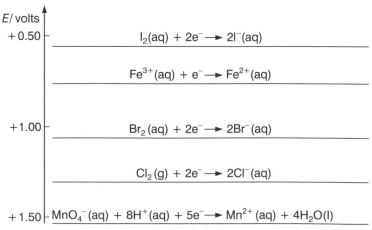

2 Electrons flow to the positive electrode of a cell. This will be the $MnO_4^-(aq)/Mn^{2+}(aq)$ half-cell, which has an electrode potential of $+1.51V$.

A reduction reaction will occur in the positive half-cell.
$$MnO_4^-(aq) + 8H^+(aq) + 5e^- \rightarrow Mn^{2+}(aq) + 4H_2O(l)$$
The other half-cell must supply electrons. Oxidation occurs.
$$Fe^{2+}(aq) \rightarrow Fe^{3+}(aq) + e^-$$

3 The number of electrons supplied and accepted must be equal. Therefore the reaction in the $Fe^{3+}(aq)/Fe^{2+}(aq)$ half-cell must occur five times each time one $MnO_4^-(aq)$ ion is reduced.
$$5Fe^{2+}(aq) \rightarrow 5Fe^{3+}(aq) + 5e^-$$
The overall equation is
$$MnO_4^-(aq) + 8H^+(aq) + 5Fe^{2+}(aq) \rightarrow Mn^{2+}(aq) + 4H_2O(l) + 5Fe^{3+}(aq)$$

Try some of the problems at the end of this section.

What exactly can we predict?

Are we limited to making predictions about redox reactions occurring in half-cells?

No. We can use the electrode potentials to make predictions about the feasibility of redox reactions whether or not we have physically arranged the reagents into two half-cells with electrodes. This makes electrode potentials useful for making predictions about *any* redox reactions.

Are we able to predict for certain whether a particular reaction will happen or not?

No. We are able to use electrode potentials to say whether it is *possible* for a reaction to happen. But we know nothing about the *rate* of the reaction. So a reaction which we have predicted as possible may not actually occur in practice because it is too slow.

Can we *make* reactions happen?

We may predict that a change is feasible, mix the reagents and find that nothing in fact happens. The rate of the reaction is so slow that no change is observable. But we can sometime change reaction rates.

If a reaction is slow it means that the activation enthalpy for the reaction must be very high. If we want the reaction to happen faster we could look for a catalyst to provide an alternative route for the reaction with a lower activation enthalpy, and so increase the rate.

However, if we predict from the electrode potentials that the reaction is *not* possible, no catalyst in the world is going to make it happen. Electrons will not flow spontaneously from a positive potential to a less positive one.

What else could we try to make reactions happen?

We must remember that if we use standard electrode potentials we are referring to reactions occurring in aqueous solution under *standard conditions*, at 298 K, and 1 atm pressure. Under different conditions, the electrode potential will be different.

If we predict that a reaction is not feasible under standard conditions, we could try changing the conditions, in order to alter the values of the electrode potentials.

Electrode potentials may vary with the concentration of the ions and molecules involved in the cell reaction (see **Section 9.4**). For example, if hydrogen or hydroxide ions are involved, pH changes will change electrode potentials, and reactions may become possible.

What happens in plants?

If electrons in ions or molecules are excited to higher energy levels, the electrode potentials change. Light can provide the energy to excite the ions or molecules.

When sunlight shines on plants some of it is absorbed by chlorophyll molecules. The *excited* chlorophyll molecules are able to cause reduction reactions which are not otherwise possible.

PROBLEMS FOR 9.3

1 Use the chart in Figure 14 to help you answer this question.
   a What reactions can occur if the following half-cells are connected so that electrons can flow?
      i $I_2(aq)/2I^-(aq)$ and $Cl_2(g)/2Cl^-(aq)$
      ii $Br_2(aq)/2Br^-(aq)$ and $MnO_4^-(aq)/Mn^{2+}(aq)$
      iii $Br_2(aq)/2Br^-(aq)$ and $I_2(aq)/2I^-(aq)$
   b Write a balanced equation for the overall reaction in each case.

2 Look at the standard electrode potentials for the following half-reactions.
   $Sn^{2+}(aq) + 2e^- \rightarrow Sn(s)$ $-0.14\,V$
   $Fe^{3+}(aq) + e^- \rightarrow Fe^{2+}(aq)$ $+0.77\,V$
   $2Hg^{2+}(aq) + 2e^- \rightarrow Hg_2^{2+}(aq)$ $+0.92\,V$
   $Cl_2 + 2e^- \rightarrow 2Cl^-(aq)$ $+1.36\,V$
   Construct an electrode potential chart and use it to predict which of the following reactions can occur.
   a $2Fe^{2+}(aq) + 2Hg^{2+}(aq) \rightarrow 2Fe^{3+}(aq) + Hg_2^{2+}(aq)$
   b $Sn(s) + 2Hg^{2+}(aq) \rightarrow Sn^{2+}(aq) + Hg_2^{2+}(aq)$
   c $2Cl^-(aq) + Sn^{2+}(aq) \rightarrow Cl_2(g) + Sn(s)$
   d $2Fe^{2+}(aq) + Cl_2(aq) \rightarrow 2Fe^{3+}(aq) + 2Cl^-(aq)$

3 Here are two half-cells:

Half-cell	E°/V
$Cl_2(g)/2Cl^-(aq)$	$+1.36$
$Cr_2O_7^{2-}(aq)/Cr^{3+}(aq)$	$+1.33$

   a Write equations for the two half-reactions which can occur if the half-cells are connected.
   b Write a balanced equation for the overall reaction.

4 a Construct an electrode potential chart for the following half-reactions.
      i $Fe^{3+}(aq) + e \rightarrow Fe^{2+}(aq)$
      ii $I_2(aq) + 2e^- \rightarrow 2I^-(aq)$
      iii $Sn^{4+}(aq) + 2e^- \rightarrow Sn^{2+}(aq)$
   b Predict whether reaction is possible between the following pairs
      i Fe^{3+} and I_2
      ii Sn^{2+} and I^-
      iii I^- and Fe^{3+}

5 The following table gives data for a series of half-reactions.

Half-reaction	E°/V
$2H^+(aq) + 2e^- \rightarrow H_2(g)$	0
$I_2(aq) + 2e^- \rightarrow 2I^-(aq)$	$+0.54$
$Fe^{3+}(aq) \rightarrow Fe^{2+}(aq)$	$+0.77$
$Br_2(aq) + 2e^- \rightarrow 2Br^-(aq)$	$+1.09$
$IO_3^-(aq) + 6H^+ + 5e^- \rightarrow \frac{1}{2}I_2(aq) + 3H_2O(l)$	$+1.19$
$Cl_2(g) + 2e^- \rightarrow 2Cl^-(aq)$	$+1.36$

   Use the data to decide which of the following can be oxidised by $IO_3^-(aq)$ in acidic solution:
   a $Fe^{2+}(aq)$ d $H_2(g)$
   b $Cl^-(aq)$ e $Br^-(aq)$
   c $I^-(aq)$

6 Dental fillings are often amalgams of silver, copper and tin. (An amalgam is a mixture of a metal with mercury). Saliva can act as an electrolyte. When someone with such a filling bites accidentally on a piece of aluminium foil they can experience a sharp pain.

Half-cell	E°/V
$Ag^+(aq)/Ag(s)$	$+0.80$
$Cu^{2+}(aq)/Cu(s)$	$+0.34$
$Sn^{2+}(aq)/Sn(s)$	-0.14
$Hg^{2+}(aq)/Hg(l)$	$+0.79$
$Al^{3+}(aq)/Al(s)$	-1.6

   a Suggest a reason for the pain experienced.
   b What is the maximum voltage which could be generated by these metal ion/metal systems under standard conditions?
   c Explain why the actual maximum voltage in the conditions of the mouth is likely to be less than your answer in **b**.

7 Electrode potentials can be useful when considering redox reactions of organic compounds.

   a Use the data given on the right to decide which reactions are feasible, under standard conditions, between oxygen and
      i   methane
      ii methanol
      iii methanal (HCHO)
   b Give equations for those reactions which are feasible.

Half-reaction	$E°$/V
HCOOH(aq) + 2H$^+$(aq) + 2e$^-$ $\rightarrow$ HCHO(aq) + H$_2$O(l)	+0.06
O$_2$(g) + 4H$^+$(aq) + 4e$^-$ $\rightarrow$ 2H$_2$O(l)	+1.23
CO$_2$(g) + 8H$^+$(aq) + 8e$^-$ $\rightarrow$ CH$_4$(g) + 2H$_2$O(l)	+0.17
HCHO(aq) + 2H$^+$(aq) + 2e$^-$ $\rightarrow$ CH$_3$OH(aq)	+0.23
CH$_3$OH(aq) + 2H$^+$(aq) + 2e$^-$ $\rightarrow$ CH$_4$(g) + H$_2$O(l)	+0.59

9.4 The effect of complexing on redox reactions

Standard electrode potentials, $E°$, always refer to systems at equilibrium under standard conditions. A change in the conditions changes the electrode potential. So, $E°$ values depend on the concentration of ions in solution as well as on temperature and pressure.

You can see the reason for this if you think about the equilibrium set up between the metal and its ions in solution. Take the Ni^{2+}(aq)/Ni(s) half-cell as an example (Figure 15).

The equilibrium set up between the metal and its ion is

$$Ni^{2+}(aq) + 2e^- \rightleftharpoons Ni(s)$$

If Ni^{2+}(aq) ions are removed the position of equilibrium changes to produce more ions in solution, and at the same time, more electrons. So removing Ni^{2+} makes the Ni electrode more negative with respect to the solution.

This is what happens when you add a ligand to the nickel half-cell. The ligand forms a complex with Ni^{2+} ions, removing them. So the concentration of Ni^{2+}(aq) falls, and the electrode potential becomes more negative.

The redox properties of metal ions can change a lot when the ligands are changed. Some values of electrode potentials are given in Table 9.

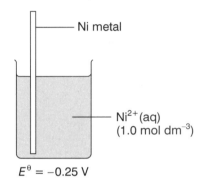

Ni metal

Ni^{2+}(aq) (1.0 mol dm^{-3})

E^θ = −0.25 V

Figure 15 The Ni^{2+}(aq)/Ni(s) half-cell

Half-reaction	$E°$/V
[Fe(H$_2$O)$_6$]$^{3+}$(aq) + e$^-$ $\rightarrow$ [Fe(H$_2$O)$_6$]$^{2+}$(aq)	+0.77
[Fe(CN)$_6$]$^{3-}$(aq) + e$^-$ $\rightarrow$ [Fe(CN)$_6$]$^{4-}$(aq)	+0.36
[Co(H$_2$O)$_6$]$^{3+}$(aq) + e$^-$ $\rightarrow$ [Co(H$_2$O)$_6$]$^{2+}$(aq)	+1.81
[Co(NH$_3$)$_6$]$^{3+}$(aq) + e$^-$ $\rightarrow$ [Co(NH$_3$)$_6$]$^{2+}$(aq)	+0.11

Table 9 Some standard electrode potentials of complex ions at 298 K

You can see from Table 9 that the effect of complexing aqueous Co^{3+} ions with NH$_3$ ligands is very large – it changes $E°$ from +1.81 V to +0.11 V. This is enough to make a major change in the redox properties of Co^{3+}. Hydrated Co^{3+} ions are such powerful oxidising agents that they can oxidise water, but the complex of Co^{3+} with NH$_3$ is stable in aqueous solution

Complexing can have an important effect on corrosion behaviour. Food cans are often made of steel coated with tin.

$$Sn^{2+}(aq) + 2e^- \rightarrow Sn(s); \quad E° = -0.14\,V$$
$$Fe^{2+}(aq) + 2e^- \rightarrow Fe(s); \quad E° = -0.44\,V$$

In cans of fruit Sn^{2+} ions form complexes with the anions of carboxylic acids in the fruit to such an extent that the electrode potential of the tin system becomes more negative than that of the iron system.

An electrochemical cell is set up between the iron and tin half-cells in the can. Tin is oxidised in preference to iron, because its electrode potential is more negative. This protects the iron from corrosion – until all the tin has reacted. (See **The Steel Story** storyline.)

PROBLEMS FOR 9.4

1 a Use the following standard electrode potentials to draw an electrode potential chart.

$$[Co(H_2O)_6]^{3+}(aq) + e^- \rightarrow [Co(H_2O)_6]^{2+}(aq);$$
$$E^\ominus = +1.81\,V$$
$$[Co(NH_3)_6]^{3+}(aq) + e^- \rightarrow [Co(NH_3)_6]^{2+}(aq);$$
$$E^\ominus = +0.11\,V$$
$$O_2(g) + 4H^+(aq) + 4e^- \rightarrow 2H_2O(l);$$
$$E^\ominus = +0.82\,V$$
$$\text{(at pH 7)}$$

b Use your chart to explain the following observations:

i When solid cobalt(III) fluoride is added to water, a colourless gas is evolved which relights a glowing splint. The resulting solution contains the pink $[Co(H_2O)_6]^{2+}(aq)$ ion.

ii When air is bubbled through a pink solution of $[Co(H_2O)_6]^{2+}(aq)$, no change in colour is observed. However on bubbling air through a yellow-brown solution of $[Co(NH_3)_6]^{2+}(aq)$, the colour rapidly changes to a dark brown.

2 a Use the following standard electrode potentials to draw an electrode potential chart.

$$[Fe(H_2O)_6]^{3+}(aq) + e^- \rightarrow [Fe(H_2O)_6]^{2+}(aq);$$
$$E^\ominus = +0.77\,V$$
$$[Fe(CN)_6]^{3-}(aq) + e^- \rightarrow [Fe(CN)_6]^{4-}(aq);$$
$$E^\ominus = +0.36\,V$$
$$I_2(aq) + 2e^- \rightarrow 2I^-(aq); \qquad E^\ominus = +0.54\,V$$

b Use your chart to predict the reactions that might occur on adding

i aqueous potassium iodide to aqueous iron(III);

ii aqueous iodine to aqueous hexacyanoferrate(II), $[Fe(CN)_6]^{4-}$.

c Explain why the reactions that you predict may not occur in practice.

the rate of the reaction means the rate at which $H_2O(l)$ and $O_2(g)$ are formed, which is the same as the rate at which $H_2O_2(aq)$ is used up. We would measure the rate of this reaction in moles of product (water or oxygen) formed per second, or moles of hydrogen peroxide used up per second. Suppose it turns out that 0.0001 mol of oxygen are being formed per second (Figure 3). The rate of the reaction is

0.0001 mol (O_2) s^{-1} or 0.0002 mol (H_2O) s^{-1} or -0.0002 mol (H_2O_2) s^{-1}

Notice that the rate in terms of moles of H_2O is twice the rate in terms of moles of O_2: this is because two moles of H_2O are formed for each mole of O_2. Notice also that the rate in terms of H_2O_2 has a *minus* sign: this is because H_2O_2 is getting used up instead of being produced.

The units for rate of reaction are usually mol s^{-1}, though they can also be mol min^{-1} or even mol h^{-1}.

Measuring rate of reaction

When you measure the speed of a car you can do so directly, using a speedometer. Unfortunately we can't measure reaction rates directly – there is no such thing as a reaction rate meter. What we have to do is to *measure the change in amount of a reactant or product in a certain time*. In practice this means measuring a property that is related to the amount of substance, such as the volume if the substance is a gas, or the colour intensity if the substance is coloured, or the pH if an acid or alkali is involved. So the procedure for measuring reaction rate is

- decide on a property of a reactant or product which you can measure, such as volume of gas or colour of solution
- measure the change in the property in a certain time
- find the rate in terms of $\dfrac{\text{change of property}}{\text{time}}$

Notice that the units of this will not be mol s^{-1}, but, say, (cm^3 of gas) s^{-1}. However, you can convert this to mol s^{-1}.

Investigating how rate depends on concentration

Now let's look more closely at the decomposition of hydrogen peroxide in solution. This reaction proceeds slowly under normal conditions, but it is greatly speeded up by catalysts. A particularly effective catalyst is the enzyme *catalase*.

You can do an experiment to investigate the enzyme-catalysed decomposition of hydrogen peroxide solution in **Activity EP6.3**, using the apparatus shown in Figure 4. The volume of oxygen is measured in the inverted burette. We will look at the kind of results you might get when you investigate how the rate of oxygen formation depends on the concentration of hydrogen peroxide solution.

We measure the total volume of oxygen given off at different times from the start of the experiment and plot a graph of this against time. This allows us to work out the rate of the reaction in (cm^3 of O_2) s^{-1}. We could convert this to mol (O_2) s^{-1}, because we know that 1 mol of oxygen occupies about 24 000 cm^3 at room temperature. But what we are interested in is *comparing* rates, and for these purposes we can use (cm^3 of O_2) without bothering to convert to moles.

Figure 5 on the next page shows a graph of the results that were obtained by starting with hydrogen peroxide of concentration 0.4 mol dm^{-3}. Notice these points about the graph in Figure 5:

- *The graph is steep at first*. The gradient of the graph gives us the rate of the reaction – the steeper the gradient, the faster the reaction. The reaction is at its fastest at the start, when the concentration of hydrogen peroxide in solution is high, before any has been used up.

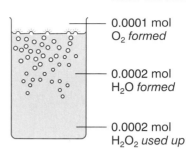

Each second

0.0001 mol O_2 *formed*

0.0002 mol H_2O *formed*

0.0002 mol H_2O_2 *used up*

$2H_2O_2 (aq) \longrightarrow 2H_2O(l) + O_2(g)$

Figure 3 The rate of reaction for the decomposition of hydrogen peroxide: an example

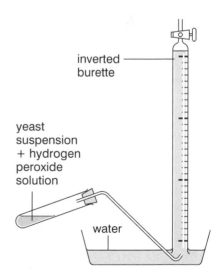

inverted burette

yeast suspension + hydrogen peroxide solution

water

Figure 4 Apparatus for investigating the rate of decomposition of hydrogen peroxide. The yeast provides the enzyme catalase

Figure 5 The decomposition of hydrogen peroxide, using a solution of concentration 0.4 mol dm⁻³. The rate is measured in terms of the volume of oxygen given off

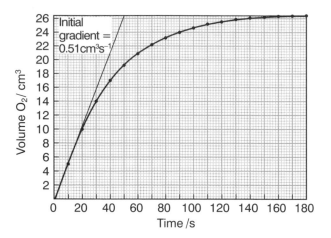

• *The graph gradually flattens out.* This is because, as the hydrogen peroxide is used up, its concentration falls. The lower the concentration, the slower the reaction. Eventually, the graph is horizontal: the gradient is zero, and the reaction has come to a stop.

The rate of the reaction at the start is called the **initial rate**. We can find the initial rate by drawing a tangent to the curve at the point t = 0, and measuring the gradient of this tangent. In the example in Figure 5, the gradient is 0.51, so the initial rate = 0.51 (cm³ of O_2) s⁻¹.

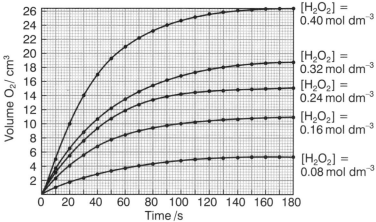

Figure 6 The decomposition of hydrogen peroxide solutions of differing concentrations

Figure 6 shows some results that were obtained when the same experiment was done using hydrogen peroxide solution of different concentrations. In each case, the concentration of the enzyme catalase was kept constant, as were all other conditions such as temperature. As you would expect, the graphs start off with differing gradients, depending on the concentration of hydrogen peroxide you started with.

Table 1 shows the initial rates of the experiments in Figure 6.

Now we are in a position to answer the question 'How does the rate of the reaction depend on the concentration of hydrogen peroxide?' Figure 7 shows the initial rates plotted against concentration of hydrogen peroxide, and you can see that it is a straight line. This means that *the rate is directly proportional to the concentration of hydrogen peroxide*. In other words

Rate ∝ [H_2O_2(aq)] (the square brackets mean 'concentration of')

or

Rate = constant × [H_2O_2(aq)]

Concentration of hydrogen peroxide at start/mol dm⁻³	Initial rate/ (cm³ of O_2(g)) s⁻¹
0.40	0.51
0.32	0.41
0.24	0.32
0.16	0.21
0.08	0.10

Table 1 Initial rates of decomposition of hydrogen peroxide

Hydrogen peroxide is not the only substance whose concentration affects the rate of this reaction. It is also affected by the concentration of the enzyme catalase, but in the series of experiments shown in Figure 6 we kept the concentration of catalase constant. However, we could do another set of experiments to find the effect of changing the concentration of catalase: this time we need to keep the concentration of hydrogen peroxide constant.

When we vary the concentration of catalase in this way, we find that the rate of the reaction is also proportional to the concentration of catalase. In other words

rate = constant × [catalase]

If we combine this with the equation involving H_2O_2, we get

rate = constant × [H_2O_2(aq)] × [catalase]

or **rate = k [H_2O_2(aq)] [catalase]**

This is called the **rate equation** for the reaction, and the constant k is called the **rate constant**. The value of k varies with temperature, so you must always say at what temperature the measurements were made when you give the rate, or the rate constant, of a reaction.

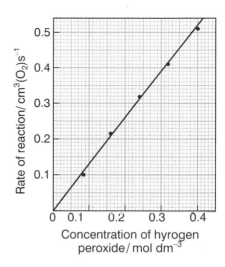

Figure 7 The initial rate of decomposition of hydrogen peroxide plotted against concentration of hydrogen peroxide

Order of reaction

We can write a rate equation for *any* chemical reaction – provided we can do an experiment first to find out how the rate depends on the concentration of the reactants. For a general reaction in which A and B are the reactants

A + B → products

the general rate equation is

rate = $k[A]^m[B]^n$

m and n are the powers to which the concentration need to be raised: they usually have values of 0, 1 or 2. m and n are called the **order of the reaction**, with respect to A and B. For example, in the hydrogen peroxide example which we met earlier,

rate = $k[H_2O_2]$ [catalase]

In this case, m and n are both equal to 1. We say that the reaction is **first order** with respect to H_2O_2 and first order with respect to catalase. The **overall order** of the reaction is given by (m + n), so in this case the reaction is **overall second order**.

Some examples will help to explain the idea of reaction order.

Example 1 *The reaction of Br radicals to form Br_2 molecules*
The equation for the reaction is
$2Br(g) \rightarrow Br_2(g)$
Experiments show that the rate equation for the formation of Br_2 is
rate = $k[Br]^2$
So this reaction is second order with respect to Br. Since Br is the only reactant involved, the reaction is also second order overall.

Example 2 *The reaction of iodide ions, I^-, with peroxodisulphate(VI) ions, $S_2O_8^{2-}$*
You may have investigated this reaction in **Activity EP6.4**. The equation for the reaction is
$S_2O_8^{2-}(aq) + 2I^-(aq) \rightarrow 2SO_4^{2-}(aq) + I_2(aq)$
Experiments show that the rate equation for the reaction is
rate = $k[S_2O_8^{2-}(aq)]$ [I^-(aq)]
So this reaction is first order with respect to $S_2O_8^{2-}$, first order with respect to I^- and second order overall.

Notice that in this second Example, the order of the reaction with respect to I⁻ is *one*, even though there are *two* I⁻ ions in the balanced equation for the reaction. This raises an important point: *you cannot predict the rate equation for a reaction from its balanced equation*. It *might* work (as in Example 1, above), but if often won't. The *only* way to find the rate equation for a reaction is by doing experiments to find the effect of varying the concentrations of reactants. This point becomes very clear if you look at the reactions in Examples 3 and 4.

Example 3   *The reaction between bromide ions, Br⁻, and bromate(V) ions, BrO_3^-.*
This reaction takes place in acidic solution:
$$BrO_3^-(aq) + 5Br^-(aq) + 6H^+(aq) \rightarrow 3Br_2(aq) + 5H_2O(l)$$
Experiments show that the rate equation for this reaction is
rate $= k[BrO_3^-][Br^-][H^+]^2$
So the reaction is first order with respect to both BrO_3^- and Br⁻, and second order with respect to H⁺. Notice that the orders do *not* correspond to the numbers in the balanced equation.

Example 4   *The reaction of propanone with iodine*
This reaction is catalysed by acid.

$$CH_3COCH_3(aq) + I_2(aq) \xrightarrow{\text{acid catalyst}} CH_3COCH_2I(aq) + H^+(aq) + I^-(aq)$$

The rate equation found by experiments is
rate $= k[CH_3COCH_3][H^+]$
(Note that the acid catalyst, H⁺, appears in the rate equation even though it is not used up.) So the reaction is first order with respect to both CH_3COCH_3 and H⁺. Notice that $[I_2]$ does not appear in the rate equation even though iodine is one of the reactants. The reaction is *zero order* with respect to I_2.

You may find it surprising that the reaction can be zero order with respect to one of the reactants: the reason is explained later, in the section on rate-determining steps.

Half-lives

To explain the important idea of half-lives, let's go back to the decomposition of hydrogen peroxide:

$$2H_2O_2(aq) \rightarrow 2H_2O(l) + O_2(g)$$

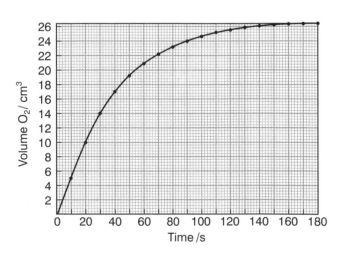

Figure 8 The decomposition of hydrogen peroxide. The rate is measured in terms of the volume of oxygen given off

The graph in Figure 8 shows the volume of oxygen produced in the decomposition of hydrogen peroxide (it is the same as Figure 5).

However, we could convert this graph to show *the amount of H_2O_2 remaining* at different times in this experiment. We can do this because we know how much H_2O_2 we started with, and we know that for every mole of O_2 produced, two moles of H_2O_2 get used up.

These calculations have been used to produce the graph in Figure 9. Notice that the graph shows that the amount of H_2O_2 *decreases* with time: this is the opposite of Figure 8, in which the amount of O_2 *increases* with time.

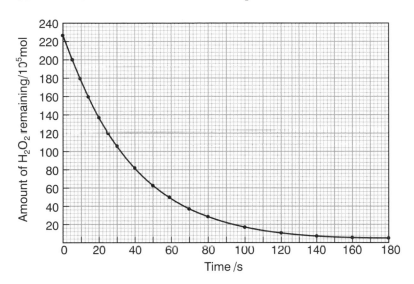

Figure 9 The same experiment as shown in Figure 8, but this time the rate of the reaction is measured in terms of the amount of H_2O_2 remaining

We can use Figure 9 to find the **half-life, $t_{\frac{1}{2}}$,** of H_2O_2 in this experiment. The half-life means *the time taken for half of the H_2O_2 to get used up*. In Figure 10 this has been done in three cases, using exactly the same graph as in Figure 9. Looking at Figure 10, you can see that to go from 200×10^{-5} mol of H_2O_2 to 100×10^{-5} mol takes 27s. In other words, starting with 200×10^{-5} mol of H_2O_2, the half-life, $t_{\frac{1}{2}} = 27$s. Starting with 100×10^{-5} mol, $t_{\frac{1}{2}}$ is again 27s. Starting with 50×10^{-5} mol, $t_{\frac{1}{2}}$ is 26s. In fact, allowing for experimental error, we find that *whatever* the starting amount $t_{\frac{1}{2}}$ is always 27s.

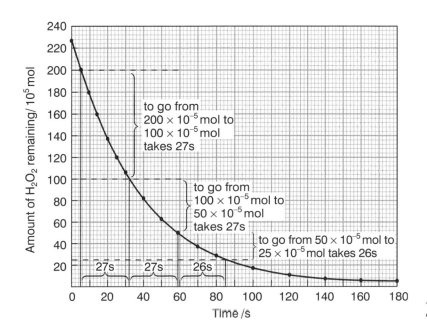

Figure 10 Finding half-lives, $t_{\frac{1}{2}}$, for the decomposition of hydrogen peroxide

The decomposition of hydrogen peroxide is a first order reaction with respect to H_2O_2, and we find the same rule for the half-lives of *all* first order reactions. *For a first order reaction, the half-life is constant, whatever the starting amount*. This characteristic gives us a useful way to decide whether a reaction is first order. Zero order and second order reactions do *not* have constant half-lives.

Radioactive decay is an important example of a first-order process. The time taken for a sample of any one isotope to decay until only half of it is left has always been the same ever since the Earth was formed (in fact, ever since the isotope was formed).

Different isotopes have different half-lives. For some, such as radium-226 ($t_\frac{1}{2}$ = 1622 years), it is thousands of years. For others it is only a fraction of a second. Both times are really quite short compared with the Earth's 4.6 billion year lifetime, but some radioactive isotopes, which were present when the Earth was formed, are still around. Uranium-235 ($t_\frac{1}{2}$ = 7×10^8 years) is an example.

This situation shows another important aspect of first order processes – they never end. There will always be some reactants left. We say a reaction is 'over' when we can no longer measure the change, not when the reactant has all gone.

Finding the order of reaction

To find the order of a reaction, you must do experiments. Most reactions involve more than one reactant, and in this case you have to do several experiments, to find the order with respect to each reactant separately. You have to control the variables so that the concentration of only one substance is changing at a time, and take all your measurements at the same temperature.

Let's look again at our trusty example of the decomposition of hydrogen peroxide in the presence of the enzyme catalase.

$$2H_2O_2(aq) \xrightarrow{\text{catalase}} 2H_2O(l) + O_2(g)$$

If we are looking at the effect of changing the concentration of hydrogen peroxide, there is no need to worry about the catalase – it is an enzyme, so it doesn't get used up, and its concentration doesn't change. But if we want to look at the effect of varying the concentration of *catalase*, we must control the concentration of hydrogen peroxide to keep it constant – otherwise we will have two variables changing at the same time. In **Activity EP 6.3**, the approach is to do several experiments, keeping the concentration of hydrogen peroxide the same each time, but varying the concentration of catalase. Another way of controlling the concentration of a reactant is to have a large excess of it, so that over the course of the experiment, the concentration does not change significantly.

Once we have collected a set of data for the effect of changing the concentration of a particular reactant, there are several methods we can use to find the order with respect to that reactant.

The initial rate method

This is the method used in the hydrogen peroxide investigation. We do several experimental 'runs' at different concentrations. For each run, we can find the initial rate graphically, as was shown in Figure 5.

Once we know the initial rate for different concentrations, we can find the order. If a graph of rate against concentration is a straight line (as in Figure 7), the reaction is first order. If a graph of rate against (concentration)^2 is a straight line, the reaction is second order. If the rate doesn't depend on concentration at all, it is zero order.

The progress curve method

A progress curve shows how the concentration of a reactant changes as the reaction progresses. A progress curve for the hydrogen peroxide decomposition was shown in Figure 9. Figure 11 shows how you can use a progress curve to find the rate of the reaction for different concentrations. The tangents show the rate of the reaction corresponding to concentrations, we can find the order (as in the initial rates method – see page 180).

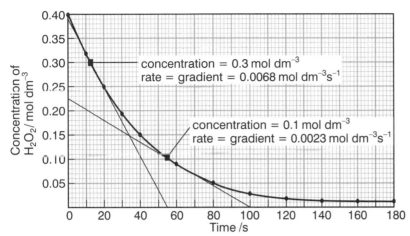

The half-lives method

We can use the progress curve to find half-lives for the reaction. If the half-lives are constant, the reaction is first-order.

Figure 11 Using a progress curve to find the rate of a reaction at different concentrations, by drawing tangents

Interpreting rate equations

Once we know the rate equation for a reaction, we can link it to the reaction mechanism. Most reaction mechanisms involve several individual steps. The rate equation gives us information about the slowest step in the mechanism – the **rate determining step**.

Rate determining steps

We have already seen that the rate equation for a reaction cannot be predicted from the balanced chemical equation. For example, consider the reaction of 2-bromo-2-methylpropane with hydroxide ions,

$$CH_3-\underset{\underset{CH_3}{|}}{\overset{\overset{CH_3}{|}}{C}}-Br \quad + \quad OH^- \quad \longrightarrow \quad CH_3-\underset{\underset{CH_3}{|}}{\overset{\overset{CH_3}{|}}{C}}-OH \quad + \quad Br^-$$

This reaction is found to be first order with respect to $(CH_3)_3CBr$ and zero order with respect to ^-OH. In other words

rate = $k[(CH_3)_3CBr]$

Why is $[OH^-]$ not involved in the rate equation? This suggests that ^-OH ions are not involved in the slow rate-determining step. The reaction cannot take place by direct reaction of 2-bromo-2-methylpropane itself with ^-OH ions. Chemists have studied this reaction in detail and they have found that it takes place in *two* steps.

First, the C–Br bond break heterolytically.

$$CH_3-\underset{\underset{CH_3}{|}}{\overset{\overset{CH_3}{|}}{C}}-Br \quad \longrightarrow \quad \left[CH_3-\underset{\underset{CH_3}{|}}{\overset{\overset{CH_3}{|}}{C}}+ \right] \quad + \quad Br^- \qquad \text{Step 1}$$

Because this step only involves $(CH_3)_3CBr$, its rate depends only on $[(CH_3)_3CBr]$, not on $[OH^-]$.

The second step involves reaction of the carbocation, $(CH_3)_3C^+$, with OH^-.

$$\left[\begin{array}{c} CH_3 \\ | \\ CH_3-C+ \\ | \\ CH_3 \end{array} \right] \quad + \quad OH^- \quad \longrightarrow \quad \begin{array}{c} CH_3 \\ | \\ CH_3-C-OH \\ | \\ CH_3 \end{array} \qquad \text{Step 2}$$

Like most ionic reactions, this process is very fast, certainly faster than Step 1. So the rate of Step 1 controls the rate of the whole reaction. That is why the rate of the reaction depends *only* on the concentration of $(CH_3)_3CBr$: the reaction is first order with respect to $(CH_3)_3CBr$, but zero order with respect to OH^-. Step 1 is said to be the rate determining step and its rate equation becomes the rate equation for the whole reaction.

The mechanism in this example involves two steps. Some simple reactions occur in a single step; other complex ones may involve more than two steps. But in every case, once you have broken the reaction down into steps, you can then write the rate equation for each step from its chemical equation. But the *overall* rate equation for the reaction can only be found by experiment.

In fact, the mechanisms for many reactions were originally found using their orders and rate equations. In the example above, the proposed mechanism for the reaction of 2-bromo-2-methylpropane with hydroxide ions was deduced from the experimentally determined rate equation.

The mechanism of enzyme-catalysed reactions

Our theories about the mechanism of enzyme-catalysed reactions were deduced from studies of the order of the reactions.

If you have done **Activities EP6.3** and **EP6.5**, you will know that when the substrate concentration is low, the rate equation for the reaction

$$\underset{\text{substrate}}{S} \quad \xrightarrow{\text{enzyme}} \quad \underset{\text{products}}{P}$$

is

rate $= k[E][S]$ where $[E]$ is the concentration of the enzyme.

From this, we deduce that the rate-determining step must involve one enzyme molecule and one substrate molecule.

$$E + S \quad \xrightarrow[\text{step}]{\text{rate-determining}} \quad \underset{\substack{\text{enzyme-substrate} \\ \text{complex}}}{ES}$$

The steps that follow this are faster:

$$ES \quad \xrightarrow{\text{fast}} \quad \underset{\substack{\text{enzyme-product} \\ \text{complex}}}{EP} \quad \xrightarrow{\text{fast}} \quad E + P$$

However, if the substrate concentration is high, the rate equation becomes

rate $= k[E]$

because, with more than enough substrate around, the first step is no longer the rate-determining one. All the enzyme active sites are occupied, and $[ES]$ is constant. What matters now is how *fast* ES or EP can break down to form the products. The rate of breakdown of ES depends on its concentration, which is effectively $[E]$ since almost all the enzyme is present as ES.

Energy barriers

Why are some steps in a reaction slow and other steps fast? One reason is that the steps have different energy barriers or activation enthalpies.

When the energy barrier is big, few pairs of molecules have enough energy to pass over it and the rate of conversion of reactants into products is slow. Figure 12 compares the enthalpy profiles for reactions with large and small activation enthalpies.

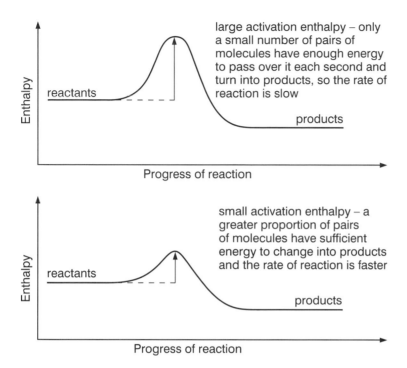

large activation enthalpy – only a small number of pairs of molecules have enough energy to pass over it each second and turn into products, so the rate of reaction is slow

small activation enthalpy – a greater proportion of pairs of molecules have sufficient energy to change into products and the rate of reaction is faster

Figure 12 The rate of a reaction depends on the size of the activation enthalpy

The enthalpy profile will be different for a reaction involving two steps. There will be two activation enthalpies – one for each step. In the case of the reaction of 2-bromo-2-methylpropane with hydroxide ions, Step 1 (the rate determining step) has the larger activation enthalpy (Figure 13). Usually, the chemicals in the middle of a reaction mechanism – the **intermediates** – are at a higher energy than the reactants or products, because they have unusual structures or bonding – like the carbocation $(CH_3)_3C^+$.

In any reaction with several steps, the rate-determining step will be the one with the largest activation enthalpy.

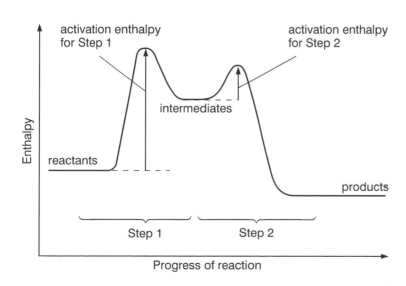

Figure 13 For a reaction involving two steps, there are two activation enthalpies

PROBLEMS FOR 10.2

1 Use the rate equations for the following reactions to write down the order of the reaction with respect to each of the reactants (and catalyst where present).

a The elimination of hydrogen bromide from bromoethane.

$$CH_3CH_2Br + OH^- \rightarrow CH_2{=}CH_2 + Br^- + H_2O$$
rate $= k[CH_3CH_2Br]$

b The acid catalysed hydrolysis of methyl methanoate.

$$HCOOCH_3 + H_2O \xrightarrow{\ H^+(aq)\ } HCOOH + CH_3OH$$
rate $= k[HCOOCH_3] [H^+]$

c The hydrolysis of urea, NH_2CONH_2, in the presence of the enzyme urease.

$$NH_2CONH_2(aq) + H_2O(l) \xrightarrow{\ urease\ } 2NH_3(aq) + CO_2(g)$$
rate $= k[NH_2CONH_2] [urease]$

d One of the propagation steps in the radical substitution of an alkane by chlorine.

$$CH_3{\cdot}(g) + Cl_2(g) \rightarrow CH_3Cl(g) + Cl{\cdot}(g)$$
rate $= k[CH_3{\cdot}] [Cl_2]$

e The formation of the World War I poison gas, phosgene, from carbon monoxide and chlorine.

$$CO(g) + Cl_2 \rightarrow COCl_2(g)$$
rate $= k[CO]^{1/2} [Cl_2]$

f The decomposition of nitrogen dioxide to oxygen and nitrogen monoxide.

$$2NO_2(g) \rightarrow 2NO(g) + O_2(g)$$
rate $= k[NO_2]^2$

2 Write down the rate equations for the following reactions.

a Experiments show that the reaction of 1-chlorobutane with aqueous sodium hydroxide is first order with respect to 1-chlorobutane and first order with respect to hydroxide ion.

$$CH_3CH_2CH_2CH_2Cl + OH^- \rightarrow$$
$$CH_3CH_2CH_2CH_2OH + Cl^-$$

b The hydrolysis of sucrose, $C_{12}H_{22}O_{11}$, is first order with respect to sucrose and first order with respect to acid catalyst, $H^+(aq)$.

$$C_{12}H_{22}O_{11} + H_2O \rightarrow 2C_6H_{12}O_6$$

3 The lipase-catalysed hydrolysis of triacetin (a natural oil) was followed using an initial rate method. An oil/water emulsion of triacetin was held at a constant temperature and pH 8. The initial rate was determined by titration with $0.0100\ mol\ dm^{-3}$ sodium hydroxide. Detergent was added to the mixture to hold the triacetin as an emulsion. Triacetin is a triester of ethanoic acid and propane-1,2,3-triol. It has the same basic structure as most fats and oils.

The results obtained are shown in Table 2.

a Write an equation for the hydrolysis of triacetin using full structural formulae.

b Determine the order of reaction with respect to
 i the enzyme, lipase
 ii the substrate, triacetin.

c Write the overall rate equation for the reaction.

4 The initial rate method was used to investigate the reaction

$$2H_2(g) + 2NO(g) \rightarrow 2H_2O(g) + N_2(g)$$

Table 3 shows the result of some studies which were made at 973 K.

[H₂] /$10^{-2}\ mol\ dm^{-3}$	[NO] /$10^{-2}\ mol\ dm^{-3}$	rate /$10^{-6}\ mol\ dm^{-3}\ s^{-1}$
2.0	2.50	4.8
2.0	1.25	1.2
2.0	5.00	19.2
1.0	1.25	0.6
4.0	2.50	9.6

Table 3

a What is the order of reaction with respect to
 i $H_2(g)$?
 ii $NO(g)$?

b Write down the rate equation for this reaction.

c Calculate a value for the rate constant for this reaction at 973 K.

Run	Volume emulsion/cm³	Volume detergent/cm³	Volume lipase/cm³	Volume water/cm³	Initial rate /$10^{-4}\ mol\ dm^{-3}\ min^{-1}$
A	15	15	2	0	5.25
B	10	20	2	0	3.40
C	5	25	2	0	1.80
D	15	15	1.5	0.5	3.84
E	15	15	1	1	2.51

Table 2

5 When cyclopropane gas is heated it isomerises to propene gas.

The following data were obtained by heating cyclopropane in a sealed container. The temperature was 700 K and the initial pressure was 0.5 atm.

Time/10^3 s	0	13	36	56	83	108
% cyclopropane	100	84	63	49	35	25

a What type of isomerism is shown by this pair of compounds?

b What would be the pressure in the reaction vessel at the end of the reaction?

c Plot a graph of % cyclopropane against time.

d Use your graph to measure some half-life values, and so determine the order of the reaction with respect to cyclopropane.

6 The radioactive isotope ^{13}N decays by β-emission with a half-life of 10 minutes.

a Write a nuclear equation for the decay of ^{13}N.

b Use the half-life to calculate how long it would take for the amount of ^{13}N in a sample to fall to the following fractions of its initial value: 0.50, 0.25, 0.125, 0.0625, 0.03125.

c Use your answers to part **b** to plot a decay curve for ^{13}N.

10.3 *The effect of temperature on rate*

How can I dissolve this sugar more quickly?
How can I get this cake to cook more quickly?
How can I get this glue to set more quickly?

The answer to all these questions is likely to involve raising the temperature. But why do processes like these go faster at higher temperatures? Whatever the reason, it is a very important effect, and the chemical industry depends heavily on it. Without it, there would be no Haber process for making ammonia, for example.

If you've measured the rate of reactions at different temperatures, you'll know that temperature has a large effect. In fact, for many reactions the rate is roughly *doubled* by a temperature rise of just 10 °C.

The collision theory of reactions

You met the basic ideas of the collision theory in **Section 10.1**. In this section we will take these ideas a little further.

Think of the reaction between nitrogen and hydrogen to make ammonia;

$$N_2(g) + 3H_2(g) \rightleftharpoons 2NH_3(g)$$

The collision theory says that reaction can only occurs when N_2 and H_2 collide. The more frequently they collide, the faster the reaction. If you increase the pressure, for example, the N_2 and H_2 molecules become closer together, so they collide and react more often. That's one of the reasons the Haber Process uses high pressures, to make ammonia more quickly.

What about the effect of temperature? Think about what happens to the molecules when you raise the temperature. They move faster, so they collide more frequently. We can work out *how much* more frequently they collide, because the average speed of molecules is proportional to the square root of the temperature. So if we increase the temperature from, say

300 K to 310 K, we would expect the average speed of the molecules to increase by a factor of $(310/300)^{1/2}$, which is about 1.016. Yet we know that the rate actually increases by much more than this.

Clearly there is more to the effect of temperature than simply making the particles collide more frequently. What matters is not just *how frequently* they collide, but also *with how much energy*. Unless the molecules collide with a certain minimum energy, they just bounce off one another and stay unreacted. In fact, this is what happens to most of the molecules most of the time. In the reaction between N_2 and H_2 at 300 K, only 1 in 10^{11} collisions results in a reaction! Even at 800 K (a temperature used in the Haber process) only 1 in 10^4 collision results in a reaction.

So the collision theory says that *reactions occur when molecules collide* **with a certain minimum energy**. *The more frequent these collisions, the faster the reaction.* This 'certain minimum energy' is the **activation enthalpy**. It is the energy that is needed to start breaking the bonds in the reacting molecules so the reaction can begin.

The distribution of energies

At any temperature, the speeds – and therefore the kinetic energies – of the molecules in a substance are spread over a wide range. It is like the walking speeds of people in a street: at any moment, some are moving slowly, some fast, and the majority are moving at moderate speeds. Molecules are similar: some have high kinetic energies, many have medium energies and some have low energies.

This distribution of kinetic energies in a gas at a given temperature is shown in Figure 14. It is called the Maxwell-Boltzmann distribution. (There is more about this distribution in **Section 4.4**.)

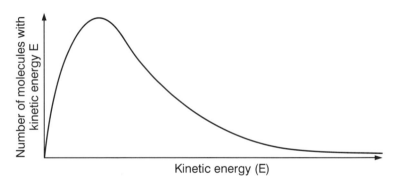

Figure 14 Distribution curve for molecular kinetic energies in a gas at a particular temperature

As the temperature increases, more molecules move at higher speeds and have higher kinetic energies. Figure 15 shows how the distribution of energies changes when you increase the temperature by 10 °C, from 300 K to 310 K. You can see that there is still a spread of energies, but now a greater proportion of molecules have higher energies.

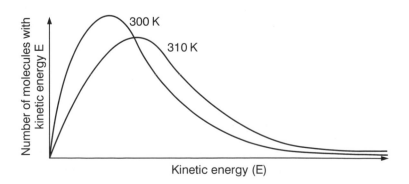

Figure 15 Distribution curves for molecular kinetic energies at 300 K and 310 K

Now let's look at the significance of this for reaction rates. We shall take as our example a reaction whose activation enthalpy, E_a, is $+50\,kJ\,mol^{-1}$, a value typical of many reactions. We need to think about how many molecules have energy greater than $50\,kJ\,mol^{-1}$, because these are the ones that can react. Of course, each individual molecule would have energy far less than $50\,kJ$: this is the energy possessed by 6×10^{23} of them.

Figure 16 shows the number of molecules with energy greater than $50\,kJ\,mol^{-1}$, for the reaction at 300 K. It's given by the shaded area underneath the curve. Only those molecules with energies in the shaded area can react.

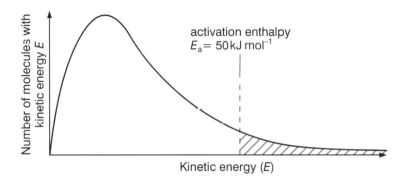

Figure 16 Distribution curve showing molecules with energy 50 kJ mol⁻¹ and above

Now look at the graph in Figure 17, which shows the curves for both 300 K and 310 K. You can see that at the higher temperature, a significantly higher proportion of molecules have energies above $50\,kJ\,mol^{-1}$ – about twice as many, in fact. This means that twice as many molecules have enough energy to react – so the reaction goes twice as fast.

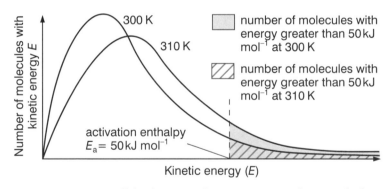

Figure 17 Distribution curves showing the effect on the proportion of molecules with energy 50 kJ mol⁻¹ and above of changing the temperature from 300 K to 310 K

We can summarise all this by saying that **reactions go faster** at higher temperatures because a larger proportion of the colliding molecules have the minimum activation enthalpy needed to react. Increasing temperature may not make much difference to the energy of each individual molecule, but it makes a big difference to the proportion of molecules with enough energy to react. In the example above, just a 10 °C increase is enough to place twice as many molecules above the minimum activation enthalpy of $+50\,kJ\,mol^{-1}$.

Temperature and rate constants

The rate constant for a reaction is k in the expression

$$\text{rate} = k[A]^a[B]^b$$

For many reactions, the rate is roughly doubled by a 10 °C temperature rise – even though everything else stays the same. Since the concentrations are not changed, the equation above shows that it must be the value of k, the rate constant, which has doubled. Changing the temperature *almost always* changes the value of k – though the 10 °C rule is only a rough one. In fact, it only works exactly for reactions with an activation enthalpy of $+50\,kJ\,mol^{-1}$, and for a temperature rise from 300 K to 310 K. Many reactions have an activation enthalpy of about this value, and a lot of chemistry is done at about 300 K, so the rule is a reasonable rough guide. In the case of the reaction of nitrogen with hydrogen in the Haber Process, the activation enthalpy is rather greater than $+50\,kJ\,mol^{-1}$, so the rate is less than doubled for a 10 °C rise.

PROBLEMS FOR 10.3

1 Use the ideas in this section to explain why
   a sugar dissolves faster at higher temperatures;
   b cakes cook faster at higher temperatures.

2 A mixture of hydrogen and oxygen doesn't react until it is ignited by a spark. Then it explodes. The mixture also explodes if you add some powdered platinum.
   a The energy of a spark is tiny, yet it is enough to ignite any quantity of hydrogen/oxygen mixture, large or small. Suggest an explanation for this.
   b Explain why platinum makes the hydrogen/oxygen reaction occur at room temperature.

3 Explain why, above a certain temperature, enzyme-catalyzed reactions actually go *more slowly* if the temperature is raised.

4 Use the **collision theory** to explain
   a Why coal burns faster when it is finely powdered than when it is in a lump.
   b Why nitrogen and oxygen in the atmosphere do not normally react to form nitrogen oxides.
   c Why reactions between two solids take place very slowly.
   d Why flour dust in the air can ignite with explosive violence.

5 The collision theory assumes that the rate of a reaction depends on
   A the rate at which reactant molecules collide with one another
   B the proportion of reactant molecules that have enough energy to react once they have collided.
   Which, out of A and B, explains each of the following observations?
   a Reactions in solution go faster at higher concentration.
   b Solids react faster when their surface area is higher.
   c Catalysts increase the rate of reactions.
   d Increasing the temperature increases the rate of a reaction.

6 Catalytic converters in car exhausts are designed to remove pollutant gases such as CO and NO_x. Yet the converters do not work effectively until the car engine has warmed up.

   Use the ideas in this section to suggest a reason.

10.4 *Catalysis*

A **catalyst** is a substance which alters the rate of a reaction without itself undergoing any permanent change.

Catalysts are not used up in chemical reactions and you can always get them back at the end. They are not changed chemically, though sometimes they may be changed *physically*. For example, the surface of a solid catalyst may crumble or become roughened. This suggests that the catalyst is taking some part in the reaction, but is being regenerated.

Only small amounts of a catalyst are usually needed. The catalyst does not affect the *amount* of product formed, only the *rate* at which it is formed.

Another interesting feature of catalysts is that they are often specific for one particular reaction. This is particularly true of biological catalysts, called **enzymes**. These are complex protein molecules which affect one particular biochemical reaction strongly, but leave a similar reaction almost unaffected. There is more about enzymes in the **Engineering Proteins** storyline.

A catalyst does not appear in the overall equation for a reaction.

Types of catalysts

If the reactants and catalyst are in the same physical state (for example, both are in aqueous solution), the reaction is said to involve **homogeneous catalysis**. Enzyme-catalysed reactions in cells take place in aqueous solution and are examples of this type of catalysis.

In contrast, many important industrial processes involve **heterogeneous catalysis**, where the reactants and the catalyst are in different physical states. This usually involves a mixture of gases or liquids reacting in the presence of a solid catalyst.

How do catalysts work?

In a chemical reaction, existing bonds in the reactants must first stretch and break. Then new bonds can form as the reactants are converted to products.

Bond breaking is an endothermic process. A pair of reacting molecules must have enough energy between them to pass over the activation enthalpy barrier before reaction can occur.

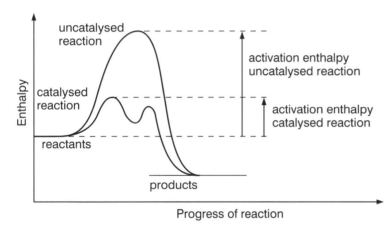

Figure 18 Shows the enthalpy profiles for an uncatalysed and a catalysed reaction.

If the enthalpy barrier is very high, relatively few pairs of molecules will have enough energy to overcome it and react to form the products – so the reaction is slow.

Catalysts speed up reactions by providing an alternative pathway for the breaking and remaking of bonds. In this case, the enthalpy barrier is lower and more pairs of molecules can pass over to form products. This means the reaction proceeds more quickly.

Heterogeneous catalysts

When a solid catalyst is used to increase the rate of a reaction between gases or liquids, the reaction occurs on the surface of the solid. Figure 19 on page 190 shows an example.

It is important that the catalyst has a large surface area for contact with reactants. For this reason, solid catalysts are used in a finely divided form or as a fine wire mesh. Sometimes the catalyst is supported on a porous material to increase its surface area and prevent it from crumbling. This happens in the catalytic converters fitted to car exhaust systems.

Zeolites (see **Developing Fuels** storyline) are widely used in industry as heterogeneous catalysts, for example, in the cracking of petroleum fractions.

You can read about the use of transition metals and their compounds as heterogeneous catalysts in **Section 11.6**.

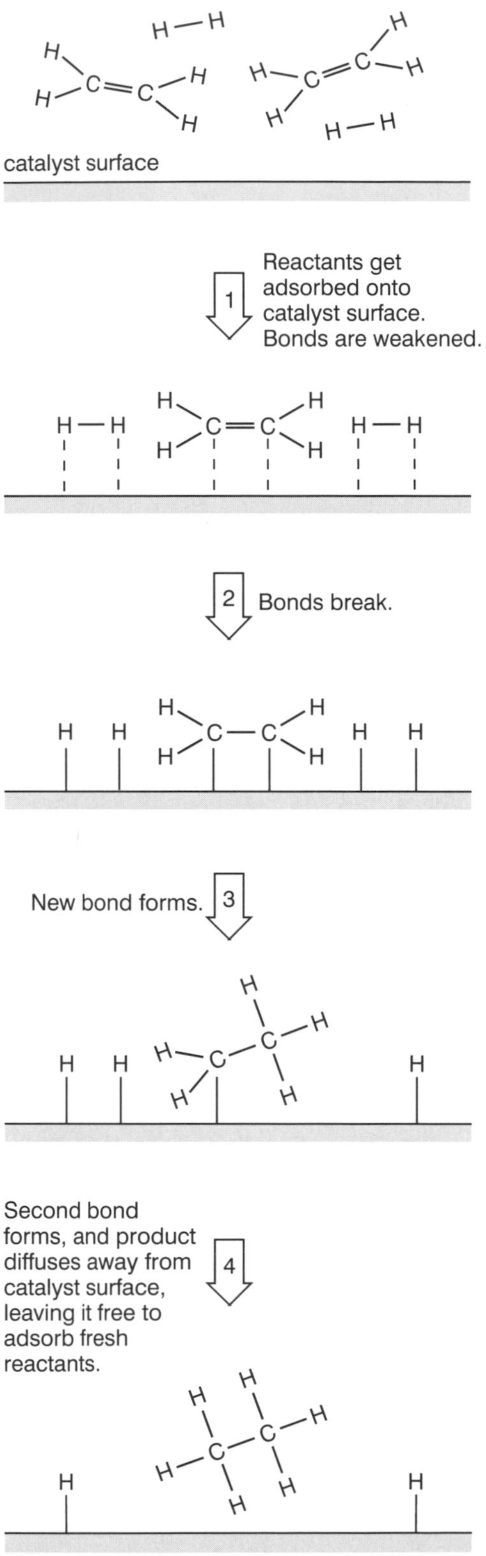

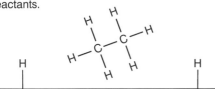

catalyst surface

Reactants get adsorbed onto catalyst surface. Bonds are weakened. **1**

2 Bonds break.

New bond forms. **3**

Second bond forms, and product diffuses away from catalyst surface, leaving it free to adsorb fresh reactants. **4**

Figure 19 An example of heterogeneous catalysis. The diagrams show a possible mechanism for nickel catalysing the reaction between ethene and hydrogen to form ethane

Homogeneous catalysts

Homogeneous catalysts normally work by forming an intermediate compound with the reactants, which then breaks down to give the products. There is more about this in **Section 11.6**, and in the **Engineering Proteins** storyline, where the mechanism of enzyme catalysis is described.

Catalyst poisoning

Catalysts can be **poisoned** so that they no longer function properly. In fact, many substances which are poisonous to humans operate by blocking an enzyme catalysed reaction.

In heterogeneous catalysis, the 'poison' molecules are adsorbed more strongly to the catalyst surface than the reactant molecules. This is the reason why leaded petrol cannot be used in cars fitted with a catalytic converter.

PROBLEMS FOR 10.4

1 Name a catalyst involved in each of the following industrial processes. In each case, state whether the process involves homogeneous or heterogeneous catalysis.
   a Reforming gasoline fractions to produce high octane petrol components.
   b Catalytic cracking of long-chain hydrocarbons.
   c Oxidation of CO and unburnt petrol in a car exhaust.

2 The activation enthalpy for the decomposition of hydrogen peroxide to oxygen and water is $+36.4\,kJ\,mol^{-1}$ in the presence of an enzyme catalyst and $+49.0\,kJ\,mol^{-1}$ in the presence of a very fine colloidal suspension of platinum.

   How will the rate of this decomposition differ for the two catalysts at room temperature? Explain your answer.

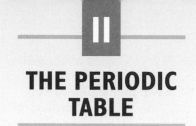

THE PERIODIC TABLE

II.1 *Periodicity*

The modern Periodic Table

Figure 1 shows a modern form of the Periodic Table.

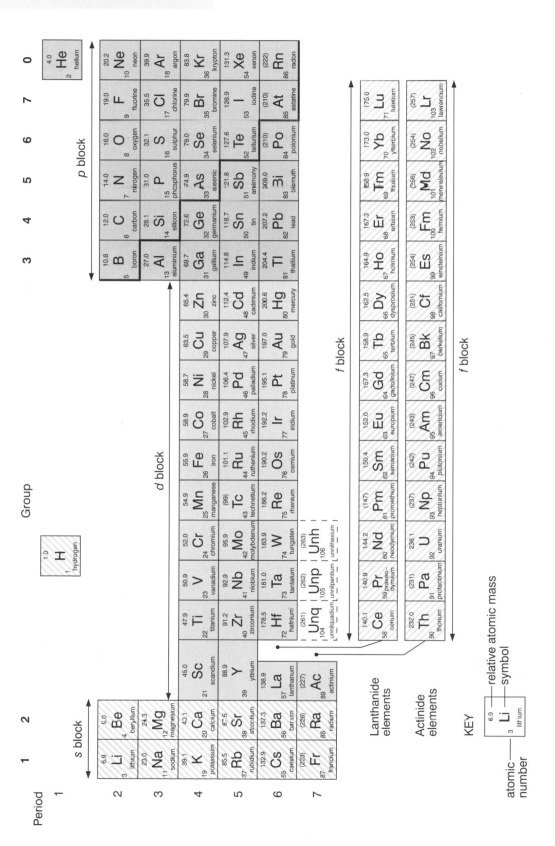

Figure 1 The modern Periodic Table

The elements appear in the table almost in order of increasing relative atomic mass – but not quite. Some pairs of elements are 'out of order', eg Tc (A_r = 128) comes before I (A_r = 127). It is **atomic number** – the number of protons in the nucleus – which really determines the place of an element in the Periodic Table.

Figure 2 shows how, with the exception of hydrogen, the elements can be organised into four **blocks**.

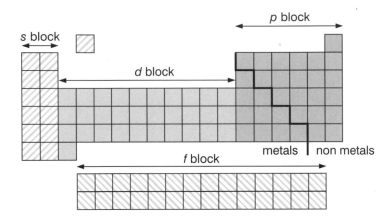

Figure 2 Blocks in the Periodic Table

Elements in the same block show general similarities. For example: all the non-metals are in the *p* block; many of the reactive metals (like sodium, potassium and strontium) are in the *s* block.

The elements in a vertical column or **Group** show more specific similarities. You may already know some of the common features of the elements of Group 1 (the alkali metals), Group 7 (the halogens) and Group 0 (the noble gases).

Horizontal rows in the Periodic Table are called **Periods**. Since they cut across the Groups, there are fewer common features among the elements of a Period. After hydrogen and helium, there are two *short Periods* (Li to Ne and Na to Ar) and four *long Periods* (K to Kr etc). There are 8 elements in a short Period, but 18 in the first long Period because of the inclusion of 10 *d* block elements.

Elements change from metallic to non-metallic across a Period. They become increasingly metallic *down* a Group – and increasingly non-metallic *up* a Group.

Physical properties and the Periodic Table

When we arrange the elements in order of atomic number, we see the patterns in physical properties which Mendeléev noted. The occurrence of periodic patterns is called **periodicity**.

Figure 3 Variation of molar atomic volume of elements with atomic number

Molar atomic volumes for elements 1–56 are plotted in Figure 3. Molar atomic volume is the volume occupied by 1 mole of the solid or liquid element.

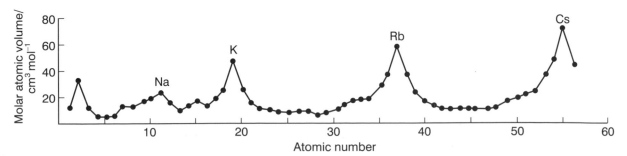

Melting points are plotted in Figure 4. This graph shows the variation of melting point with atomic number for elements up to Ba. Notice the 'peaks' which correspond to the Group 4 elements, and the pattern of melting points for the *d* block elements.

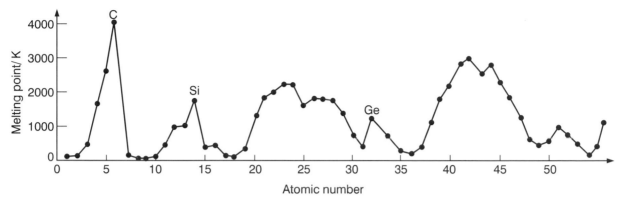

Figure 4 Variation of melting points of elements with atomic number

Periodicity in chemical formulae

Figure 5 shows the variation in the formulae of the chlorides of the elements of Periods 2 and 3. The vertical axis represents the number of moles of chlorine atoms which combine with 1 mole of atoms of the element: for example, CCl_4 corresponds to 4 moles of Cl atoms per mole of C atoms. Figure 6 is a similar plot for the oxides of the elements of Periods 2 and 3. For example, Al_2O_3 corresponds to 1.5 moles of O atoms per mole of Al atoms. Both plots show the pattern in chemical formulae that occur across the Periodic Table.

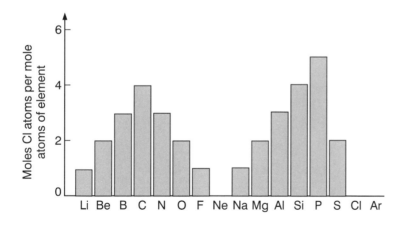

Figure 5 Periodicity in the formulae of chlorides (where more than one chloride exists, the highest is shown here)

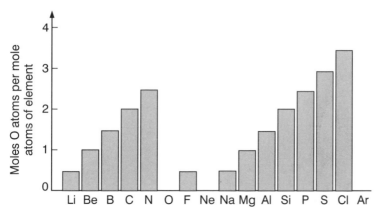

Figure 6 Periodicity in the formulae of oxides (where more than one oxide exists, the highest is shown here)

PROBLEMS FOR II.I

1. In which block of the Periodic Table are each of the following elements found?
   a. tungsten, W
   b. antimony, Sb
   c. rubidium, Rb
   d. holmium, Ho.

2. Look at Figure 3.
   a. What is meant by the *molar atomic volume* of an element?
   b. What do the elements at the *peaks* of the molar atomic volume graph have in common?
   c. Explain why this graph provides evidence for periodicity.

3. Look at Figure 4. Concentrate on the first 20 elements.
   a. What elements are at the *peaks* of the melting point graph? What elements are at the *troughs*?
   b. Explain why this graph provides evidence for periodicity.

4. Look at Figure 5.
   a. Work out the formulae of the chlorides shown for each of the elements in the chart.
   b. Describe the pattern shown by the formulae of the chlorides as you go across the Periods of the Periodic Table.

5. Look at Figure 6.
   a. Work out the formulae of the oxides shown for each of the elements in the chart.
   b. Describe the pattern shown by the formulae of the oxides as you go across the Periods of the Periodic Table.

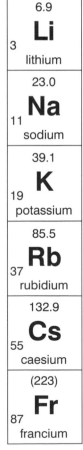

Figure 7 The elements of Group 1

II.2 *The s block: Groups 1 and 2*

The *s* block contains two groups of reactive metals. Group 1 metals (Figure 7) are also called the **alkali metals**. Group 2 metals (Figure 8) are also called the **alkaline earth metals**. These Groups illustrate two trends that also apply to other Groups in the Periodic Table:

- *elements become more metallic as you go down a Group.* For this reason, the most reactive metals in Groups 1 and 2 are to be found at the bottom of each Group.
- *elements become less metallic as you go across a Period from left to right.* For this reason, the Group 1 metals are more reactive than the Group 2 metals in the same Period.

Physical properties

Although we talk about the *s* block elements as being typical metals, this really only applies to their *chemical* properties. Physically these metals do not compare well with the familiar metals of the *d* block, such as iron, copper and chromium. The *s* block metals tend to be soft, weak metals with low melting points, and the metals themselves have few uses. However, the *compounds* of the *s* block elements are very important.

Chemical reactions

Group 1 and 2 metals are all very reactive, and for this reason they are never found in nature in the native, uncombined state. However, compounds of *s* block metals are very common throughout nature. Indeed, a lot of the earth beneath your feet is made from compounds of *s* block metals such as magnesium and calcium.

Like all Groups in the Periodic Table, Groups 1 and 2 show patterns of reactivity as you go down the Group. There are *similarities* between the reactions of the elements within a Group, but also *differences* which show up as patterns, or trends. The *similarities* are because the elements in a particular Group all have similar arrangements of electrons in their atoms (see **Section 2.3**). The *differences* are because, as you go down the Groups, the size of the atoms increases. We can illustrate these similarities and trends by looking at Group 2 in a little detail.

Some chemical properties of Group 2 elements and their compounds

We will not include the element at the bottom of the Group, radium, which is radioactive and only occurs on Earth in tiny quantities.

We will also leave out beryllium, which is difficult to work with because its compounds are very poisonous. This leaves us with Mg, Ca, Sr and Ba, which you will have already met in experimental work.

The elements of Group 2 are all reactive. They form compounds containing ions with a 2+ charge, such as Mg^{2+} and Ca^{2+}.

Reactions of the elements with water

All the elements react with water to form hydroxides and hydrogen, with a steady increase in the vigour of the reaction as you move down the Group. Magnesium reacts only slowly, even when the water is heated. Barium reacts rapidly, giving a steady stream of hydrogen. (Even so, none of the elements reacts as vigorously with water as the Group 1 elements such as sodium and potassium do.)

The general equation, using M to represent a typical Group 2 metal reacting with water, is

$$M(s) + 2H_2O(l) \rightarrow M(OH)_2(aq) + H_2(g)$$

Oxides and hydroxides

The general formula of the oxides is MO and of the hydroxides is $M(OH)_2$.

In water, the oxides and hydroxides form alkaline solutions (though they do not dissolve in water completely). This is typical of metal oxides and hydroxides, in contrast to non-metals, whose oxides are usually acidic. The most strongly alkaline oxides and hydroxides are those at the bottom of the Group.

As you would expect, the oxides and hydroxides react with acids to form salts. For example,

$$MO(s) + 2HCl(aq) \rightarrow MCl_2(aq) + H_2O(l)$$
$$M(OH)_2(s) + H_2SO_4(aq) \rightarrow MSO_4(aq) + 2H_2O(l)$$

This neutralising effect is used by farmers when they put lime (calcium hydroxide) on their fields to neutralise soil acidity.

Effect of heating carbonates

The general formula of Group 2 carbonates is MCO_3.

When you heat the carbonates, they decompose, forming the oxide:

$$MCO_3(s) \rightarrow MO(s) + CO_2(g)$$

The carbonates become more difficult to decompose as you go down the Group. For example, magnesium carbonate is quite easily decomposed by heating in a test tube over a Bunsen burner flame, but calcium carbonate needs much stronger heating directly in the flame before it will decompose. We say that the **thermal stability** of calcium carbonate is greater than that of magnesium carbonate.

The decomposition of calcium carbonate (limestone) is an important process, used to manufacture calcium oxide (quicklime).

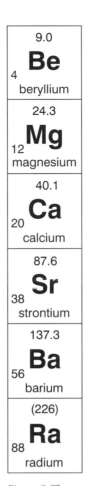

| 9.0 |
| **Be** |
| 4   beryllium |

| 24.3 |
| **Mg** |
| 12   magnesium |

| 40.1 |
| **Ca** |
| 20   calcium |

| 87.6 |
| **Sr** |
| 38   strontium |

| 137.3 |
| **Ba** |
| 56   barium |

| (226) |
| **Ra** |
| 88   radium |

Figure 8 The elements of Group 2

Solubilities of compounds

The solubilities of compounds of Group 2 elements show clear trends as you go down the Group. The solubilities of the hydroxides and carbonates are summarised in Table 1.

Solubility of hydoxides Here the pattern is for the hydroxides to become *more* soluble as you go down the Group. Thus, as Table 1 shows, $Mg(OH)_2$ is much less soluble than $Ba(OH)_2$. This pattern is repeated for most Group 2 compounds where the anion has a single charge (1–).

Solubility of carbonates Here the pattern is for the carbonates to become *less* soluble as you go down the Group. As Table 1 shows, $MgCO_3$ is more soluble than $CaCO_3$. This general patter is repeated for most Group 2 compounds where the anion has a double charge (2–).

Notice that the figures in Table 1 do not show a perfect trend – for example the pattern in solubilities of hydroxides is uneven between Sr and Ba. This usually happens with patterns in the Periodic Table – they are rarely perfect, because there are so many factors operating to decide the properties of elements and their compounds. When we describe a pattern such as 'carbonates become less soluble as you go down a Group' we are stating a rule that generally works and is useful for predicting properties. But, as always in science, predictions can only be confirmed by doing experiments, or looking up the results of experiments done by others.

	Solubility /mol per 100 g water
Hydroxides	
$Mg(OH)_2$	0.000 02
$Ca(OH)_2$	0.0016
$Sr(OH)_2$	0.033
$Ba(OH)_2$	0.024
Carbonates	
$MgCO_3$	0.000 13
$CaCO_3$	0.000 013
$SrCO_3$	0.000 007
$BaCO_3$	0.000 009

Table 1 Solubilities of Group 2 hydroxides and carbonates

PROBLEMS FOR II.2

1   Draw up a table like the one below and fill it in to show the trends in properties down Group 2.

Element	Trend in reactivity with water	Trend in thermal stability of carbonate	Trend in pH of hydroxide in water	Trend in solubility of hydroxide	Trend in solubility of carbonate
Mg					
Ca					
Sr					
Ba					

2   Predict the trend in solubility of the sulphates as you go down Group 2.

3   On the basis of what you know about Group 2, predict the following concerning the elements of Group 1.
    a   The element which reacts most vigorously with water.
    b   The products of the reaction of this element with water. Write an equation for the reaction.
    c   The element with the least thermally stable carbonate.

4   Most of the compounds of Group 1 elements are very soluble in water. The chlorides of Group 2 elements are very soluble in water, but their carbonates and sulphates are generally not very soluble.
    a   Use this information, and your own knowledge, to say whether each of the following compounds of Group 1 and 2 elements normally occurs
    A on land (in rocks)   B in the sea
    C in both places.

    i calcium carbonate   iv calcium sulphate
    ii magnesium carbonate v sodium chloride
    iii magnesium chloride   vi potassium chloride.
    b   Give reasons for your answer to part **a**.
    c   For as many as possible of the compounds in part **a**, give the common name of the compound by which it is known when found in the sea or in rocks.

5   Predict the following properties of radium, Ra.
    a   How it reacts with water.
    b   The thermal stability of its carbonate. Write an equation for the reaction which occurs when the carbonate decomposes.
    c   The solubility (in mol per 100 g of water) of its carbonate.

11.3 *The p block: Group 4*

General survey of the Group

The elements in Group 4 are:

carbon   [He] $2s^2 2p^2$
silicon   [Ne] $3s^2 3p^2$
germanium   [Ar] $3d^{10} 4s^2 4p^2$
tin   [Kr] $4d^{10} 5s^2 5p^2$
lead   [Xe] $4f^{14} 5d^{10} 6s^2 6p^2$

In some Groups, like Groups 1 and 7, elements within the Group resemble one another quite closely. The Group 4 elements, though, differ considerably from one another. Carbon is a non-metal. Silicon and germanium are semiconductors. Tin and lead are metals.

You can see that all the elements have the same outer electron configuration, s^2p^2. As a result they all form compounds in which they have oxidation state +4. Examples are CO_2, $SiCl_4$, and PbO_2.

In addition, tin and lead form compounds in which they are in oxidation state +2. In these compounds only the outer p electrons are involved in bonding. The stability of the +2 oxidation state relative to the +4 state increases down the Group (Figure 10). Elements at the top of the Group form few compounds in the +2 state.

Compounds in which the Group 4 element is in the +4 oxidation state are generally covalently bonded. Thus the tetrachlorides are all covalently bonded liquids – including $PbCl_4$, which is a little surprising since lead is normally considered to be a metallic element.

On the other hand, compounds in which the element is in the +2 state are mainly ionic in character. Thus, for example, $PbCl_2$ is a crystalline solid with a melting point of 501 °C – very different from $PbCl_4$.

The nature of the *oxides* changes as you go down the Group, and also changes according to whether the element is in the +2 or +4 oxidation state. In the +4 state, the oxides are acidic at the top of the Group (eg CO_2), but become amphoteric further down the Group (eg PbO_2). In the +2 state the oxides (if they exist) tend to be more basic. Thus PbO shows mainly basic character.

A simple way of looking at the +2 and +4 states is to think of the elements showing more metallic character in the +2 state, and more non-metallic character in the +4 state.

Carbon and silicon

Carbon and silicon are both very important elements. All living things are made up of compounds containing carbon, and silicon is a major component of the Earth's crust. Carbon exists in two main allotropic forms – graphite and diamond. Both of these have giant covalent structures. Recently a third allotropic form has been discovered; a variety of *fullerenes* are now known in which carbon atoms are joined together in large molecules shaped like balls (Figure 11).

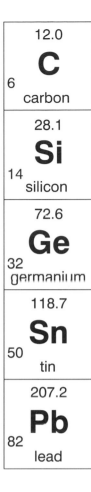

12.0	
C	
6	
carbon	
28.1	
Si	
14	
silicon	
72.6	
Ge	
32	
germanium	
118.7	
Sn	
50	
tin	
207.2	
Pb	
82	
lead	

Figure 9 The elements of Group 4

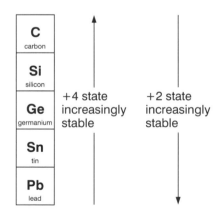

Figure 10 The relative stabilities of the +2 and +4 oxidation states in Group 4

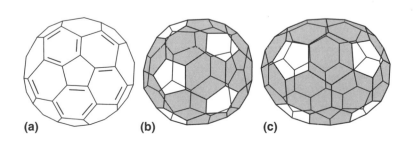

(a)   **(b)**   **(c)**

Figure 11 The fullerenes are an allotropic form of carbon.
a *shows C_{60}, named buckminsterfullerene, which is shaped like a round ball*
b *is another way of representing C_{60}, with the 5-membered rings white and the 6-membered rings shaded darker*
c *shows C_{70}, which is shaped like a rugby ball*

Carbon has an exceptional ability to bond to itself to form chains and rings of atoms, and to form multiple bonds with other carbon atoms. The huge variety of compounds of carbon is dealt with in organic chemistry. Unlike carbon, silicon does not readily bond with itself to form chains and rings.

Both carbon and silicon can be oxidised on heating to form oxides:

$$C(s) + O_2(g) \rightarrow CO_2(g)$$
carbon dioxide

$$Si(s) + O_2(g) \rightarrow SiO_2(s)$$
silicon(IV) oxide

Carbon also forms carbon monoxide, CO, a product of incomplete combustion.

Carbon and silicon oxides

We breathe CO_2 and walk on SiO_2. Both are essential for life on Earth: one is used in photosynthesis, and the other produces the soil which supports growing plants.

You could hardly imagine two more physically different substances. CO_2 is a gas at room temperature, with a boiling point of 195 K, and SiO_2 is a hard solid which melts at 1883 K. (Sand, for example, is impure quartz.)

The reason for this dramatic difference in physical properties is the difference in bonding between carbon and oxygen on the one hand, and silicon and oxygen on the other. The small size of the carbon atom makes it possible for carbon to form double bonds with oxygen so that carbon dioxide is composed of individual molecules

$$O{=}C{=}O$$

Intermolecular forces between the carbon dioxide molecules are weak. Little energy is needed to separate individual molecules in the solid and liquid phases to form a gas, so CO_2 is a gas at room temperature. It freezes at $-78\,°C$ to a white solid (dry ice). It is soluble in water, giving an acidic solution.

Silicon atoms are larger than carbon atoms and they normally bond to four oxygens. Silica, SiO_2, is an extended network of SiO_4 units in which the central silicon is covalently bonded to each of four oxygen atoms. Each Si atom has a half-share in four oxygen atoms (Figure 12).

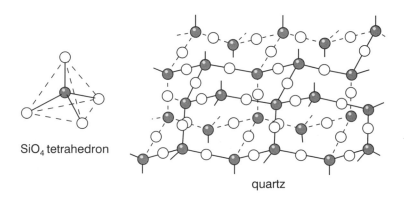

SiO$_4$ tetrahedron

quartz

Figure 12 The structure of silicon(IV) oxide, (quartz). Sand is an impure form of quartz

Because of its extended network structure, silicon(IV) oxide is insoluble in water, and has high melting and boiling points (1883 K and 2503 K respectively). The reason for these properties is the strong covalent bonding which exists throughout the SiO_2 structure. Considerable energy is needed to break bonds within the structure so that a very high temperature is needed to melt silicon(IV) oxide.

Note that CO_2 and SiO_2 are named differently. CO_2 is called carbon dioxide because this name describes its simple molecules, made up of one carbon and two oxygen atoms. SiO_2 is sometimes called silicon dioxide, but this can be misleading because it implies that the substance contains simple molecules.

PROBLEMS FOR II.3

1 Explain in your own words why CO_2 is a gas at room temperature and pressure, while SiO_2 is solid.

2 The change in the relative stabilities of the +2 and +4 oxidation states is particularly dramatic between tin and lead. This is demonstrated in their redox behaviour. Tin(II) is a useful reducing agent, whilst lead(IV) oxide is a powerful oxidising agent.
  a Refer to the Data Sheets and give one or two examples of substances which could be oxidised by PbO_2.
  $$PbO_2(s) + 4H^+(aq) + 2e^- \rightarrow Pb^{2+}(aq) + 2H_2O(l);$$
  $$E^\circ = +1.46\,V$$
  b Refer to the Data Sheets to give an example of a substance which could be reduced by $Sn^{2+}(aq)$.
  $$Sn^{4+}(aq) + 2e^- \rightarrow Sn^{2+}(aq); \quad E^\circ = +0.15\,V$$

3 Draw dot-cross diagrams (showing the outer electrons only) to show the bonding in
  a $PbCl_4$ **b** $PbCl_2$

4 Silane, SiH_4, is analogous to methane, CH_4.
  a Write an equation, including state symbols, to represent the combustion of silane.
  b When the air supply is plentiful, methane burns with a blue, smokeless flame. What would you expect to observe when silane burns in a plentiful air supply?
  c Give the formula of the silicon compound that is analogous to ethane.

5 For each of the following oxides, say whether you would expect it to be **A** acidic **B** basic **C** amphoteric:
  a CO_2 **b** SnO_2 **c** PbO_2 **d** PbO.

II.4 *The* p *block: Group 5*

Group 5 of the Periodic Table (Figure 13) is a typical Group in the *p* block. At the top are non-metals: nitrogen and phosphorus. At the bottom are metalloids (metallic elements with some non-metal character): antimony (Sb) and bismuth (Bi).

The electronic structures of the Group 5 elements are shown below.

nitrogen	$[He]\ 2s^2 2p^3$
phosphorus	$[Ne]\ 3s^2 3p^3$
arsenic	$[Ar]\ 3d^{10} 4s^2 4p^3$
antimony	$[Kr]\ 4d^{10} 5s^2 5p^3$
bismuth	$[Xe]\ 4f^{14} 5d^{10} 6s^2 6p^3$

Atoms of these elements can form three covalent bonds by sharing the three unpaired electrons. This gives compounds in which the oxidation state of the Group 5 element is +3 or –3.

Each atom also has a lone pair of electrons. These enable the atoms to form dative bonds. When they do this the Group 5 elements can form compounds in which their oxidation state is +5. You will see later how nitrogen does this.

The most important members of Group 5 are nitrogen and phosphorus. Both are constituent elements in living things, and both are essential for healthy plant growth. Although nitrogen gas is all around us in the air, we have problems getting it into a form which can be used in agriculture.

Nitrogen

Nitrogen gas may be abundant, but it is exceptionally unreactive. Thousands of litres of it pass unreacted through your lungs every day. It even passes through the hot cylinders of a motor car mostly unreacted.

14.0
N
7
nitrogen
31.0
P
15
phosphorus
74.9
As
33
arsenic
121.8
Sb
51
antimony
209.0
Bi
83
bismuth

Figure 13 The elements of Group 5

The low reactivity of the N_2 molecule arises from the strong triple bond holding the atoms together (Figure 14).

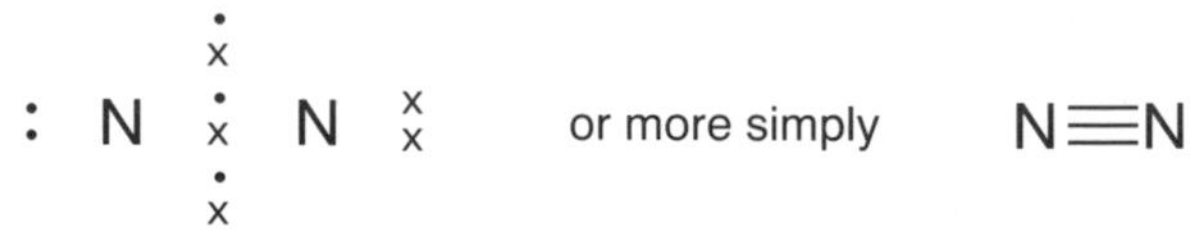

$$: N \overset{\bullet}{\underset{\bullet}{\overset{\times}{\underset{\times}{\times}}}} N \overset{\times}{\times} \qquad \text{or more simply} \qquad N{\equiv}N$$

Figure 14 The bonding in the N_2 molecule

Bond enthalpy of N≡N: + 945 kJ mol⁻¹
Compare with the bond enthalpy of a single N—N bond: + 158 kJ mol⁻¹

Before nitrogen can react, the triple bond between the atoms must be broken, or partly broken. The bond enthalpy of N≡N is very large, +945 kJ mol⁻¹, and most reactions of molecular N_2 have a high activation enthalpy and require high temperature and catalysts to make them occur. That's why the Haber process for making ammonia needs such high temperatures. During thunderstorms, the highly energetic lightning flash can provide enough energy to make nitrogen react with oxygen to form nitrogen oxides.

Once nitrogen has reacted, though, it can form many compounds. The most important of these are *ammonia*, *nitrogen oxides* and *nitrates*. All are involved in *the nitrogen cycle*.

Ammonia

Ammonia is nitrogen hydride. The bonding of the ammonia molecule is shown in Figure 15. Notice that the lone pair of electrons on the N atom is not involved in the bonding, so it is available to form dative covalent bonds.

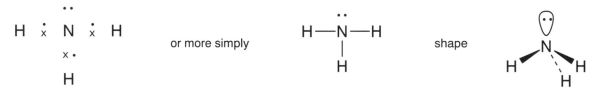

Figure 15 The bonding in the NH_3 molecule

Ammonia readily forms dative covalent bonds to H⁺ ions, which means ammonia acts as a base, forming the ammonium ion. Figure 16 shows the bonding in the ammonium ion.

Ammonia also forms dative bonds to transition metal ions, which makes it a good ligand in complexes.

Figure 16 The bonding in the NH_4^+ ion

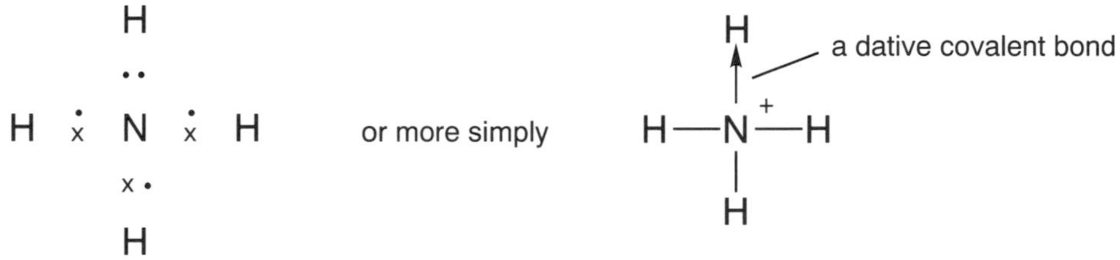

However, all four bonds are equivalent, so we normally represent the ammonium ion as

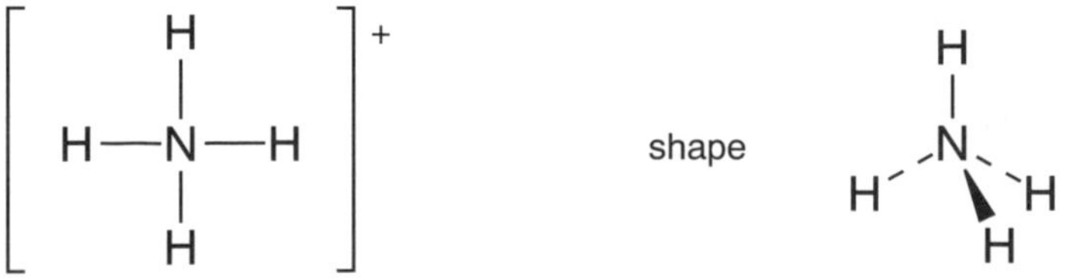

Nitrogen oxides

Nitrogen forms many different oxides, all of them gases. The most important ones are shown in Table 2.

Name and formula	Appearance	Where it comes from
nitrogen monoxide, NO	colourless gas, turns to brown NO_2 in air	combustion processes, especially vehicle engines
		thunderstorms
		formed in the soil by denitrifying bacteria
nitrogen dioxide, NO_2	brown gas (toxic)	from oxidation of NO in atmosphere
dinitrogen oxide, N_2O	colourless gas	formed in the soil by denitrifying bacteria

Table 2 Oxides of nitrogen

Nitrates

Two kinds of nitrate ions are involved in the nitrogen cycle: nitrate(III), NO_2^- and nitrate(V), NO_3^-. Notice that they are both named as nitrates, but they are distinguished from one another by showing the oxidation number of the nitrogen. Figure 17 shows the bonding in the nitrate(III) and nitrate(V) ions.

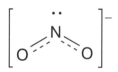   or more simply

The charge is delocalised over the two N—O bonds, which are equivalent.

nitrate(III)

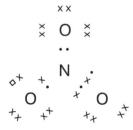

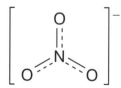   or more simply

By using its lone pair to form a dative bond to an oxygen atom, the nitrogen increases its oxidation number to +5.

nitrate(V)

Figure 17 The bonding in the NO_2^- and NO_3^- ions

Name and formula	Commonly called	Properties
nitrate(V) ion, NO_3^-	nitrate	oxidising agent
		nitrates(V) are very soluble in water.
nitrate(III) ion, NO_2^-	nitrite	reducing and oxidising agent
		toxic

Table 3 Nitrates

PROBLEMS FOR II.4

1 Nitrogen, N_2, at the top of Group 5, is very unreactive. The next member of Group 5, phosphorus, is highly reactive: white phosphorus, P_4, catches fire spontaneously in air. Figure 18 shows the shape of the P_4 molecule.

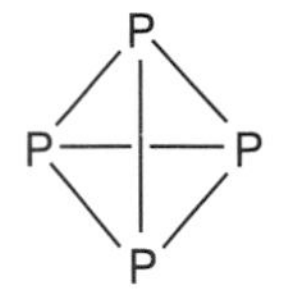

Figure 18 The P_4 molecule

Suggest an explanation for the difference in reactivity between these two close neighbours in Group 5.

2 Look at the following pairs of nitrogen compounds. In each case
  a give the oxidation state of N in each of the two compounds,
  b say whether the conversion of the first compound to the second involves: **A** oxidation **B** reduction **C** neither oxidation nor reduction

i	NO	NO_2	**v**	N_2	NH_3
ii	NO_2^-	NO_3^-	**vi**	NO_2^-	NO
iii	N_2O	NH_3	**vii**	NO_3^-	NO_2
iv	NH_4^+	NH_3			

3 Nitric(V) acid is manufactured by the catalytic oxidation of ammonia. This produces NO, which is further oxidised and then dissolved in water to produce HNO_3:

$$4NH_3(g) + 5O_2(g) \rightarrow 4NO(g) + 6H_2O(g)$$
(reaction 1)

$$2NO(g) + O_2(g) \rightarrow 2NO_2(g)$$
(reaction 2)

$$3NO_2(g) + H_2O(l) \rightarrow 2HNO_3(aq) + NO(g)$$
(reaction 3)

  a The final stage of this process, reaction 3, produces NO as well as HNO_3. How would the NO be dealt with?
  b Name one compound, useful in agriculture, that is manufactured using nitirc acid.
  c **i** Starting with 1000 kg of ammonia, what is the maximum mass of HNO_3 that could be produced?
    ii Give two reasons why the mass of nitric(V) acid actually produced will be less than your answer in **i**.
  d What particular environmental protection measures might be needed in a plant using this manufacturing process?

19.0	
F	
9	
fluorine	
35.5	
Cl	
17	
chlorine	
79.9	
Br	
35	
bromine	
126.9	
I	
53	
iodine	
(210)	
At	
85	
astatine	

Figure 19 The elements of Group 7

II.5 *The p block: Group 7*

What are the halogens?

The *halogens* are the elements in Group 7 of the Periodic Table (Figure 19). All halogen atoms have seven electrons in the outer shell. The halogen are the most reactive Group of non-metals, and none of them is found naturally in the element form. They are all found in compounds as *halide* ions (the singly negatively charged ion, eg Br^-). Calcium fluoride and sodium chloride are naturally occurring halides. Iodine is also found as sodium iodate where it is in the form of iodate(V) ions, IO_3^-.

Fluorine and chlorine are the most abundant halogens, bromine occurs in smaller quantities, iodine is quite scarce and astatine is an artificially produced, short-lived, radioactive element.

All the halogen elements occur as *diatomic molecules*, eg F_2 and Br_2. The two atoms are linked by a single covalent bond (Figure 20).

shared pair of electrons

Figure 20 Covalent bonding in the I_2 molecule

In compounds, a halogen atom can complete its outer shell of electrons by
- gaining an electron from a metal to form a halide ion in an ionically bonded compound (Figure 21a)
- sharing an electron from another atom in a covalently bonded compound (Figure 21b).

a Forming an ion (Br^-)   **b** Forming a molecule (HBr)

Figure 21 How a halogen can complete its outer shell during compound formation

In both cases, the halogen has an oxidation state of –1 in the compound. It is also possible for halogens (other than fluorine) to expand their outer shell of electrons so it holds more than eight electrons. This makes it possible for the halogens to reach higher oxidation states such as +5 or +7. For example, in the chlorate(V) ion, chlorine has expanded its outer shell to hold 12 electrons (Figure 22).

Physical properties of the halogens

Some properties of the halogens are shown in Table 4.

	Fluorine	Chlorine		Bromine		Iodine
isotopes and abundances	^{19}F 100%	^{35}Cl 75%	^{37}Cl 25%	^{79}Br 50%	^{81}Br 50%	^{127}I 100%
melting point/K	53	172		266		386
boiling point/K	85	238		332		456
solubility at 293 K/g per 100 g of water	reacts with water	0.6		3.6		0.02

Table 4 Some physical properties of the halogens

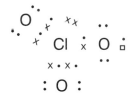

Figure 22 In the ClO_3^- ion, chlorine has expanded its outer shell to hold 12 electrons

The halogens are much more soluble in organic solvents such as hexane than they are in water. The appearances of their solutions are shown in Table 5.

Chemical properties of the halogens

The halogens are a reactive Group of elements. They tend to remove electrons from other elements: they are oxidising agents. The elements at the top of the Group are the most reactive, and the strongest oxidising agents.

Halogen	In water	In hexane
chlorine	pale-green	pale green
bromine	red-brown	red
iodine	brown	violet

Table 5 Colours of halogens in solution

Reactions with metals

With many metals (eg the *s* block) the halogens react to form compounds containing halide ions, eg potassium bromide KBr, calcium chloride $CaCl_2$. In all these compounds the halogen is in oxidation state –1.

Reactions with non-metals

With non-metals, and some *p* block and transition metals, halogens form molecular compounds containing covalent bonds, eg carbon tetrachloride CCl_4, tin tetrachloride $SnCl_4$, phosphorus trichloride PCl_3. In most of these compounds the halogen is in oxidation state –1.

Halogens also react with one another – to form interhalogen compounds, eg ICl, BrF_3. In these compounds, the less reactive of the two halogens is in a positive oxidation state.

Reactions with other halide ions

If you add a solution containing chlorine to a solution of iodide ions there is a chemical reaction. The solution (which would be almost colourless if nothing happened) turns brown and iodine is produced.

$$Cl_2(aq) + 2K^+I^-(aq) \rightarrow 2K^+Cl^-(aq) + I_2(aq)$$

We can simplify this equation, because K^+ ions stayed unchanged, so they can be left out:

$$Cl_2(aq) + 2I^-(aq) \rightarrow 2Cl^-(aq) + I_2(aq)$$

A similar thing happens if bromine solution is added to iodide ions.

$$Br_2(aq) + 2I^-(aq) \rightarrow 2Br^-(aq) + I_2(aq)$$

Notice that we are using **ionic equations** here. Full chemical formulae have not been used for some of the reagents: eg I^-. That's because the equations represent *general* reactions: chlorine will react in the same way with any aqueous solution of iodide ions, whether they come from NaI, KI, CaI_2, and so on. So although the ionic equation is shorter than the full chemical equation, it tells us more – it tells us, for example, how chlorine reacts with a whole range of iodides.

These are redox reactions. Bromine, for example, oxidises I^- (oxidation state –1) to iodine (oxidation state 0). In the process, Br_2 (oxidation state 0) is reduced to Br^- (oxidation state –1). In general, a more reactive halogen will oxidise the halide ions of a less reactive one.

The reactions are not reversible. Iodine will not liberate bromine from potassium bromide solution because iodine is less reactive than bromine.

Reactions of halide ions

Oxidation to form halogens

Halide ions are oxidised to halogens, provided a strong enough oxidising agent is available. Using X^- to represent a general halide:

$$2X^- \rightarrow X_2 + 2e^- \text{ (removed by oxidising agent)}$$

The oxidising agent might be another halogen (see above) or a strong oxidising agent such as potassium manganate(VII) or concentrated sulphuric acid. For example, potassium manganate(VII) oxidises chloride ions to chlorine. This is a useful way to prepare chlorine in the laboratory.

Reactions with silver ions

Silver halides are precipitated when a solution of silver ions is added to a solution containing Cl^-, Br^- or I^- ions. The general equation is

$$Ag^+(aq) + X^-(aq) \rightarrow AgX(s)$$

The appearance of the silver halides are

    silver chloride   white
    silver bromide   cream
    silver iodide   yellow

AgCl and AgBr are decomposed by light to produce silver and the halogen. The decomposition is the basis of most photographic processes.

Silver halides dissolve in sodium thiosulphate solution. The reaction is used to clear unreacted silver halide crystals from films in the fixing process.

PROBLEMS FOR 11.5
---

1 a Make a copy of Table 4, adding a column for astatine, At. Astatine only exists as artificial radioactive isotopes, so you will not be able to record isotopic abundances. Astatine-210 is one of 23 known isotopes.

From the trends in physical properties down Group 7, predict approximate values for the solubility in water, melting and boiling points of astatine.

b Use your knowledge of the colours and the reactions of the halogens to predict the changes (If any) that you might *observe* when the following reagents are mixed. Where a reaction is predicted, write a balanced ionic equation with state symbols.

  i Aqueous bromine water is added to aqueous sodium astatide, NaAt(aq).

  ii Aqueous hydroastatic acid, HAt(aq), is added to aqueous iodine.

  iii Aqueous silver nitrate is added to aqueous sodium astatide.

  iv Hot astatine vapour, $At_2(g)$ is passed over heated metallic sodium.

2 Chlorine is used to manufacture a wide range of chemicals. One of the most familiar of these is the bleach used as a powerful domestic disinfectant. Household bleach is an aqueous solution of sodium chloride and sodium chlorate(I), NaClO, in a one to one mole ratio. It is produced by dissolving chlorine gas in cold dilute aqueous sodium hydroxide.

a Write a balanced equation, with state symbols, for the reaction of chlorine with sodium hydroxide to form bleach.

b i In the reaction in part **a** which element(s) change oxidation states?

  ii Write down these oxidation state changes and identify which involves an oxidation and which a reduction.

c The active component of bleach is the sodium chlorate(I). This unstable compound cannot be isolated as a solid. Bleach left in sunlight slowly decomposes releasing oxygen, O_2.

  i Write an equation for this decomposition including state symbols.

  ii Which element(s) change oxidation states?

  iii Write down these oxidation state changes and identify which involves an oxidation and which a reduction.

3 Some river water is believed to have been contaminated with potassium iodide.

a Suggest some simple test tube experiments which would enable you to show that iodide ions were present in the water. Describe how you would carry out the tests and give the observations that you expect if iodide ions were present in significant concentration.

b How might you make a quantative measurement of the concentration of iodide ion in $g\,dm^{-3}$? Give an outline of an experiment, together with an indication of how the iodide ion concentration could be calculated.

11.6 *The* d *block: characteristics of transition metals*

The *d* block consists of three horizontal series in Periods 4, 5, and 6 respectively (see Figure 2 on page 192). Each series contains ten elements. Their chemistry is quite different from that in other parts of the Periodic Table. It results from the special electronic configurations of *d* block elements and the energy levels associated with their electrons. Differences between elements within a Group in the *d* block are less sharp than in the *s* and *p* block, and similarities across the Period are greater, so that we can discuss the *d* block as a collection of elements with many features in common.

Electronic configuration

Across the first tow of the *d* block (the ten elements Sc to Zn) each element has one more proton in the nucleus and one more electron than the previous element. Each 'additional' electron enters the 3*d* shell. Remember this is *not* the outermost shell, because the outer 4*s* orbital has already been filled. The electronic arrangements of the atoms of the first row of the *d* block are shown in Figure 23. Only the two outer-sub-shells are shown because the elements all have an identical core of electrons. The core is the electronic arrangement of the noble gas argon, Ar, $1s^2 2s^2 2p^6 3s^2 3p^6$.

Element	Atomic number	Electronic arrangement						
					3d			4s
Sc	21	[Ar]	↑					↑↓
Ti	22	[Ar]	↑	↑				↑↓
V	23	[Ar]	↑	↑	↑			↑↓
Cr	24	[Ar]	↑	↑	↑	↑	↑	↑
Mn	25	[Ar]	↑	↑	↑	↑	↑	↑↓
Fe	26	[Ar]	↑↓	↑	↑	↑	↑	↑↓
Co	27	[Ar]	↑↓	↑↓	↑	↑	↑	↑↓
Ni	28	[Ar]	↑↓	↑↓	↑↓	↑	↑	↑↓
Cu	29	[Ar]	↑↓	↑↓	↑↓	↑↓	↑↓	↑
Zn	30	[Ar]	↑↓	↑↓	↑↓	↑↓	↑↓	↑↓

building up of inner 3*d* sub–shell   outer shell

Figure 23 Arrangement of electrons in the ground state of elements of the first row of the d-*block. [Ar] represents the electronic configuration of argon*

These elements have essentially the same *outer* electronic arrangement as each other, in the same way as the elements in a vertical Group all have the same outer shell structure. Moreover, unlike the elements in a Group, they do not differ by a complete electron shell but by having one more electron in the *inner, incomplete* 3*d* sub-shell.

Why are chromium and copper different?

Look carefully at Figure 23. The electronic configurations of Cr and Cu don't fit the pattern of building up the 3*d* sub-shell.

In the ground state of an atom electrons are always arranged to give the lowest total energy. Because of their negative charge, electrons repel one another, so a lower total energy is obtained with electrons singly in orbitals than if they are paired in an orbital.

You might expect Cr to have the electronic arrangement

[Ar] | ↑ | ↑ | ↑ | ↑ | | | ↑↓ |

(continuing the pattern of the previous elements). However, its electronic arrangement is in fact

[Ar] | ↑ | ↑ | ↑ | ↑ | ↑ | | ↑ |

Figure 24

because this has a lower energy by avoiding having a pair of electrons in the same orbital.

The energies of the $3d$ and $4s$ orbitals are very close together in Period 4. At chromium the orbital energies are such that putting one electron into each $3d$ and $4s$ orbital gives a lower energy than having two in the $4s$ orbital (Figure 24). At copper putting two electrons into the $4s$ orbital would give a higher energy than filling the $3d$ orbitals.

Why are transition metals special?

The characteristic properties of transition metals, such as coloured compounds and variable oxidation states, are due to the presence of an inner incomplete d sub-shell. Electrons from both the inner d sub-shell and outer s sub-shell can be involved in compound formation.

Not all d block elements have an incomplete d sub-shell. For example, zinc has the electronic configuration [Ar] $3d^{10}4s^2$ in which the d sub-shell is full. Zinc does not show typical transition metal properties. Also, its ion Zn^{2+} ([Ar]$3d^{10}$) is not a typical transition metal ion.

Similarly, scandium forms the Sc^{3+} ion which has a closed shell electronic configuration like argon. This ion, the only one formed by scandium, has no $3d$ electrons and is unlike transition metal ions in its properties.

For this reason, a transition metal is defined as *an element which forms at least one ion with a partially filled sub-shell of* d *electrons*. In the first row of the d block only the eight elements from titanium to copper are classed as transition metals.

Note that, when d block elements form ions, it is the s electrons that are lost first. So ions of d block elements contain only d electrons in the outer shell.

What are transition elements like?

Transition elements are all metals. They are similar *to each other* but show distinct differences from s block metals such as sodium or magnesium, and p block metals such as aluminium and tin.

Some of the physical properties of chromium, iron and cobalt, together with those of sodium and magnesium, are listed in Table 6 and illustrate this point.

Property	Element				
	Cr	**Fe**	**Co**	**Na**	**Mg**
metallic (atomic) radius/nm	0.13	0.13	0.13	0.19	0.16
melting point/K	2163	1813	1773	371	923
boiling point/K	2753	3273	3173	1165	1383
density/g cm^{-3}	7.2	7.9	8.9	0.97	1.74

Table 6 Physical properties of some d *and* s *block metals*

Transition elements are dense metals with high melting and boiling points. They tend to be hard and durable, with high tensile strength and good mechanical properties. These properties are a result of strong **metallic bonding** between the atoms in the metal lattice.

Metallic bonding

Strong forces exist between the separate atoms in a metal and these are known as metallic bonds. They can be explained by using a model in which the outer shell electrons of the metal move randomly throughout a lattice of regularly spaced positive ions. The moving electrons are sometimes described as a 'sea' or 'cloud' of moving and fluctuating negative charge (Figure 25).

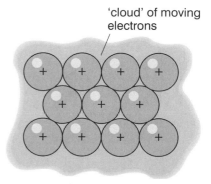

'cloud' of moving electrons

Figure 25 A model of metallic bonding

Each positively charged ion is attracted to the 'cloud' of negative electrons and vice versa. These electrostatic attractions bind the entire metal crystal together as a single unit. In this model, one particular electron does not belong to any one metal ion but is attracted to all the positive ions in the lattice. These electrons are said to be *delocalised* or spread out over the lattice.

The strength of metallic bonding depends on several factors including the number of electrons per atom available for delocalisation in this way. Thus magnesium (two outer shell electrons) has stronger metallic bonding than sodium (one outer shell electron) and this is reflected in the higher melting point and boiling point of magnesium.

Transition metals can release electrons for bonding from both the outer and inner shells. For example, an element such as iron from the first row of the *d* block can use both 3*d* and 4*s* electrons, forming strong metallic bonds. Sodium and magnesium, on the other hand, can only use their 3*s* electrons.

We use transition metals to make things such as cars, cooking utensils and tools and to construct buildings and bridges. Their properties make them more suitable for this than *s* block metals like sodium.

The metals are sometimes used pure, but more often they are used in the form of **alloys** – particularly with iron. This is important in engineering because the properties of an alloy can be controlled as desired. Transition metals form a wide range of alloys with each other. Their atoms are often similar in size and behaviour and so the lattice structure may not by altered greatly as a result of substituting one atom for another. Even so, alloying modifies the properties and usually makes the metal harder and less malleable.

The effect of alloying on metal properties

The bonds between the atoms in a metal are strong but are not directed between particular atoms. When a force is applied to a metal crystal, the layers of atoms can 'slide' over one another. This is known as **slip**. After slipping the atoms settle once again into a close-packed structure. Figure 26 shows the positions of the atoms before and after slip has taken place.

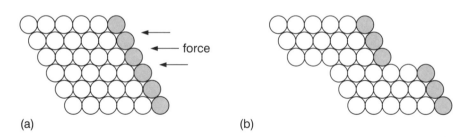

*Figure 26 The arrangement of metal atoms in a crystal **a** before and **b** after slip has taken place. The shaded circles represent the end row of atoms*

(a)   (b)

This is why metals can be hammered into different shapes or drawn into wires without breaking, ie they are **malleable** and **ductile.**

In an alloy, differently sized atoms interrupt the orderly arrangement of atoms in the lattice and make it more difficult for the layers to slide over one another (Figure 27). The metal becomes harder and less malleable.

Figure 27 The arrangement of metal atoms in an alloy

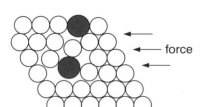

The open circles represent atoms of iron, and the black circles are the larger atoms of a metal added to make an alloy.

Smaller atoms, often non-metals such as carbon or nitrogen, can be taken into the metal lattice and fit into the holes between the metal atoms. This happens in some types of steel. The carbon atoms distort the lattice, making slip between layers more difficult. In some steels, the carbon forms crystals of iron carbide which are very hard. The regions of iron carbide in the softer iron make the steel very strong.

Characteristic chemical properties of transition elements

As with physical properties, the chemical properties of transition metals are very different from those of s block metals. The typical chemical properties can be summarised as follows:

1 formation of compounds in a variety of oxidation states
2 catalytic activity of the elements and their compounds
3 strong tendency to form complexes
4 formation of coloured compounds

Now let's look at these special properties in more detail. **1** and **2** are dealt with in this section. **3** and **4** are covered in **Section 11.7** which deals with complex formation.

Variable oxidation states

Transition elements show a great variety of oxidation states in their compounds. By comparison, s block metals are limited to oxidation numbers of +1 (for Group 1) and +2 (for Group 2). The reason for this can be seen by comparing successive ionisation enthalpies for an s block metal (calcium) and a d block metal (vanadium) as shown in Table 7.

| | Ionisation enthalpies/kJ mol^{-1} | | | |
	$\Delta H_i(1)$	$\Delta H_i(2)$	$\Delta H_i(3)$	$\Delta H_i(4)$
Ca [Ar] $4s^2$	+590	+1150	+4940	+6480
V [Ar] $3d^3 4s^2$	+648	+1370	+2870	+4600

Table 7 Successive ionisation enthalpies for calcium and vanadium

$\Delta H_i(1)$ $\Delta H_i(2)$ $\Delta H_i(3)$ and $\Delta H_i(4)$ are enthalpy changes for the following reactions:

$$M(g) \rightarrow M^+(g) + e^-; \quad \Delta H_i(1)$$
$$M^+(g) \rightarrow M^{2+}(g) + e^-; \quad \Delta H_i(2)$$
$$M^{2+}(g) \rightarrow M^{3+}(g) + e^-; \quad \Delta H_i(3)$$
$$M^{3+}(g) \rightarrow M^{4+}(g) + e^-; \quad \Delta H_i(4)$$

These enthalpy changes are called the **first**, **second**, **third** and **fourth ionisation enthalpies** respectively and are a measure of the energy needed to remove a particular electron from the atom or ion.

Both calcium and vanadium always lose the 4s electrons. For calcium, the first and second ionisation enthalpies are relatively low since these correspond to removing two s electrons from the outer shell. There is then a sharp increase to the third and fourth ionisation enthalpies since it is much more difficult to remove further electrons from the filled 3p sub-shell.

For vanadium, there is a more gradual increase in successive ionisation energies as first the 4s and then the 3d electrons are removed. You can see from Table 7 that removing a third electron from vanadium does not involve a big jump in ionisation enthalpy as it does for calcium.

Figure 28 summarises the most common oxidation states of the *d* block elements scandium to zinc. The most important oxidation states are in boxes.

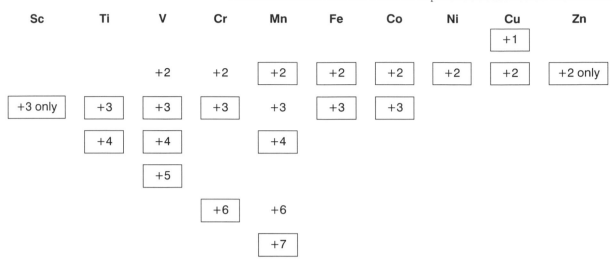

Sc	Ti	V	Cr	Mn	Fe	Co	Ni	Cu	Zn
								+1	
	+2	+2	+2	+2	+2	+2	+2	+2	+2 only
+3 only	+3	+3	+3	+3	+3	+3			
	+4	+4		+4					
		+5							
			+6	+6					
				+7					

Figure 28 Oxidation states shown by elements in the first row of the d-block. The most important oxidation states are in boxes

The oxidation state +1 is only important in the case of copper. For all the other metals, the sum of the first two ionisation energies is low enough for two electrons to be removed.

Except for scandium and titanium, all the elements show the oxidation state +2 in which the $4s$ electrons have been lost by ionisation.

All the elements except nickel, copper and zinc show an oxidation state of +3.

The number of oxidation states shown by an element increases from Sc to Mn. In each of these elements, the highest oxidation state is equal to the total number of $3d$ and $4s$ electrons in atoms of the metal.

After Mn, there is a decrease in the number of oxidation states shown by each element. The highest oxidation states become lower and progressively less stable. It seems that the increasing nuclear charge binds the d electrons more strongly in these elements, making them harder to remove.

In general, the lower oxidation states are found in simple ionic compounds; for example, compounds containing Cr^{3+}, Mn^{2+}, Fe^{3+}, Cu^{2+} ions. The metals in their higher oxidation states are usually bound covalently to an electronegative element such as O or F, often in an anion. For example,

VO_3^- vanadate(V) ion
MnO_4^- manganate(VII) ion

Simple ions with high oxidation states such as V^{5+} and Mn^{7+} are not formed because the charge density would be too high.

Stability of oxidation states

The change from one oxidation state of a metal to another is a redox reaction. Thus, the relative stability of different oxidation states can be predicted by looking at the standard electrode potentials ($E^\ominus$ values) for these reactions.

The general trends which emerge are

- higher oxidation states become less stable relative to lower ones on moving from left to right across the series;
- compounds containing metals in high oxidation states tend to be oxidising agents (eg MnO_4^- manganate(VII) ions) whereas compounds with metals in low oxidation states are often reducing agents (eg V^{2+} and Cr^{2+} ions);

- the relative stability of the +2 state with respect to the +3 state increases across the series. For elements early in the series, the +2 state is highly reducing. Thus, solutions of V^{2+} and Cr^{2+} ions are strong reducing agents. Later in the series, the +2 state is stable and the +3 state highly oxidising; Co^{3+}, for example, is a strong oxidising agent, and Ni^{3+} and Cu^{3+} do not exist in aqueous solution.

Catalytic activity

A catalyst alters the rate of a chemical reaction without being used up in the process. Transition metals and their compounds are effective and important catalysts both in industry and in biological systems.

Chemists believe that catalysts offer a new reaction pathway which has a lower activation enthalpy barrier than that of the uncatalysed reaction (see **Section 10.4**).

Once again, it is the availability of 3*d* as well as 4*s* electrons and the ability to change oxidation state which are among the vital factors in making transition metals such good catalysts.

Heterogeneous catalysis

In heterogeneous catalysis, the catalyst is in a different phase from the reactants. In the case of transition metals, this usually means a solid metal catalyst with reactants in the gas phase or liquid phase.

Transition metals can use the 3*d* and 4*s* electrons of atoms on the metal surface to form weak bonds to reactants. Once the reaction has occurred on the surface, these bonds can break to release the products.

An important example is the reaction of hydrogenation of alkenes, which is catalysed by nickel or platinum. A suggested mechanism for this reaction is given in Figure 19 in **Section 10.4**.

Homogeneous catalysis

In homogeneous catalysis, the catalyst is in the same phase as the reactants. In the case of transition metals, this usually means the reaction is in the aqueous phase, the catalyst being an aqueous transition metal ion.

Homogeneous catalysis usually involves the transition metal ion forming an intermediate compound with one or more of the reactants, which then breaks down to form the products. An example is the reaction in **Activity SS3.2**, between 2,3-dihydroxybutanoate ions and hydrogen peroxide.

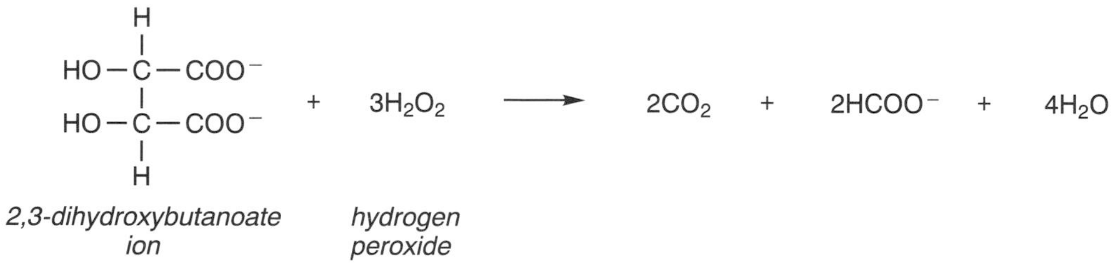

$$
\begin{array}{c}
\text{H} \\
| \\
\text{HO}-\text{C}-\text{COO}^- \\
| \\
\text{HO}-\text{C}-\text{COO}^- \\
| \\
\text{H}
\end{array}
\quad + \quad 3H_2O_2 \longrightarrow 2CO_2 \quad + \quad 2HCOO^- \quad + \quad 4H_2O
$$

2,3-dihydroxybutanoate ion   *hydrogen peroxide*

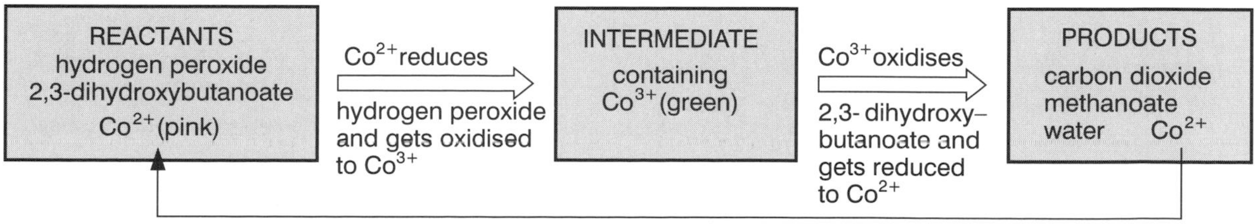

| REACTANTS hydrogen peroxide 2,3-dihydroxybutanoate Co^{2+} (pink) | Co^{2+} reduces hydrogen peroxide and gets oxidised to Co^{3+} | INTERMEDIATE containing Co^{3+} (green) | Co^{3+} oxidises 2,3- dihydroxy–butanoate and gets reduced to Co^{2+} | PRODUCTS carbon dioxide methanoate water Co^{2+} |

Figure 29 A suggested mechanism for the catalytic action of Co^{2+}. The reaction mixture turns from pink to green, then back to pink

Figure 29 shows a suggested mechanism for this catalysis. Figure 30, on page 212, shows the corresponding energy profile.

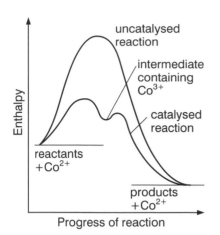

Figure 30 The energy profiles for the catalysed and uncatalysed reactions

Transition metal ions are particularly effective catalysts in redox reactions. This is because they can readily move from one oxidation state to another, in the way that Co readily moves between the +2 and +3 states in the reaction on the previous page.

***

PROBLEMS FOR 11.6

1   Use Figure 23 to write out the electron configuration for each of the elements in the series scandium to zinc (eg Sc [Ar] $3d^1 4s^2$).

2   **a**   Write the electronic configurations of the following ions:

    i Cu^{2+} **iii** Fe^{3+} **v** Cr^{3+}

    ii Cu^+ **iv** V^{3+} **vi** Ni^{2+}

   b   Explain why Cu^{2+} behaves as a typical transition metal ion, but Cu^+ does not.

3   Look at the information in Table 6.

   a   What do you notice about the size of the three transition metal atoms compared with those of sodium and magnesium?

   b   How do the melting and boiling points of transition metals compare with those of non-transition metals?

   c   How do the densities of s and d block metals compare?

   d   Write out the electronic configurations of the elements
Cr, Fe, Co, Na and Mg (eg Na $1s^2 2s^2 2p^6 3s^1$).

   e   Suggest why certain properties are common to transition metals but different from the properties of s block metals like sodium and magnesium.

4   Look at Figure 31 which shows E° values for the elements titanium to cobalt for the reaction
$M^{3+}(aq) + e^- \rightarrow M^{2+}(aq)$
The more positive the value of E°, the more likely is the aqueous M^{3+} ion to become reduced to the M^{2+} ion.

The E° values steadily become more positive across the series, but this is interrupted by what appear to be an abnormal values for manganese and iron. Thus Mn^{2+} is unexpectedly stable with respect to Mn^{3+}, whereas Fe^{3+} is unexpectedly stable with respect to Fe^{2+}.

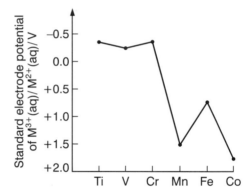

Figure 31 Standard electrode potentials for $M^{3+}(aq) + e^- \rightarrow M^{2+}(aq)$

Draw out 'electrons in a box' arrangements for the Mn^{2+}, Mn^{3+}, Fe^{2+} and Fe^{3+} ions. Use these to suggest a reason for the anomalies in Figure 31.

5   Use the given values of standard electrode potentials to explain the following observations.

   a   In acidic solution, aqueous iron(II) is slowly oxidised by air to iron(III) whereas aqueous manganese(II) is not oxidised to manganese(III).
$Fe^{3+}(aq) + e^- \rightarrow Fe^{2+}(aq)$;

$$E^\circ = +0.77\,V$$

$Mn^{3+}(aq) + e^- \rightarrow Mn^{2+}(aq)$;

$$E^\circ = +1.49\,V$$

$O_2(g) + 4H^+(aq) + 4e^- \rightarrow 2H_2O(l)$;

$$E^\circ = +1.23\,V$$

   b   Red copper(I) oxide dissolves in dilute sulphuric acid to give a blue solution of copper(II) sulphate and a red-brown precipitate of copper metal.

     $Cu^{2+}(aq) + e^- \rightarrow Cu^+(aq)$; $E^\circ = +0.15\,V$

     $Cu^+(aq) + e^- \rightarrow Cu(s)$; $E^\circ = +0.52\,V$

11.7 *The* d *block: complex formation*

What are complexes?

A complex consists of a central metal atom or ion surrounded by a number of negatively charged ions or neutral molecules possessing a lone pair of electrons. These surrounding anions or molecules are called **ligands**.

A complex may have an overall positive charge, a negative charge or no charge at all. For example,

$[Fe(H_2O)_6]^{3+}$
$[NiCl_4]^{2-}$
$Ni(CO)_4$

If a complex is charged, it is called a **complex ion**. The overall charge is the sum of the charge on the central metal ion and the charges on the ligands.

For $[NiCl_4]^{2-}$,　charge $= (2+) + 4(1-) = 2-$

In reality the charges on a complex ion are delocalised over the whole ion.

Bonding in complexes is complicated. It usually involves electron pairs from the ligand being shared with the central ion. Ligands are thus *electron donors*.

The number of bonds from ligands to the central ion is known as the **coordination number** of the central ion. The most common coordination numbers are six and four, but two does occur, particularly in complexes of Ag(I) and Cu(I).

Shapes of complexes

The shape of a complex depends on its coordination number. (There is more about the shapes of molecules in **Section 3.3**).

Complexes with *coordination number* 6 usually have an *octahedral* arrangement of ligands around the central metal ion (Figure 32). This is the most common coordination number.

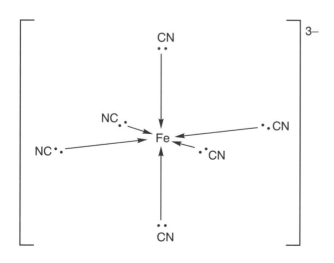

octahedral complex of Fe(III)
coordination number 6

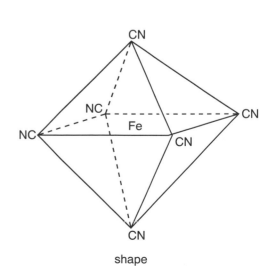

shape

Figure 32 Octahedral complex of Fe(III). Coordination number 6

Complexes with *coordination number 4* usually have a *tetrahedral* arrangement of ligands around the central metal ion (Figure 33). Some four-coordinate complexes have a *square planar* structure (Figure 34).

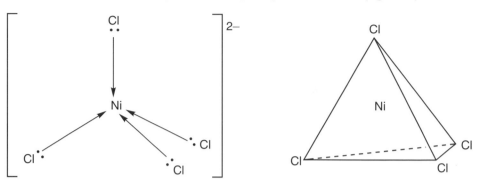

Figure 33 Tetrahedral complex of Ni(II). Coordination number 4

tetrahedral complex of Ni(II) coordination number 4

shape

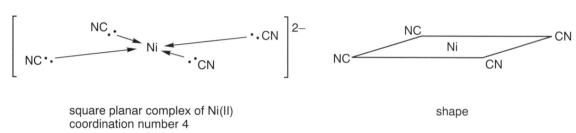

square planar complex of Ni(II) coordination number 4

shape

Figure 34 Square planar complex of Ni(II). Coordination number 4

Complexes with *coordination number 2* usually have a *linear* arrangement of ligands (Figure 35).

Table 8 gives a summary of the shapes of complexes with some examples.

$$\left[H_3N: \longrightarrow Ag \longleftarrow :NH_3 \right]^+$$

Figure 35 Linear complex of Ag(I). Coordination number 2

Table 8 The shapes of complexes.

Coordination number	Shape of complex	Examples
6	octahedral	$[Fe(CN)_6]^{3-}$ $[Ni(NH_3)_6]^{2+}$
4	tetrahedral	$[NiCl_4]^{2-}$
4	square planar	$[Ni(CN)_4]^{2-}$
2	linear	$[Ag(NH_3)_2]^+$

Most complexes have regular shapes like the ones in Figures 32 to 35 but some important examples do not. For example, in the $[Cu(H_2O)_6]^{2+}$ ion, four of the water ligands are held more strongly than the remaining two, so that four of the copper-oxygen bonds are shorter. This produces a distorted octahedral arrangement.

Naming complex ions

The systematic naming of complex ions is based on four rules.

1 Give the number of ligands of the same type around the central cation using the Greek prefixes mono-, di-, tetra-, penta- and hexa-.
2 Identify the ligands (in alphabetical order if more than one type) using the ending -o for anions, eg F⁻ fluoro, CL⁻ chloro, CN⁻ cyano, ⁻OH hydroxo. Neutral ligands keep their name except H_2O aqua, NH_3 ammine.
3 Name the central metal ion. Use the English name for a positively charged or neutral complex. If the overall charge of the complex is negative, use

the Latinised name for the cation and add the suffix -*ate*, ie cuprate for copper, ferrate for iron, zincate for zinc etc.

4 Indicate the oxidation number of the central metal in brackets.
For example,

$[Cr(H_2O)_6]^{3+}$ hexaaquachromium(III) ion
$[Fe(CN)_6]^{2-}$ hexacyanoferrate(II) ion
$[CuCl_4]^{2-}$ tetrachlorocuprate(II) ion

Why are some complexes more stable than others?

Different ligands form complexes with different stabilities. For example, ammonia ligands displace water ligands when added to blue aqueous copper(II) ions. The complex is more stable with NH_3 ligands than with H_2O ligands.

$$[Cu(H_2O)_6]^{2+}(aq) + NH_3(aq) \rightleftharpoons [CuNH_3(H_2O)_5]^{2+}(aq) + H_2O(l)$$
$$[CuNH_3(H_2O)_5]^{2+}(aq) + NH_3(aq) \rightleftharpoons [Cu(NH_3)_2(H_2O)_4]^{2+}(aq) + H_2O(l)$$
(and so on)

The water is displaced in a series of steps until a deep blue solution of $[Cu(NH_3)_4(H_2O)_2]^{2+}(aq)$ is formed. Each step is reversible and an equilibrium mixture is formed containing both the starting materials and products for that step.

The stability of a complex can be expressed in terms of the equilibrium constants for the ligand displacement reactions (see **Section 7.2** for an explanation of equilibrium constant). These are known as **stability constants**. Usually an **overall stability constant K_{stab}** is given rather than the stepwise stability constants for the individual ligand displacement steps. Like all equilibrium constants, stability constants vary with temperature.

For the *overall* reaction described above, the equation is

$$[Cu(H_2O)_6]^{2+}(aq) + 4NH_3(aq) \rightleftharpoons [Cu(NH_3)_4(H_2O)_2]^{2+}(aq) + 4H_2O(l)$$

$$K_{stab} = \frac{[Cu(NH_3)_4(H_2O)_2]^{2+}(aq)]}{[[Cu(H_2O)_6]^{2+}(aq)]\,[NH_3(aq)]^4} = 1 \times 10^{12}\,dm^{12}\,mol^{-4} \text{ at } 298\,K$$

The stability constants of complexes can be extremely large or extremely small, and so the values are often given on a logarithmic scale. In the copper example above, $\lg K_{stab} = 12$.

Notice that because water is also the solvent, its concentration is virtually constant and it is omitted from the equilibrium expression.

Stability constants can be used to compare the stability of any two ligands, but the values usually quoted give the stability of a complex relative to the simple aqueous ion where the ligand is water. *The higher the value of the stability constant, the more stable the complex.*

Table 9 shows some values of $\lg K_{stab}$.

Complex ion	lg K_{stab}
$[Ni(NH_3)_6]^{2+}$	8.6
$[CuCl_4]^{2-}$	5.6
$[Fe(SCN)(H_2O)_5]^{2+}$	3.9
$[Ni(en)_3]^{2+}$	18.3
$[Ni(edta)]^{2-}$	19.3
$[Cu(en)_2]^{2+}$	20.0

Table 9 The stability constants of some complexes. The structures of en and edta^{4-} are shown in Figures 36 and 37

Polydentate ligands

Some ligand molecules, like NH_3 and H_2O, can only bond to a metal ion through a single atom or ion and are **monodentate** ligands. Others can bond through more than one, and are **polydentate** ligands. For example, the ethanedioate ion and 1,2-diaminoethane, shown in Figure 36, are **bidentate** ('two-toothed') ligands because they can form *two* bonds with a metal ion by using pairs of electrons from two oxygen or nitrogen atoms.

The metal ion (for example, an M^{2+} ion) is held in a five-membered ring as if it was in the claws of a crab. The ring is called a **chelate ring** after the Greek word *chele* meaning crab.

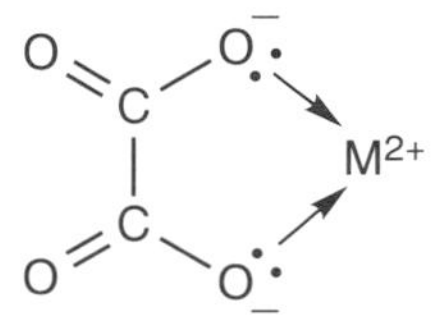

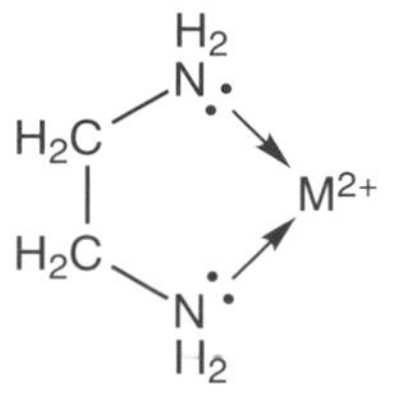

Figure 36
a *The ethanedioate ion acting as a bidentate ligand*
b *The 1,2-diaminoethane ligand (en for short)*

Edta^{4-} forms six bonds to metal ions and so is a **hexadentate** ligand (see Figure 37a). Edta^{4-} complexes via the two nitrogen atoms and four oxygen atoms of the COO$^-$ groups. Edta^{4-} acts as a kind of cage and traps the metal atom inside as shown for the complex with nickel in Figure 37b. (Edta is an abbreviation for **e**thylene**d**iamine**tetra**cetic acid, which is the acid from which the ion edta^{4-} is produced.)

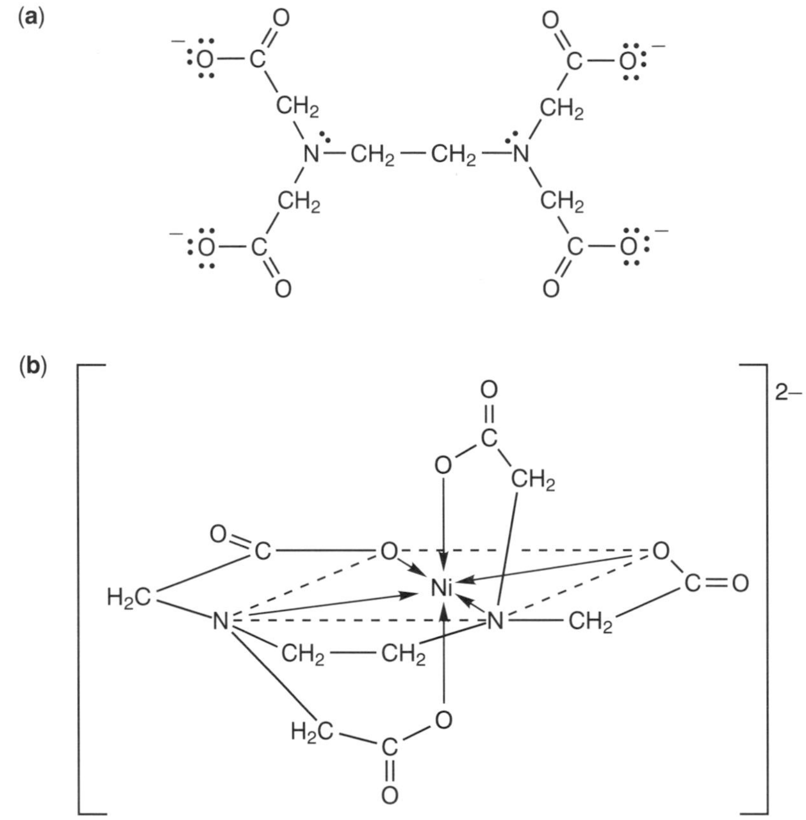

(a)

(b)

Figure 37
(a) *Edta^{4-}, a hexadentate ligand*
(b) *The nickel-edta complex ion*

Stability constants for complexes with polydentate ligands are strikingly large. Look at Table 9. The reason that these complexes are so stable in aqueous solution is the entropy increase which occurs as they are formed. In the case of edta^{4-} a single ligand can replace up to six water molecules.

$$[Ni(H_2O)_6]^{2+}(aq) + edta^{4-}(aq) \rightarrow [Ni(edta)]^{2-}(aq) + 6H_2O(l)$$

The reaction increases the number of separate particles present, and so increases the entropy of the system. This favourable entropy change for the forward reaction shifts the equilibrium towards the products. (See **Section 4.3** for more about entropy.)

Coloured compounds

Compounds of *d* block transition metals are frequently coloured both in the solid state and in solution. You can get an idea of the great range of colours shown by transition metal ions in aqueous solution by looking at Table 10.

Table 10 Colours of some transition metal ions in aqueous solution

Ion	Outer electrons	Colour	Ion	Outer electrons	Colour
Ti^{3+}	$3d^1$	purple	Fe^{3+}	$3d^5$	yellow
V^{3+}	$3d^2$	green	Fe^{2+}	$3d^6$	green
Cr^{3+}	$3d^3$	violet	Co^{2+}	$3d^7$	pink
Mn^{3+}	$3d^4$	violet	Ni^{2+}	$3d^8$	green
Mn^{2+}	$3d^5$	pale pink	Cu^{2+}	$3d^8$	blue

The intensity of colour varies greatly. For example, MnO_4^- is an intensely deep purple, but Mn^{2+} is very pale pink. The colour of transition metal compounds can often be related to the presence of unfilled or partly filled d orbitals (and so to unpaired electrons) in the metal ion. When white light falls on a substance some may be absorbed, some transmitted and some reflected (**Section 6.4**). If light in the visible region of the spectrum is absorbed, the compound will appear coloured.

What happens when a d block compound absorbs light?

Light is only absorbed by an atom when its energy matches the energy gap between two energy states in the atom. If it does, an electron is promoted from an orbital of lower to one of higher energy. So the atom or ion absorbing the radiation changes from its ground state to an excited state.

For a simple metal ion in the gas phase, the five d orbitals all have the same energy. In a complex, however, the metal ion is surrounded by negative ligands or by polar molecules with lone pairs of electrons. The presence of these ligands affects the electrons in the d orbitals of the central metal ion. Orbitals close to the ligands are pushed to slightly higher energy levels than those further away. As a result the five d orbitals are split into two groups at different energy levels.

This splitting of the $3d$ sub-shell in the presence of an octahedral arrangement of ligands is illustrated for the Ti^{3+} ion in Figure 38.

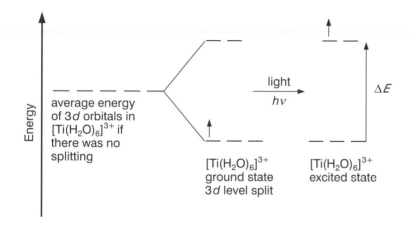

Figure 38 Relative energy levels for the five $3d$ orbitals of the hydrated Ti^{3+} ion

When light passes through a solution of $[Ti(H_2O)_6]^{3+}$ ions, a photon light may be absorbed. The energy of this photon corresponds to excitation of an electron from a low energy d orbital to a high energy d orbital.

The frequency (v) of the light absorbed depends on the energy difference between these two levels, ΔE. ($\Delta E = hv$) where h is Planck's constant). For most d block transition metals the size of ΔE is such that the light absorbed falls in the visible part of the spectrum.

The absorption spectrum for the $[Ti(H_2O)_6]^{3+}$ ion is shown in Figure 39.

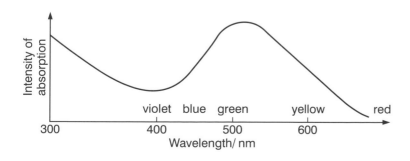

Figure 39 The absorption spectrum of the hydrated Ti^{3+} ion, $[Ti(H_2O)_6]^{3+}$

$[Ti(H_2O)_6]^{3+}$ is a simple example because the ion contains only one d electron. The situation is more complicated when there are several d electrons and many more transitions are possible between d levels. However, the basic principles are the same.

The colour of a transition metal complex depends on

- the number of d electrons present in the transition metal ion;
- the arrangement of ligands around the ion, since this affects the splitting of the d sub-shell;
- the nature of the ligand, since different ligands have a different effect on the relative energies of the d orbitals in a particular ion.

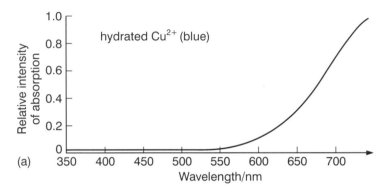

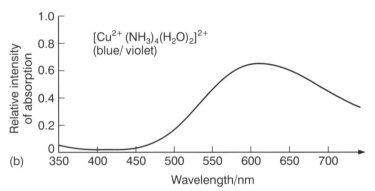

Figure 40 The absorption spectra for
a *hydrated Cu^{2+} and*
b *[Cu(NH$_3$)$_4$(H$_2$O)$_2$]$^{2+}$*
In the ammonia complex, the ammonia ligands cause a bigger splitting of d orbital energies. This means that light of higher energy, and lower wavelength, is absorbed

For example, NH_3 ligands cause a larger difference than H_2O ligands in splitting d orbital energies. The blue colour of hydrated Cu^{2+} ions, for example, changes to deep blue/violet when NH_3 is added (see Figure 40).

$$4NH_3(aq) + [Cu(H_2O)_6]^{2+} \rightarrow [Cu(NH_3)_4(H_2O)_2]^{2+} + 4H_2O(l)$$
blue   deep blue/violet

You can see from the following complexes of chromium that the colour also depends on the number of each kind of ligand present:

$[Cr(H_2O)_6]^{3+}$ $[Cr(H_2O)_5Cl]^{2+}$ $[Cr(H_2O)_4Cl_2]^+$
violet   green   dark green

Finally, the colour of complexes depends not only on what the ligands are but on the way the ligands are arranged around the metal ion. For example, the two geometric isomers of $[Co(NH_3)_4Cl_2^-]^+$ are different colours (**Section 3.5** deals with geometric isomerism). Figure 41 shows the structures of the two isomers.

Some ions of d block elements are colourless, because it is not possible for d electrons to move between energy levels. For example, Zn^{2+} has a full set of 10 $3d$ electrons, so the $3d$ orbitals are full up and movement is impossible. So Zn^{2+} compounds are white.

There is more about the chemistry of colour in **Section 6.6**.

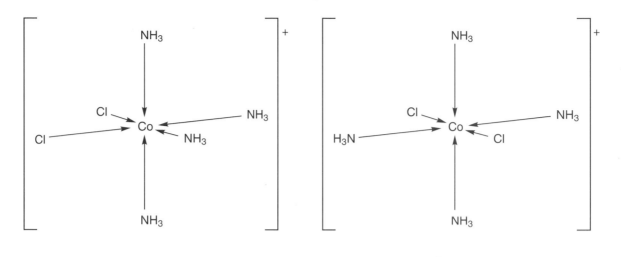

cis-isomer: violet   trans-isomer: green

Figure 41 The cis *and* trans *isomers of [Co(NH$_3$)$_4$Cl$_2$]$^+$ have different colours*

PROBLEMS FOR II.7

1 Give the coordination number for each of the following complex ions
   a [Ag(H$_2$O)$_2$]$^+$
   b [CoCl$_4$]$^{2-}$
   c [Co(CN)$_6$]$^{3-}$
   d [Cr(H$_2$O)$_5$OH]$^{2+}$

2 Write down the formula of the following complex ions. Make sure you include the charge on the ion.
   a hexaaquamanganese(II) ion
      (ligand = H$_2$O; metal ion = Mn^{2+})
   b tetraamminezinc(II) ion
      (ligand = NH$_3$; metal ion = Zn^{2+})
   c hexafluoroferrate(III) ion
      (ligand = F$^-$; metal ion = Fe^{3+})
   d pentaaquamonohyroxochromium(III) ion
      (ligands = H$_2$O and OH$^-$; metal ion = Cr^{3+})

3 What is the oxidation state of the central metal ion of each of the complexes **a** to **d** in Problem **1**?

4 Give systematic names for the following complex ions:
   a [V(H$_2$O)$_6$]$^{3+}$
   b [Fe(CN)$_6$]$^{4-}$
   c [CoCl$_4$]$^{2-}$
   d [Ag(NH$_3$)$_2$]$^+$
   e [Cr(H$_2$O)$_4$Cl$_2$]$^+$

5 a Explain why titanium(IV) oxide (the white pigment in white paint) is not coloured.
   b Explain why compounds of Sc^{3+}, Zn^{2+} and Cu$^+$ are not coloured.

6 Titanium(IV) chloride dissolves in concentrated hydrochloric acid to give the ion [TiCl$_6$]$^{2-}$.
   a What is the coordination number of the titanium ion?
   b What is the oxidation number of the titanium ion?
   c Suggest a name for [TiCl$_6$]$^{2-}$.
   d Draw a likely structure for this ion.

7 Copper forms a complex with 1,2-diaminoethane (en) and with a similar bidentate ligand, 1,3-diaminopropane (pn),
   $$NH_2CH_2CH_2CH_2NH_2$$
   The lg K_{stab} values for these two complexes are shown below

	[Cu(en)$_2$]$^{2+}$	[Cu(pn)$_2$]$^{2+}$
lg K_{stab}	20.03	17.17

   Look at models of the two complexes of copper. What size of chelate ring seems to lead to the more stable structure?

8 Classify the following ligands as mono-, bi- or polydentate ligands:
   a H$_2$O
   b ![benzene ring with two OH groups]
   c porphyrin ring which complexes with iron(II) in haemoglobin (see **The Steel Story** Storyline Section **SS3**)
   d ![benzene ring with OH and COO$^-$ groups]

9 Complexes formed by edta^{4-} involve pairs of electrons on nitrogen and oxygen atoms in the same way as complexes formed by NH_3 and H_2O. Explain why stability constants of edta complexes in aqueous solution are generally so much larger than those of corresponding complexes with NH_3 and H_2O.

10

> **Problems inside a can of rhubarb**
>
> The canning of rhubarb presents food tech-nologists with a considerable challenge. The juice from rhubarb is one of the most corrosive possible environments inside a food can and the metal has to be protected by an internal coat of unreactive lacquer. Even with this coating, tins of rhubarb have only a relatively short shelf life of about 6 months.

Rhubarb contains ethanedioic acid. This is a weak dicarboxylic acid which dissociates to give a solution which is acidic and contains ethanedioate ions:

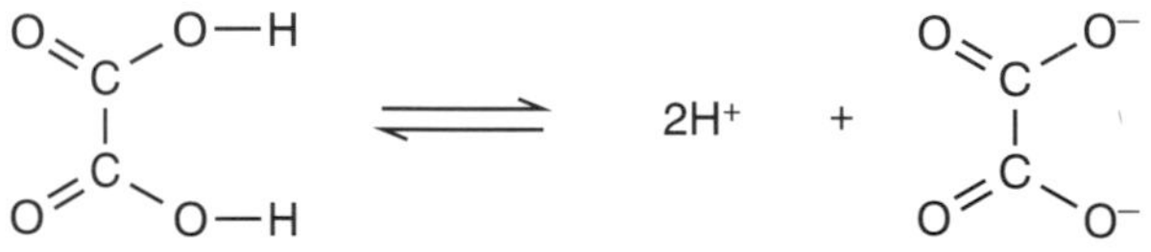

As well as ethanedioate ions in rhubarb the anions of three other carboxylic acids are found in a number of different fruits. Their structures are shown below.

Naturally-occurring carboxylic acids such as these are often still called by their old names. (The old name of ethanedioic acid is oxalic acid.) You can check their systematic names which are given in brackets below the traditional names.

When fruits are put into cans, the anions of these acids form soluble complexes with any Sn^{2+} ions. The removal of Sn^{2+} in this way encourages the formation of more Sn^{2+} ions, and hence the oxidation of the tin coating inside the can. The more stable the complex the more rapid is the consequent corrosion.

J. C. Sherlock and S. C. Britton from the Tin Research Institute have carried out experiments to estimate stability constants for Sn^{2+}/fruit acid complexes. Their results are as follows:

Acid	lg K_{stab}
tartaric	6.66
malic	7.43
citric	7.65
oxalic	12.39

a Which acid forms the most stable complex with Sn^{2+} ions?

b Which acid would you expect to lead to the most rapid corrosion of the tin?

c Extending their investigations Sherlock and Britton sealed pieces of tin in solutions of the four acids. After 35 days the tin content of each solution was measured by atomic absorption spectroscopy.

Acid	A	B	C	D
Dissolved tin (ppm)	8.5	20.3	5.5	8.5

Work out which acid of A, B, C, and D is which.

d Discuss the implications of the information given here for corrosion in tins of apples, grapes, oranges and rhubarb.

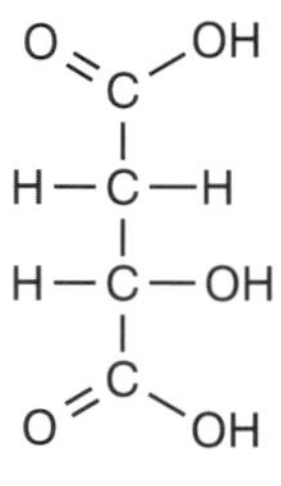

malic acid
(2-hydroxybutanedioic acid)

(common in apples and berries)

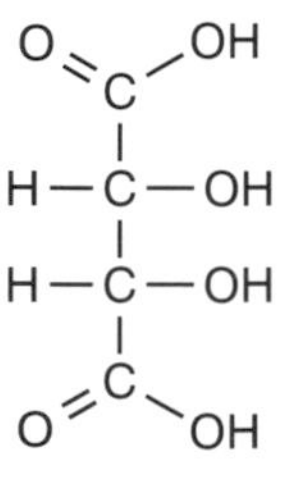

tartaric acid
(2,3-dihydroxybutanedioic acid)

(one of its salts is found in grapes)

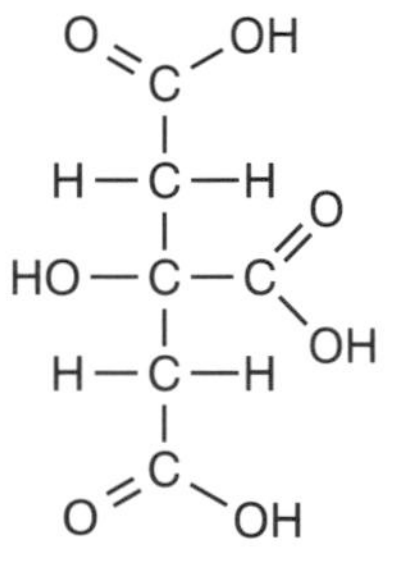

citric acid
(3-hydroxy-3-carboxypentanedioic acid)

(common in citrus fruits)

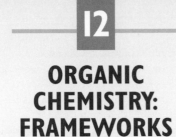

12

ORGANIC CHEMISTRY: FRAMEWORKS

Figure 1 *The electron structure of carbon*

Figure 2 *Covalent bonds in methane*

n	Molecular formula	Name
1	CH_4	methane
2	C_2H_6	ethane
3	C_3H_8	propane
4	C_4H_{10}	butane
5	C_5H_{12}	pentane

Table 1 *Some simple alkanes*

The name of all alkanes ends in **-ane**

12.1 *Alkanes*

Many carbon compounds are found in living organisms which is why their study got the name **organic chemistry**. Today, organic chemistry includes all carbon compounds whatever their origin, except CO, CO_2 and the carbonates, which traditionally are included in inorganic chemistry.

Only carbon can form the diverse range of compounds necessary to produce the individuality of living things.

Why carbon?

Carbon's electron structure is shown in Figure 1. This electron structure makes it the first member of Group 4 in the centre of the Periodic Table, and is responsible for its special properties.

A carbon atom has 4 electrons in its outer shell. It could achieve a stable noble gas electron arrangement by losing or gaining 4 electrons; but this is too many electrons to lose of gain. The resulting carbon ions would have charges of +4 or –4, respectively, and would be too highly charged. So when carbon forms compounds, the bonds are *covalent* rather than ionic.

In methane (CH_4), for example, the carbon atom achieves a stable octet of 8 electrons by *sharing* its outer electrons with four hydrogen atoms (Figure 2). Four C–H covalent bonds are formed.

Carbon forms strong covalent bonds with itself to give *chains* and *rings* of its atoms joined by C–C covalent bonds. This property is called **catenation** and leads to the limitless variety of organic compounds possible.

Each carbon atom can form 4 covalent bonds, so the chains may be straight or branched and can have other atoms or groups substituted in them.

Hydrocarbons

Chemists cope with the vast number of organic compounds by dividing them into groups of related compounds.

Hydrocarbons contain only carbon and hydrogen atoms and are represented by the molecular formula C_xH_y. There are different types of hydrocarbons. For example.

CH_4 (methane)	is an alkane
C_2H_4 (ethene)	is an alkene
C_6H_6 (benzene)	is an aromatic hydrocarbon (or arene)

Hydrocarbons are relatively unreactive – this is particularly true of alkanes and arenes. They form the unreactive framework of organic compounds. But when you attach other groups, such as OH, to the hydrocarbon framework, its properties are modified. So we can think of organic compounds as having hydrocarbon **frameworks**, with **modifiers** attached. A modifier such as the hydroxyl group, OH, is also called a **functional group**. In this chapter we are concerned with frameworks: modifiers are dealt with in Chapter 13.

Alkanes

Alkanes are **saturated** hydrocarbons. 'Saturated' means that they contain the maximum amount of hydrogen possible, with no double or triple bonds between carbon atoms.

The general molecular formula of the alkanes is C_nH_{2n+2} where n = 1, 2, 3 ... . Table 1 shows the names and formulae of some simple alkanes.

Look at the names in Table 1. The first part indicates the number of carbon atoms present. *Alk-* is the general term for *meth-*, *eth-*, *prop-* and so on. The second part *-ane* shows that the molecules are saturated.

A series of compounds related to each other in this way is called a **homologous series**. All the members of the series have the same general molecular formula and each member differs from the next by a —CH_2— unit. All the compounds in a series have similar chemical properties, so chemists can study the properties of the group rather than those of individual compounds. However, physical properties such as melting point, boiling point and density change gradually in the series as the number of carbon atoms in the molecules increases.

Finding the formulae of alkanes

Section 1.1 explains the difference between empirical and molecular formulae. The molecular formula is more useful, because it tells you how many atoms of each type are present in a molecule of the compound. The molecular formula may or may not be the same as the empirical formula for the compound, which gives only the *simplest ratio* of the different types of atoms present.

molecular formula = m × empirical formula

where **m** = 1, 2, 3, ...

The empirical formula of a compound is worked out from its percentage composition. For a hydrocarbon, the percentage composition is easily found by burning a known mass in oxygen and measuring the amounts of carbon dioxide and water formed. This is called **combustion analysis** and can be performed automatically by a machine.

Example 0.100 g of a hydrocarbon **X** on complete combustion gave 0.309 g CO_2 and 0.142 g H_2O. Calculate the empirical formula of the compound **X**.

Answer First, calculate the masses of C and H in 0.100 g of the compound.
44 g CO_2 contains 12 g C
∴ mass of C in 0.100 g **X** = (12/44) × 0.309 g = 0.0843 g
18 g H_2O contains 2 g H
∴ mass of H in 0.100 g **X** = (2/18) × 0.142 g = 0.0158 g

	C	:	H
ratio by mass	0.0843	:	0.0158
ratio by moles	0.00703	:	0.0158
simplest ratio (divide by smaller)	1	:	2.25
whole number ratio	4	:	9

∴ Empirical formula of **X** is C_4H_9

Once we have the empirical formula, we can work out the molecular formula provided we know the relative molecular mass of the compound.

Example The relative molecular mass of **X** was found to be 114 by using a mass spectrometer. Find the molecular formula of X.

Answer Empirical formula of **X** is C_4H_9.
But M_r (C_4H_9) = 57. This is half of 114. So the molecular formula of **X** must be $(C_4H_9)_2 = \mathbf{C_8H_{18}}$

You can practice this kind of calculation in problems 1 to 5 at the end of this section.

Structure of alkanes

Figure 2 shows a dot-cross formula for methane. It shows all the *outer* electrons in each atom and how electrons are shared to form the covalent bonds. For larger molecules, dot-cross formulae are rather cumbersome and the shared electron pairs can be replaced by lines representing the covalent bonds (Figure 3).

For ethane

molecular formula

C_2H_6

empirical formula

CH_3

Figure 3

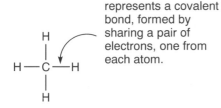

represents a covalent bond, formed by sharing a pair of electrons, one from each atom.

This is called a **structural formula** or sometimes a **graphical formula**. It shows the way atoms are bound together in the molecule.

Look at Table 2. It gives the full structural formulae and also shortened formulae, for some alkanes. Note that each carbon atom is bonded to four other atoms. Each hydrogen atom is bonded to one other atom.

Name	Molecular formula	Structural formula	Shortened structural formula
methane	CH_4	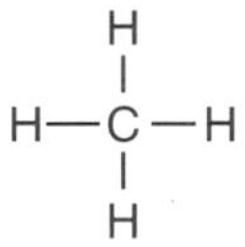 H–C–H (with H above and below)	CH_4
ethane	C_2H_6	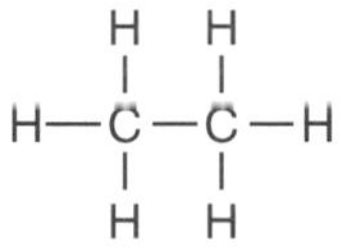 H–C–C–H (with H's)	$CH_3–CH_3$ or CH_3CH_3
propane	C_3H_8	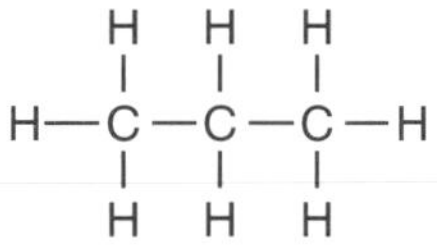 H–C–C–C–H (with H's)	$CH_3–CH_2–CH_3$ or $CH_3CH_2CH_3$

Table 2 *Structural formulae of alkanes*

Branched alkanes

It is possible to have alkanes with straight or branched chains. So it is often possible to draw more than one structural formula for a given molecular formula. There is often a straight-chain compound and one or more branched-chain compounds with the same molecular formula.

For C_4H_{10}, there are two possible structural formulae.

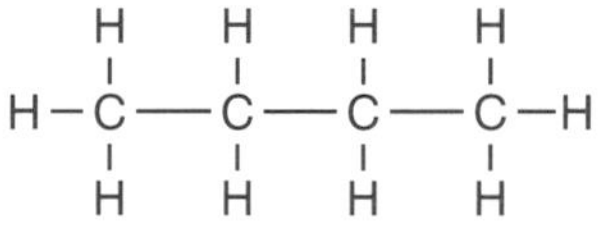

butane and *methylpropane*

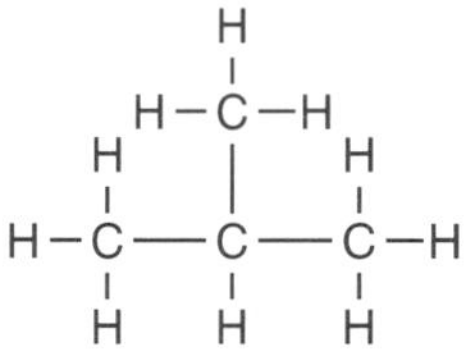

Alkyl group	Formula
methyl	$CH_3–$
ethyl	$CH_3CH_2–$
propyl	$CH_3CH_2CH_2–$
butyl	$CH_3CH_2CH_2CH_2–$
pentyl	$CH_3CH_2CH_2CH_2CH_2–$

Table 3 *Some common alkyl groups*

These two compounds are **isomers** because they have the same molecular formulae but different structural formulae. There is more about isomerism in **Section 3.4**.

Note how the branched-chain isomer is named. It is regarded as being formed from the straight-chain alkane, propane, by attaching a —CH_3 side group to the second carbon atom. It is therefore called methylpropane.

The –CH_3 group is just methane with a hydrogen atom removed so that it can join to another atom. It is called a **methyl group**.

Side groups of this kind are called **alkyl groups**. They have the general formula C_nH_{2n+1} and are often represented by the symbol **R** (see Table 3).

An alkyl group R has the general formula C_nH_{2n+1}

Cycloalkanes

As well as open-chain alkanes, it is also possible for alkane molecules to form rings.

These molecules are called **cycloalkanes** and have the general formula C_nH_{2n}. They have two less hydrogen atoms than the corresponding alkane, because there are no CH_3— groups at the ends of the chain. Table 3 shows some different ways of representing cycloalkanes. The **skeletal formula** shows only the shape of the carbon framework.

Cycloalkane	Shortened structural formula	Skeletal formula
cyclopropane C_3H_6	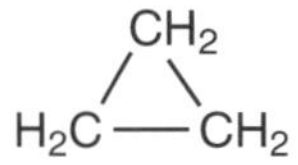 $\begin{array}{c} CH_2 \\ H_2C - CH_2 \end{array}$	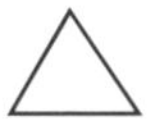 △
cyclobutane C_4H_8	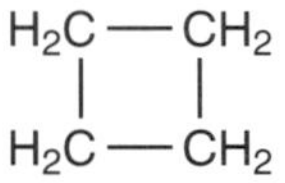 $\begin{array}{c} H_2C - CH_2 \\ H_2C - CH_2 \end{array}$	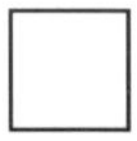 □
cyclohexane C_6H_{12}	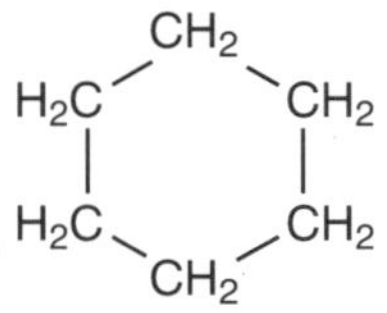 $\begin{array}{c} CH_2 \\ H_2C \quad CH_2 \\ H_2C \quad CH_2 \\ CH_2 \end{array}$	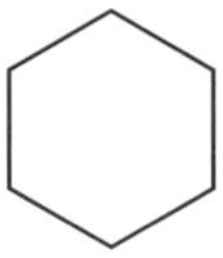 ⬡

Table 3 Cycloalkanes

Shapes of alkanes

Representing structures in a 2-dimensional way on paper can give a misleading picture of what the molecule looks like.

The pairs of electrons in the covalent bonds repel one another and so arrange themselves round the carbon atom as far apart as possible. Thus, the C—H bonds in methane are directed so they point towards the corners of a regular tetrahedron (Figure 4). The carbon atom is at the centre of the tetrahedron, and the H—C—H bond angles are 109° (109° 28' to be precise). See Figure 5.

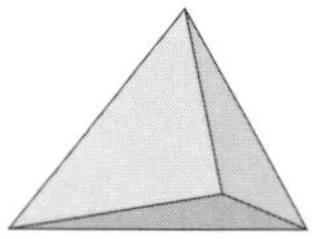

Figure 4 A regular tetrahedron

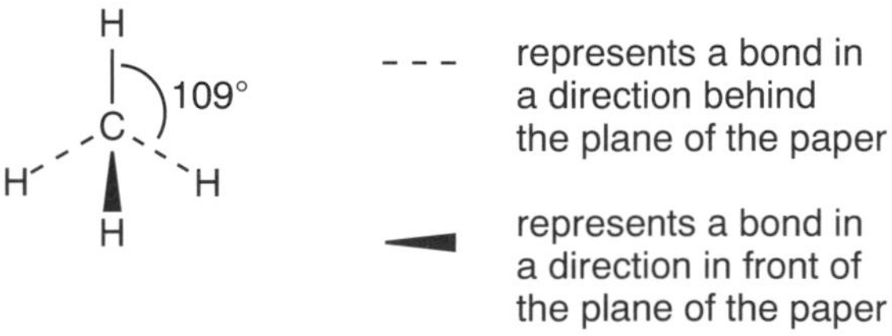

Figure 5 The 3-dimensional shape of methane

— represents a bond in the plane of the paper

- - - represents a bond in a direction behind the plane of the paper

◄ represents a bond in a direction in front of the plane of the paper

The structure of ethane in three dimensions is shown in Figure 6. Each carbon atom is at the centre of a tetrahedral arrangement.

Figure 6 The 3-dimensional shape of ethane

a simpler way of drawing ethane which shows the shape less accurately

Figure 7 shows the 3-dimensional structure of butane. You can see that hydrocarbon chains are not really 'straight', but a zigzag of carbon atoms. All the bond angles are 109°.

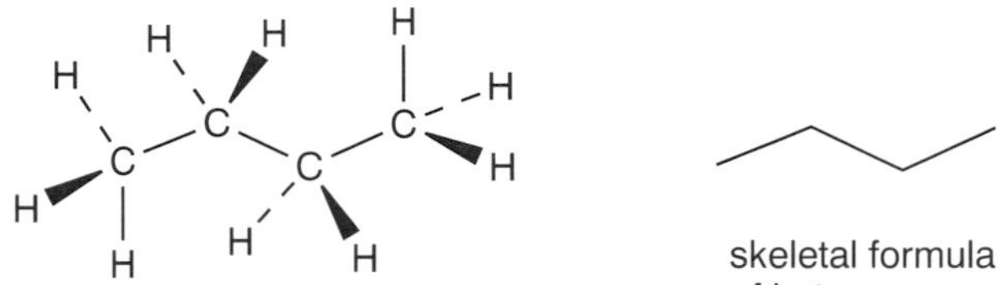

skeletal formula
of butane

Figure 7 The 3-dimensional shape of butane

The shape of a hydrocarbon chain is often represented by a skeletal formula, also shown in Figure 7.

It is a good idea to build models of some of these compounds. You can do this in **Developing Fuels, Activity DF4.1**.

What are alkanes like?

Whether an alkane is solid, liquid or gas at room temperature depends on the size of its molecules. The first members of the series (C_1–C_4) are colourless gases. Higher members (C_5–C_{16}) are colourless liquids and the remainder are white waxy solids.

Alkanes mix well with each other, but do not mix with water. The alkanes and water form two separate layers. This is because alkanes contain non-polar molecules, but liquids like water and methanol contain polar molecules which attract each other and prevent the alkane molecules mixing with them. (There is more about polarity of molecules in **Section 5.3**.)

Chemical reactions of alkanes

Alkanes are very unreactive towards many laboratory reagents. They are unaffected by polar reagents such as acids, alkalis, metals and oxidising agents such as potassium manganate(VII).

When they do react, it is usually in the gas phase and energy must be supplied to get the reaction started.

Oxidation of alkanes

Alkanes do not react with air at room temperature, but, if heated, they burn readily to give carbon dioxide and water. The reaction has a high activation enthalpy and energy must be supplied to overcome this. The bonds in the alkane and in O_2 must first be broken before the new bonds in CO_2 and H_2O can form.

Alkanes must be vaporised before they will burn and so the less volatile alkanes ignite less readily. This is why petrol ignites so much more easily than oil.

Once started, the combustion is very exothermic which is why alkanes are such good fuels.

$$CH_4(g) + 2O_2(g) \rightarrow CO_2(g) + 2H_2O(l); \quad \Delta H^{\ominus}_c = -890\,kJ\,mol^{-1}$$
$$C_7H_{16}(g) + 11O_2(g) \rightarrow 7CO_2(g) + 8H_2O(l); \quad \Delta H^{\ominus}_c = -4817\,kJ\,mol^{-1}$$

If the supply of air is limited, the combustion may be incomplete and produce CO and C (soot) along with partially oxidised hydrocarbons.

Cracking alkanes

When alkanes are heated to high temperatures in the absence of air, they break up to form smaller molecules. Such reactions are known as **cracking**. They can be carried out at a lower temperatures in the presence of a catalyst. When the molecule splits, one of the products is an **alkene** and contains a C=C double bond,

$$\underset{decane}{C_{10}H_{22}} \rightarrow \underset{octane}{C_8H_{18}} + \underset{ethene}{CH_2{=}CH_2}$$

There are many ways an alkane can break down and a complex mixture of products is obtained. The alkanes often rearrange to give branched rather than straight-chain products.

PROBLEMS FOR 12.1

1 Write down the empirical formula for each of the following hydrocarbons:
  a C_3H_8 b $C_{20}H_{42}$ c C_6H_6

2 A hydrocarbon contains 85.7% C and 14.3% H by mass. Its relative molecular mass is 28.
  a Find its empirical formula.
  b Suggest a possible molecular formula for this compound.

3 A hydrocarbon contains 92.3% C. Its relative molecular mass is 78. Find
  a its empirical formula
  b its molecular formula.

4 A compound contains C, H and O. Combustion analysis shows its composition is 40% C, 6.7% H and 53.3% O by mass. Its relative molecular mass is 180. Find
  a its empirical formula
  b its molecular formula.

5 A hydrocarbon contains 82.8% by mass of carbon. Work out its empirical formula. Its relative molecular mass is found to be 58. What is its molecular formula?

6 Which of the following formulae represent alkanes?
  a CH_2 d C_7H_{16}
  b C_2H_6 e C_9H_{20}
  c C_5H_{10} f $C_{13}H_{26}$

7 Below are the names of alkanes corresponding to values of n between 6 and 10.

n	Name of alkane
6	hexane
7	heptane
8	octane
9	nonane
10	decane

What is the molecular formula of each of the following alkanes?
  a heptane
  b octane
  c decane

8 Draw 'dot-cross' formulae for ethane and propane.

9 Draw out the full structural formula of
  a $CH_3CH_2CH_2CH_2CH_3$
  b $CH_3(CH_2)_8CH_3$
  c

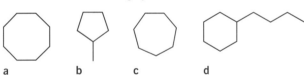

  [sometimes written $CH_3CH(CH_3)CH(CH_3)CH_3$]

10 There are two different compounds with the molecular formula C_4H_{10}. Draw out their structures.

11 Draw skeletal formula for
  a cyclopentane
  b methylcyclohexane
  c propylcyclopropane

12 Name the following hydrocarbons:

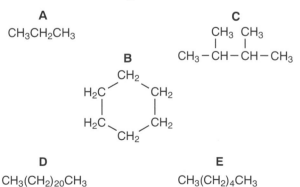

   a   b   c   d

13 Draw the skeletal formula for
  a decane $C_{10}H_{22}$
  b heptane C_7H_{16}
  c methylpropane

14 Look at the following alkanes:

A
$CH_3CH_2CH_3$

B

$$\begin{array}{ccc} & CH_2 & \\ H_2C & & CH_2 \\ | & & | \\ H_2C & & CH_2 \\ & CH_2 & \end{array}$$

C

$$\begin{array}{c} CH_3 \;\; CH_3 \\ | \quad\;\; | \\ CH_3-CH-CH-CH_3 \end{array}$$

D
$CH_3(CH_2)_{20}CH_3$

E
$CH_3(CH_2)_4CH_3$

  a Which alkane(s) would you expect to be solid at 298 K?
  b Which alkane would you expect to be the most volatile?

c Which alkane does *not* fit the general formula C_nH_{2n+2}?

d Which two alkanes are isomers?

e Give an example of an alkane which might have been formed by cracking another of the alkanes shown above. (Identify both alkanes).

f For which alkane(s) are there no isomeric compounds?

15 Write balanced equations for each of the following reactions.

a Pentane, C_5H_{12}, burns in a plentiful supply of air to form carbon dioxide and water.

b Pentane burns in a limited supply of air to form carbon monoxide and water.

16 When an alkane is cracked, each molecule forms at least two new molecules.

a What conditions are needed to cause cracking reactions in alkanes?

b Write the structural formula of three pairs of compounds that could be formed by cracking heptane, C_7H_{16}.

12.2 Alkenes

What are alkenes?

Ethene is the simplest example of a class of hydrocarbons called **alkenes**. It has the structure

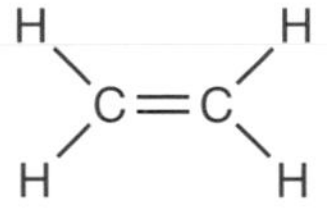

Alkenes are distinguished from other hydrocarbons by the presence of the C=C double bond.

Examples of other members of the alkene family are

propene	$CH_3CH{=}CH_2$
but-1-ene	$CH_3CH_2CH{=}CH_2$
but-2-ene	$CH_3CH{=}CHCH_3$
pent-1-ene	$CH_3CH_2CH_2CH{=}CH_2$

As with the alkanes, the boiling points of alkenes increase as the number of carbon atoms increases. Ethene, propene and the butenes are gases. After that they are liquids and eventually solids.

Notice how the alkenes are named. Take but-1-ene as an example. The *but-* part tells you it has the same hydrocarbon chain as butane. The *-ene* suffix (the end part of the name) tells you it is an alkene and has a C=C double bond. The number tells you which carbon atom in the chain is the first to be involved in the double bond, assuming that you start counting from the end of the chain which is closest to the double bond.

Note that all non-cyclic alkenes have the empirical formula CH_2. This is the same as the empirical formula of the cycloalkanes, eg cyclohexane, but the double bond makes alkenes react very differently from cycloalkanes.

There are **cycloalkenes**, like cyclohexene, which is an important intermediate in the production of some types of nylon, and there are **dienes**, like penta-1,3-diene

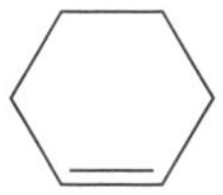

$CH_3{-}CH{=}CH{-}CH{=}CH_2$

cyclohexene *penta-1,3-diene*

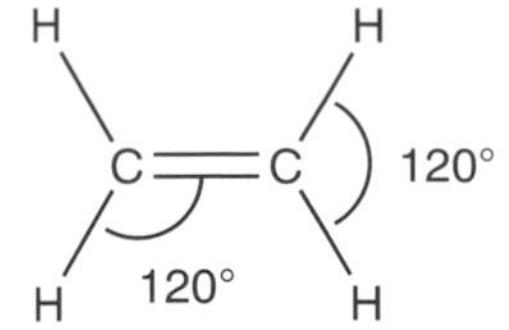

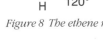

Figure 8 The ethene molecule is flat

Shapes of alkenes

Ethene is a flat molecule with the shape and approximate bond angles shown in Figure 8.

The bonds around the C=C group are arranged the same way in all alkenes. One way of explaining this is to say that this arrangement places the groups of electrons as far apart as possible. The groups of electrons constitute regions of negative charge, and repulsions between them are minimised when the angles are about 120°. (The four electrons in the double bond count as one region of negative charge.)

We can use the same general rule to predict the shapes of other molecules. For example, the tetrahedral shape of methane places the four groups of electrons around the carbon atom as far apart as possible (see **Section 12.1**) There is more about the shapes of molecules in general in **Section 3.3**.

Reactions of ethene

The four electrons in ethene's double bond give the region between the two carbon atoms a greater than usual density of negative charge. Positive ions, or molecules with a partial positive charge on one of the atoms, will be attracted to this negatively charged region. They may then go on to react by accepting a pair of electrons from the double bond. When they do this we describe them as **electrophiles**. Compare them with *nucleophiles*, which are attracted to regions of positive charge (**Section 13.1**). Most of the reactions of alkenes involve electrophiles.

Electrophilic addition reactions

Reaction with bromine

When we bubble ethene gas through bromine, the bromine becomes decolourised – this is a good general test for an alkene. The ethene reacts with bromine to form 1,2-dibromoethene

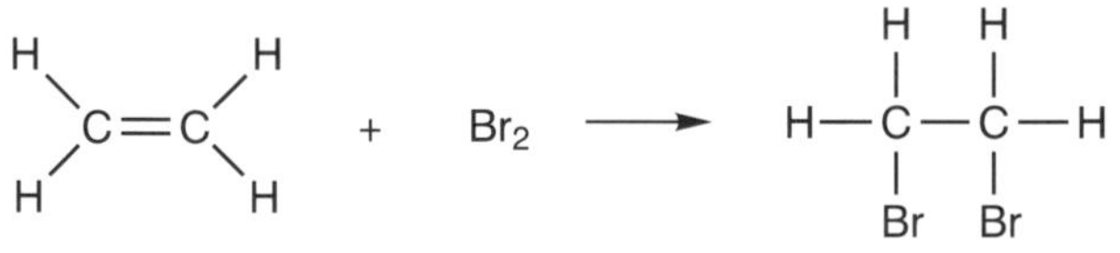

1,2-dibromoethane

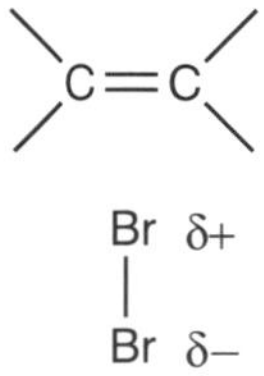

Lets look at the mechanism of this reaction. Chemists believe that the bromine molecule becomes **polarised** as it approaches the alkene.

This means that the electrons in the bromine are repelled by the alkene electrons and are pushed back along the molecule. The bromine atom nearest the alkene becomes slightly positively charged and the bromine atom furthest from the alkene becomes slightly negatively charged.

The positively charged bromine atom now behaves as an eletrophile and reacts with the alkene double bond.

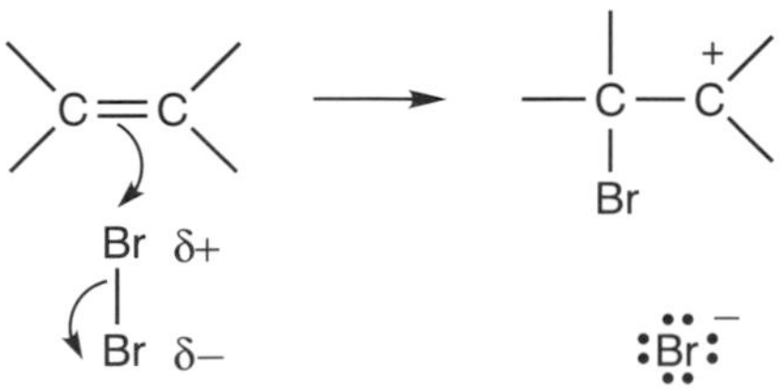

Bromine molecule being polarised by an alkene

Remember, we use curly arrows like these to represent the movement of pairs of electrons in chemical reactions.

Notice that one of the carbon atoms now has a share in only six outer electrons. It has become positively charged: it is a **carbocation**. Carbocations react very rapidly with anything which has electrons to share – such as the bromide ion. A pair of electrons moves from the Br⁻ to the positively charged carbon.

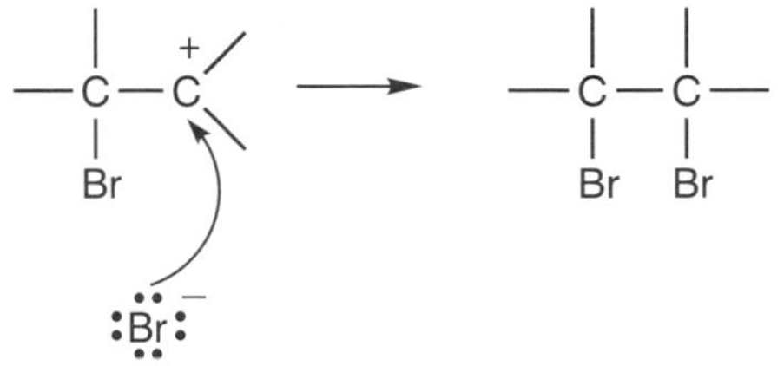

(Note that the Br⁻ could attack from *either side* of the positively charged carbon atom, though here we have shown it attacking from below.)

The overall equation for the two steps in the mechanism is

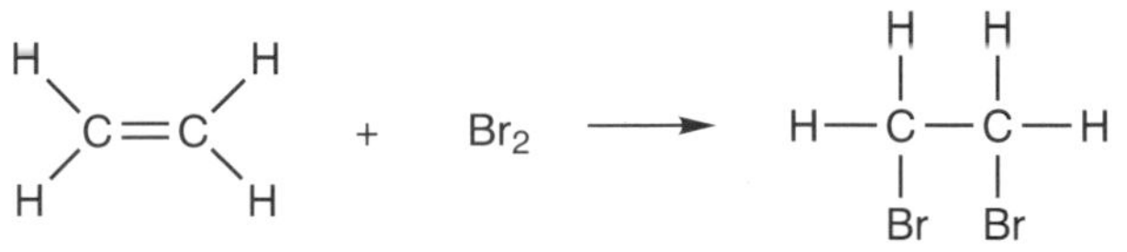

1,2-dibromoethane

This equation represents an **addition reaction**, and since the initial attack is by an electrophile, the process is called an **electrophilic addition.**

Often the test for an alkene involves *bromine water* rather than pure bromine. In this case, there is an alternative to the second stage in the reaction. Water molecules have lone pairs of electrons and can act as nucleophiles in competition with Br⁻ ions

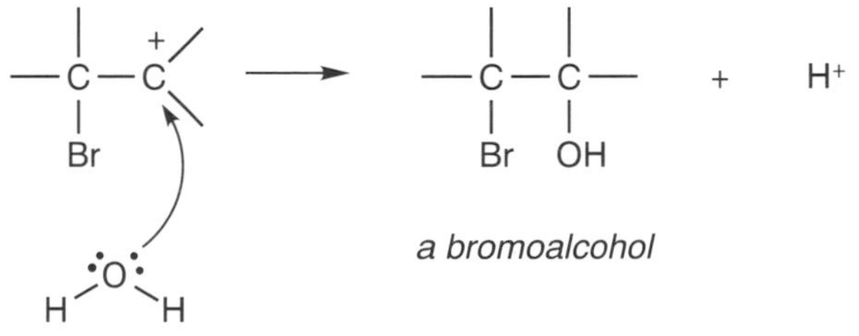

a bromoalcohol

If the bromine water is dilute, there will be many more H_2O molecules than Br⁻ ions present, and the bromoalcohol will be the main product of the reaction. This does not affect what you *see* – the bromine water is still decolourised.

Reaction with hydrogen bromide

Ethene reacts readily at room temperature with an aqueous solution of HBr. It is another example of electrophilic addition.

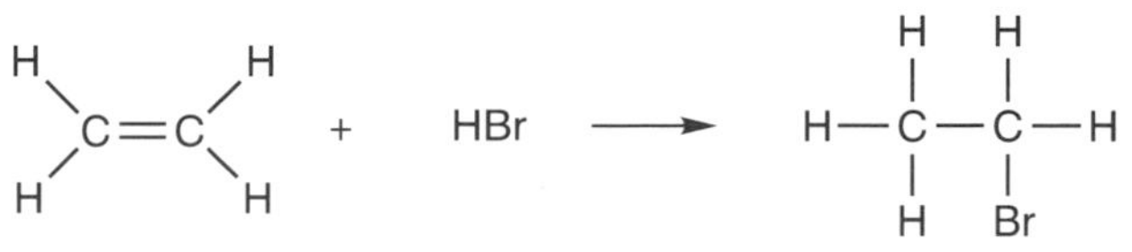

bromoethane

Alkenes also react with *gaseous* HBr but here ions are not involved and the mechanism involves a *radical* addition. (The conditions under which a reaction is carried out can be very important in determining the mechanism.)

Reaction with water

In the presence of a catalyst, ethene and water undergo electrophilic addition. The reaction is used for the industrial manufacture of ethanol. It needs a catalyst of phosphoric acid adsorbed onto silica, and the reaction is carried out at around 300 °C and 60 atm pressure.

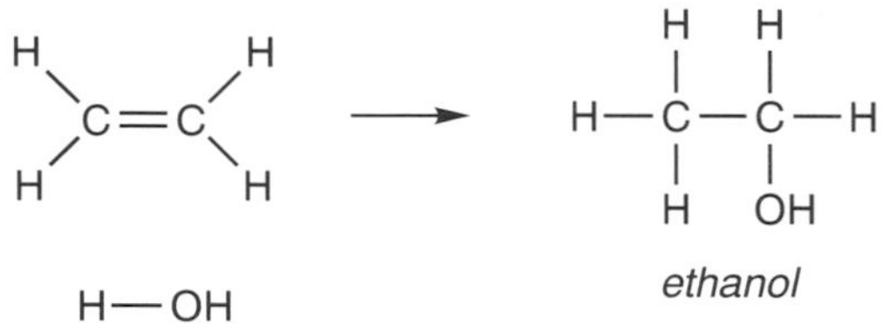

Reaction with hydrogen

This is another example of an addition reaction, but here the mechanism involves hydrogen atoms, and takes place on the surface of a catalyst. A catalyst is needed to help break the strong H—H bond and form H atoms before alkenes react with hydrogen. If a platinum catalyst is used the process takes place under normal laboratory conditions. Nickel is a cheaper but less efficient catalyst. It needs to be very finely powdered and the gases need to be at 150 °C and 5 atm pressure for **hydrogenation** to occur.

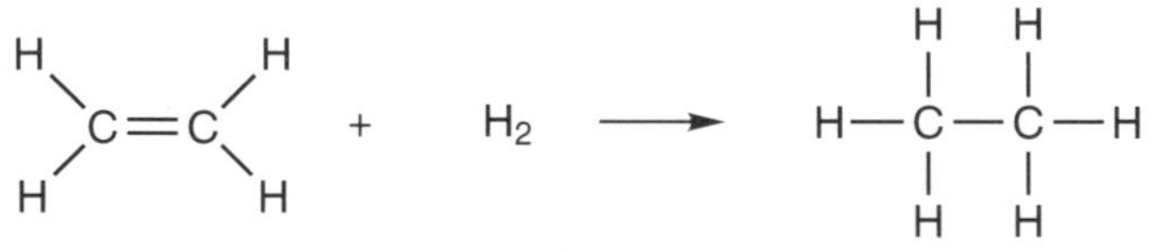

Figure 19 in **Section 10.4** suggests a possible mechanism for this reaction. Margarine manufacture is based on the addition of hydrogen to C=C bonds. Plant and animal oils are *hardened*, or made more solid and fatty like butter, by hydrogenation.

The ability of alkenes to take up other atoms or groups indicates that their bonding capacity is not fully used and they are said to be **unsaturated**. You have probably come across this term in relation to margarines *'high in polyunsaturates'*. These margarines have only had some of the C=C double bonds in the oils saturated by hydrogenation.

Addition polymerisation

Addition polymerisation is covered in detail in the **Polymer Revolution** storyline. Here we will look at the mechanism of addition polymerisation of ethene.

The reaction can be summarised as

$$CH_2{=}CH_2 \ + \ CH_2{=}CH_2 \ + \ CH_2{=}CH_2 \ \longrightarrow \ {-}CH_2{-}CH_2{-}CH_2{-}CH_2{-}CH_2{-}CH_2{-}$$

The reaction requires a catalyst. In the original ICI process, the catalyst was dioxygen, O_2, but often an organic peroxide is used. The reaction mechanism involves *radicals* (**Section 6.3**), and the catalyst provides the radicals to get the reaction started.

Let's look at the three stages of the reaction.

Initiation   This involves the creation of a radical from the catalyst. If the catalyst is an organic peroxide, the first step is for the weak O—O bond to break. If the letter R is used to represent an alkyl group, the breakdown of a peroxide can be shown as:

$$R{-}O{-}O{-}R \ \longrightarrow \ R{-}O{\bullet} \ + \ R{-}O{\bullet}$$

Propagation A radical now combines with one of the electrons from the double bond in the alkene:

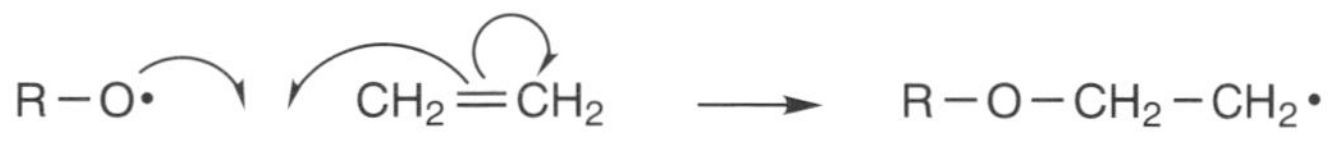

$$R-O\cdot \quad CH_2{=}CH_2 \quad \longrightarrow \quad R-O-CH_2-CH_2\cdot$$

Section 6.3 explains the significance of the half-arrows. The second electron from the broken bond remains with the other carbon atom and so the product is also a radical. Note that there is now only a single bond between the two carbon atoms.

The new radical goes on to react with more alkene:

$$R-O-CH_2-CH_2\cdot \quad + \quad CH_2{=}CH_2 \quad \longrightarrow \quad R-O-CH_2-CH_2-CH_2-CH_2\cdot$$

... and so the chain grows. The propagation stage is very rapid: chains containing up to 10 000 monomer units can be produced per second.

As the chains grow they flail around, and sometimes a growing chain may attack itself by *back-biting*. Figure 9 illustrates what happens.

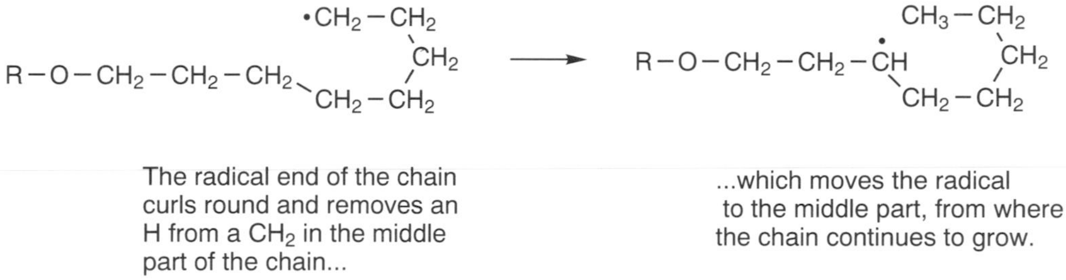

$$R-O-CH_2-CH_2-CH_2 \overset{\displaystyle \cdot CH_2-CH_2}{\underset{\displaystyle CH_2-CH_2}{\diagdown \, \underset{}{CH_2} \diagup}} \quad \longrightarrow \quad R-O-CH_2-CH_2-\overset{\displaystyle \cdot}{CH} \overset{\displaystyle CH_3-CH_2}{\underset{\displaystyle CH_2-CH_2}{\diagdown \, \underset{}{CH_2} \diagup}}$$

The radical end of the chain curls round and removes an H from a CH$_2$ in the middle part of the chain...

...which moves the radical to the middle part, from where the chain continues to grow.

The result of 'back-biting' is to create a new growing point in the middle part of the chain. A *branched* chain will now grow. This is what happens in the process used to make low-density polythene: on average there is a branch every 50 atoms.

Figure 9 'Back-biting'

Termination The reaction ends when the radicals are used up. One way this can happen is by radicals simply joining together.

$$R-O-(CH_2)_n{-}CH_2\cdot \quad + \quad R-O-(CH_2)_m{-}CH_2\cdot$$

$$\downarrow$$

$$R-O-(CH_2)_n{-}CH_2-CH_2-(CH_2)_m{-}O-R$$

PROBLEMS FOR 12.2

1 Name the alkenes with the following structures:

a $CH_3-CH{=}CH_2$

b $CH_3-CH{=}CH-CH_2-CH_2-CH_2-CH_3$

c
$\overset{\displaystyle CH_3}{\underset{\displaystyle CH_3}{\diagdown \, C{=}CH_2 \diagup}}$

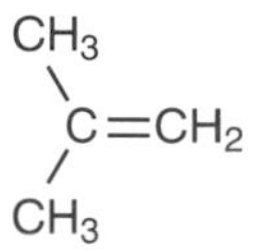

d $CH_2{=}CH-CH_2-CH{=}CH-CH_2-CH_3$

e (cyclopentene structure)

2 Draw structures for the following alkenes:
  a   pent-2-ene   c   2,3-dimethylpent-2-ene
  b   hex-3-ene   d   cyclopenta-1,3-diene

3 a Write overall equations for the reaction of
  i but-2-ene with hydrogen bromide
  ii cyclohexene with steam
  iii hexa-1,4-diene with excess hydrogen.
b In each case, suggest appropriate conditions for carrying out the reaction.

4 This question is about the substance *citronellol*. Citronellol belongs to a class of compounds called *terpenes*. Many closely related terpenes occur in nature where they are responsible for the fragrances of plants. We often convert them into more volatile ester derivatives for use in perfumes. Citronellol is found in geraniums and citronella grass. Its structure is shown below.

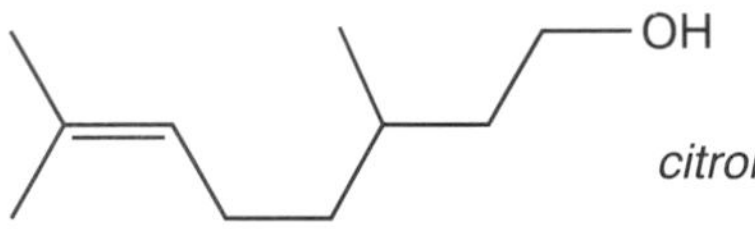

citronellol

We can think of all terpenes as being built up from the compound which is shown as **A** below.

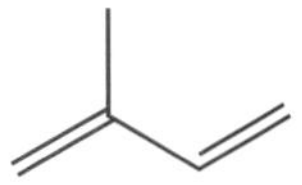

compound A

The naturally occurring compounds usually contain one or more C=C double bonds, —OH groups or C=O groups. Different plants produce different arrangements of these groups within the terpene molecules, so a large number of terpenes are found.
a Name compound **A**.
b Write down the molecular formula of compound **A**.
c Write down the molecular formula of citronellol.
d Name the functional groups present in citronellol.
e How many units of compound **A** combine to produce citronellol?
f Using molecular formulae, write equations to represent compound **A** reacting with
  i H_2
  ii HBr.

g What would you expect to see if some citronellol was shaken with a solution of bromine dissolved in an organic solvent such as hexane? Write an equation for this process using molecular formulae.

5 Copy and complete the table below to show the structure of the repeating units and the major uses of the polymers which are made from the monomers listed. The first example has been done for you.
  You may need to refer to reference books to find some of the information.

Monomer	Repeating unit	Major uses
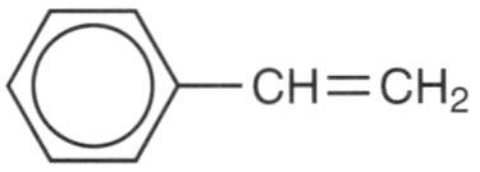 $-CH=CH_2$ phenylethene (*styrene*)	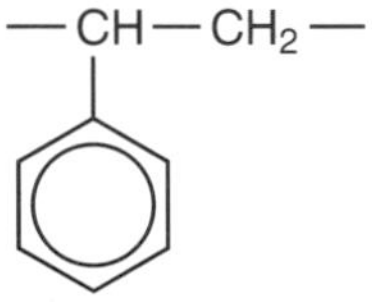	packaging cups, cartons household items
$CH_2=CHCl$ chloroethene (*vinyl chloride*)		
$CH_2=CHCN$ propenenitrile (*acrylonitrile*)		
$CF_2=CF_2$ tetrafluoroethene		
methyl 2-methylpropenoate (*methyl methacrylate*)		

12.3 *Arenes*

What's special about benzene?

Benzene is a colourless liquid – nothing special there. It has a molecular formula C_6H_6 and so it must be very unsaturated. The puzzle is that benzene does not behave like a normal unsaturated compound. It is much less reactive and has its own characteristic properties.

The reason for this is to do with the rather special cyclic structure of benzene. Look at Figure 10 which shows the structure of the benzene ring. The benzene ring is a flat hexagon (Figure 11). All the bond angles are 120° and all the carbon-carbon bonds are the same length. This length is less than for a carbon-carbon single bond, but greater than for a carbon-carbon double bond (see Table 4).

Figure 10 The structure of the benzene ring

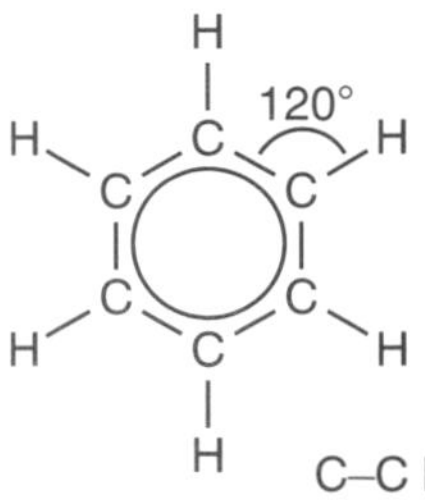

C–C bond length
0.139 nm

Try drawing a dot-cross formula to represent the bonding in a benzene ring. Each carbon atom has four outer electrons which can be used to form bonds. Three of these electrons are used to form single bonds with the two carbons next door in the ring and with a hydrogen atom. This leaves one electron on each carbon atom.

Instead of overlapping in pairs to form three separate double bonds, these remaining electrons are spread out evenly and are shared by all six carbon atoms in the ring. The spreading out of electrons in this way is called **electron delocalisation**. You can see one way of drawing a dot-cross formula for benzene in Figure 12.

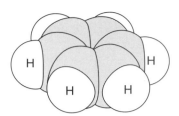

Figure 11 A diagram of a space-filling model of a benzene molecule, showing its planar shape

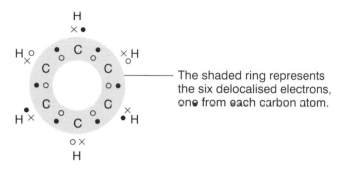

The shaded ring represents the six delocalised electrons, one from each carbon atom.

Figure 12 Dot-cross formula for benzene. The circles represent electrons from C atoms; the crosses represent electrons from H atoms

Additional evidence for the delocalised structure comes from **electron density maps**, which can be drawn up from X-ray diffraction studies (Figure 13). Figure 13 shows a uniform electron density in all the carbon-carbon bonds.

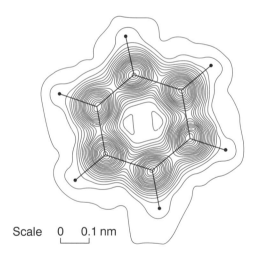

Scale 0   0.1 nm

Bond		Bond length/nm
alkane	C—C	0.154
alkene	C=C	0.134
benzene ring	C⚌C	0.139

Table 4 Some carbon-carbon bonds lengths

Figure 13 An electron density map for benzene at −3°C. The lines are like contour lines on a map: they show parts of the molecule with equal electron density

You can think of the delocalised structure of benzene as halfway between the two extreme structures:

    and

These are sometimes called Kekulé structures because they were first proposed by August Kekulé in 1865. Neither of these forms actually exists, and the delocalised arrangement is often represented by drawing a circle inside the ring.

Stability of benzene

The cyclic delocalisation of electrons in benzene makes the molecule more stable than you would expect if it had a Kekulé-type structure, with three double bonds and three single bonds. Electrons repel one another, so a system in which they are delocalised and as far apart from one another as possible will involve minimum repulsion and will be stabilised.

The stability of the benzene ring has an important effect on the way benzene reacts. Benzene tends to undergo reactions in which the stable ring is preserved. Reactions which disrupt the delocalised electron system are less favourable and need higher temperatures and more vigorous conditions. You will find out more about the way benzene reacts in **Section 12.4**.

The results of thermochemical experiments give us a way to estimate just how much more stable a benzene ring is than a hypothetical Kekulé-type structure. One way to do this is to measure the enthalpy change when benzene reacts with hydrogen to form cyclohexane:

$$\text{benzene} \quad + \quad 3H_2 \quad \longrightarrow \quad \text{cyclohexane} \qquad \Delta H^{\ominus} = -208 \text{ kJ mol}^{-1}$$

You can compare this with the value you would expect if benzene had the hypothetical Kekulé structure. This value cannot be measured of course, but you can make a reasonable estimate. The enthalpy change when *cyclohexene* reacts with hydrogen can be measured.

$$\text{cyclohexene} \quad + \quad H_2 \quad \longrightarrow \qquad \Delta H^{\ominus} = -120 \text{ kJ mol}^{-1}$$

If you assume that the three double bonds in the Kekulé structure behave independently of each other, the enthalpy change of hydrogenation would be three times the value for cyclohexene.

$$\text{Kekulé's benzene} \quad + \quad 3H_2 \quad \longrightarrow \qquad \Delta H^{\ominus} = 3 \times (-120 \text{ kJ mol}^{-1})$$
$$= -360 \text{ kJ mol}^{-1}$$

The enthalpy changes for these reactions are summarised in Figure 14.

Figure 14 Enthalpy changes for the hydrogenation of benzene and the hypothetical Kekulé structure

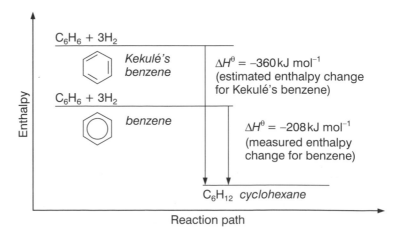

Thus, when one mole of benzene molecules are hydrogenated $(360\,kJ\,mol^{-1} - 208\,kJ\,mol^{-1}) = 152\,kJ\,mol^{-1}$ less energy is given out than would be expected if they had the Kekulé structure. So, benzene must be more stable than the Kekulé structure by $152\,kJ\,mol^{-1}$.

What are arenes?

Hydrocarbons like benzene which contain rings stabilised by electron delocalisation are called **arenes**. The ending *-ene* tells you they are unsaturated, like alkenes. The *ar-* prefix comes from *aromatic*, which means *sweet-smelling*. Arenes are sometimes called **aromatic hydrocarbons**.

The term *aromatic* was originally used to describe the characteristic fragrance (*aroma*) of some naturally occurring oils which contained benzene rings. It is now used in a much wider sense since not all arenes are sweet-smelling. Benzene itself has a rather strong unpleasant smell.

There are many arenes. Below are some examples in which hydrogen atoms on the benzene ring have been replaced by alkyl groups. Note how they are named.

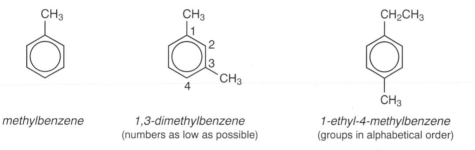

methylbenzene   1,3-dimethylbenzene   1-ethyl-4-methylbenzene
(numbers as low as possible)   (groups in alphabetical order)

Benzene rings can be 'joined together' to give **fused ring systems**, such as **naphthalene** and **anthracene**. Notice that where the rings join, they share a pair of carbon atoms. So naphthalene, with two rings, has 10 carbon atoms rather than 12.

naphthalene, $C_{10}H_8$ anthracene, $C_{14}H_{10}$

Like benzene, these molecules contain delocalised electrons, but the delocalisation extends over all the rings. It is harder to represent the delocalisation in fused rings using circles.
For example, the structure

for naphthalene would suggest that the molecule contains two separate delocalised systems with six delocalised electrons in each. This is not the case, so we draw the rings with double bonds – but remember this is just a convenient representation of a complex delocalised structure.

Compounds derived from arenes

Hydrogen atoms on the benzene ring can be replaced by different functional groups. For example,

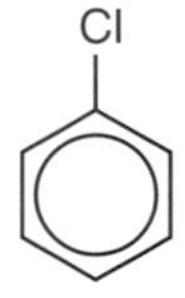

chlorobenzene    *nitrobenzene* 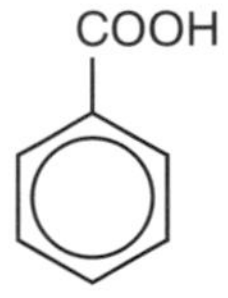   *benzoic acid*    *benzaldehyde*

The group C_6H_5— derived from the benzene ring is called a **phenyl** group. Two important aromatic compounds whose names are based on this group are **phenol** (C_6H_5OH) and **phenylamine** ($C_6H_5NH_2$).

phenol   *phenylamine*

PROBLEMS FOR 12.3

1 Name the following arenes.

a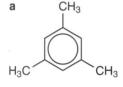

b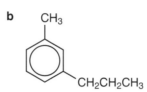

2 Draw the structural formulae of the following compounds:

  a bromobenzene

  b methyl benzoate

  c phenylethene.

3 a Kekulé first proposed the cyclic structure for benzene in 1865. His ring contained three double bonds and three single bonds. Draw out the structure of this ring and mark in the carbon–carbon bond lengths you would expect (see Table 4 on page 233). What effect would these bond lengths have on the overall shape of the ring?

  b If Kekulé's structure was correct, there should be two isomers of 1,2-dimethylbenzene. Draw out the structure of these isomers.

  c In fact, only one form of 1,2-dimethylbenzene is known. Explain why this is so.

4 The enthalpy change of formation of gaseous benzene is the enthalpy change when a mole of gaseous benzene molecules are formed from the elements carbon and hydrogen:

$$6C(s) + 3H_2(g) \rightarrow C_6H_6(g)$$

  a Assume that the benzene ring has a Kekulé structure with alternating single and double bonds. Work out a theoretical value for the enthalpy change of formation of gaseous benzene by the following stages:

    i Calculate the enthalpy change involved in producing six moles of gaseous carbon atoms from C(s), and six moles of gaseous hydrogen atoms from $H_2(g)$.

$$C(s) \rightarrow C(g); \qquad \Delta H = +715\,kJ\,mol^{-1}$$
$$\tfrac{1}{2}H_2(g) \rightarrow H(g); \qquad \Delta H = +218\,kJ\,mol^{-1}$$

    ii Use bond enthalpies to calculate the energy released when 1 mole of gaseous 'Kekulé benzene' molecules are formed from gaseous atoms of carbon and hydrogen:

$$6C(g) + 6H(g) \rightarrow C_6H_6(g)$$

Bond enthalpy of C=C (average)
$$= +612\,kJ\,mol^{-1}$$
Bond enthalpy of C—C (average)
$$= +348\,kJ\,mol^{-1}$$
Bond enthalpy of C—H (average)
$$= +413\,kJ\,mol^{-1}$$

    iii Use your answers in **i** and **ii** to estimate the total enthalpy change when a mole of gaseous 'Kekulé benzene' molecules are formed from the elements carbon and hydrogen.

  b An experimental value for the enthalpy change of formation of gaseous benzene can be found from enthalpy changes of combustion. It has a value of $+82\,kJ\,mol^{-1}$. Compare this with the value you obtained for 'Kekulé benzene' in **a iii**. What do your results suggest about the relative stabilities of 'real benzene' and 'Kekulé benzene'?

12.4 *Reactions of arenes*

The six electrons in the delocalised system in benzene do not belong to any particular carbon atom and are free to move around the ring. They are much more loosely held than electrons in normal bonds. They spread out in a cloud which extends above and below the plane of the benzene ring (Figure 15).

regions of high electron density above and below the benzene ring

Figure 15 The regions of electron density above and below the benzene ring

These regions of high electron density tend to attract positive ions, or atoms with a partial positive charge within molecules. So benzene, like alkenes, reacts with **electrophiles**. The reactions are much slower with benzene, though, because the first step in the reaction mechanism disrupts the delocalised electron system, which requires a substantial input of energy.

Alkenes react with electrophiles in **addition reactions** (see **Section 12.2**). The product is a saturated molecule. For example,

$$CH_2{=}CH_2 \quad + \quad Br_2 \quad \longrightarrow \quad CH_2Br{-}CH_2Br$$

Benzene undergoes **substitution** rather than addition reactions with electrophiles. By reacting in this way, the stable benzene ring system is kept intact.

Bromination of benzene

Benzene reacts with bromine in the presence of a catalyst, such as iron filings or iron(III) bromide. The bromine is decolourised and fumes of HBr are given off. The reaction which takes place is a substitution reaction:

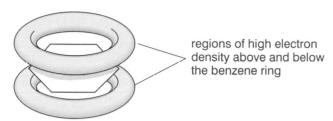

The first step involves reaction of the benzene ring with the electrophile Br^+, and so the process is called **electrophilic substitution**.

The bromine molecule becomes polarised as it approaches the benzene ring:

Br δ+
|
Br δ−

The positively charged end of the bromine molecule is now electrophilic. Even so, these polarised bromine molecules react only very slowly with the stable benzene ring. Iron filings help to speed up this reaction by making the bromine molecules more polarised. First, the iron reacts with bromine to form iron(III) bromide ($FeBr_3$):

$$2Fe + 3Br_2 \rightarrow 2FeBr_3$$

It is thought that the $FeBr_3$ helps to polarise the bromine molecule by accepting a lone pair from one of the bromine atoms. The bromine molecule becomes so polarised that it splits into Br^+ and $FeBr_4^-$:

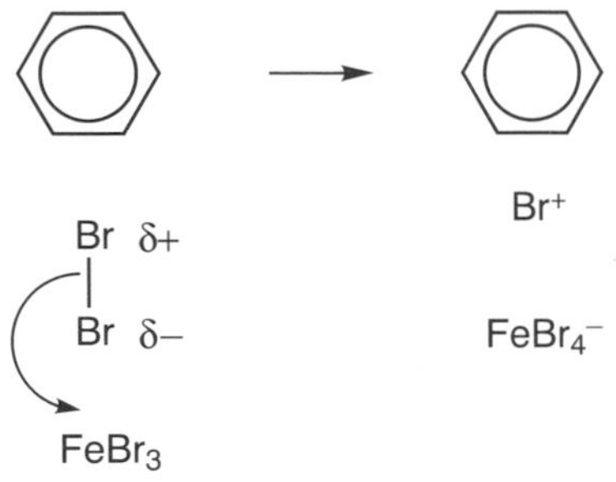

Br^+ then becomes bound to a ring carbon atom and H^+ is lost from the ring. This H^+ reacts with $FeBr_4^-$ to produce HBr and regenerate the $FeBr_3$ catalyst:

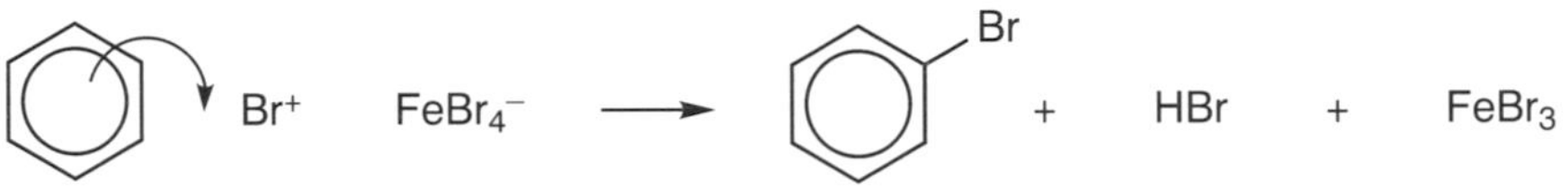

It is easy to see why addition reactions are much more difficult. If a molecule of benzene underwent addition with a molecule of bromine, the stable ring of delocalised electrons would be broken.

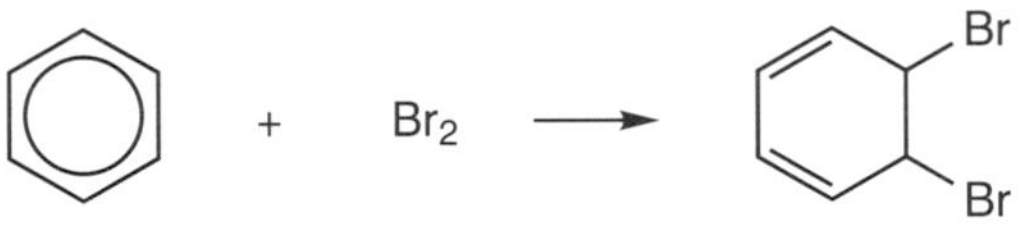

Other electrophilic substitution reactions of benzene

Benzene undergoes many substitution reactions. In every case, there has to be an electrophile which is attracted to the delocalised electron system in the benzene ring.

Nitration

Benzene reacts with a mixture of concentrated nitric acid and concentrated sulphuric acid (called a **nitrating mixture**). If the temperature is kept below 55 °C, the product is nitrobenzene:

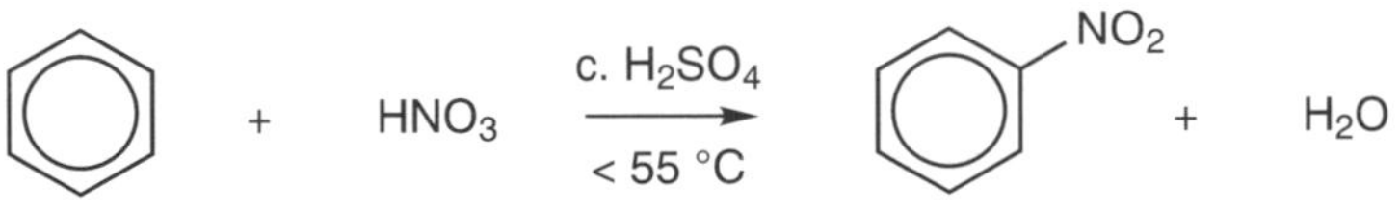

nitrobenzene

At higher temperatures, further substitution of the ring takes place to give the di– and tri-substituted compounds:

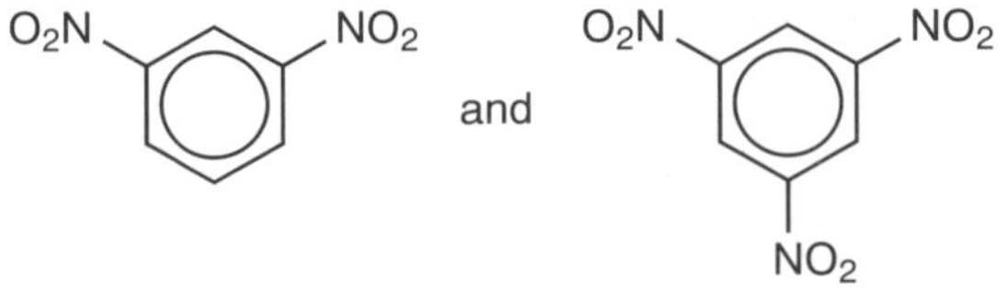

The electrophile which reacts with the benzene ring is the NO_2^+ cation. There is good evidence that this is formed in the nitrating mixture by the reaction of sulphuric acid with nitric acid:

$$HNO_3 \quad + \quad 2H_2SO_4 \longrightarrow \underbrace{NO_2^+ \quad + \quad 2HSO_4^- \quad + \quad H_3O^+}$$

nitrating mixture

Sulphonation

When benzene and concentrated sulphuric acid are heated together under reflux for several hours, benzenesulphonic acid is formed:

benzenesulphonic acid

The electrophile in this case is thought to be SO_3, which is present in the concentrated sulphuric acid. SO_3 carries a large partial positive charge on the sulphur atom and it is this atom which becomes bound to the benzene ring:

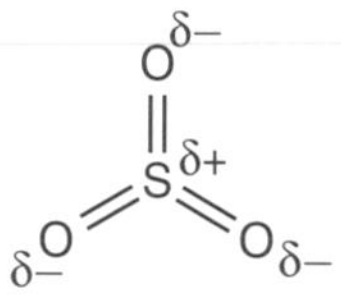

The full structural formula of benzenesulphonic acid is

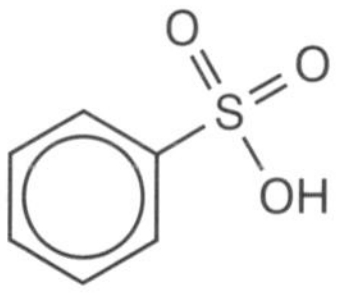

It is a strong acid and forms salts in alkaline solution:

sodium benzenesulphonate

Most solid detergents contain salts of this kind, with a long alkyl group attached to the benzene ring. The hydrocarbon part of the molecule mixes with fats, and the ionic part mixes with water.

Chlorination

A chlorine atom may be substituted onto a benzene ring in much the same way as a bromine atom. An aluminium chloride catalyst is often used.

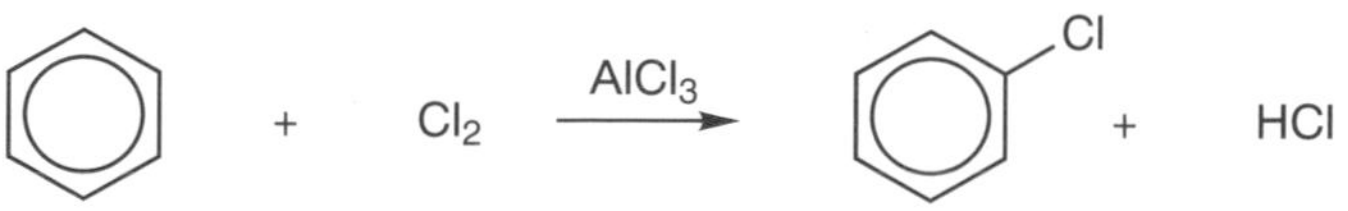

The mechanism is similar to the bromination reaction. The aluminium chloride helps to polarise the chlorine molecule. This produces the electrophile Cl^+, which reacts with the benzene ring to form chlorobenzene.

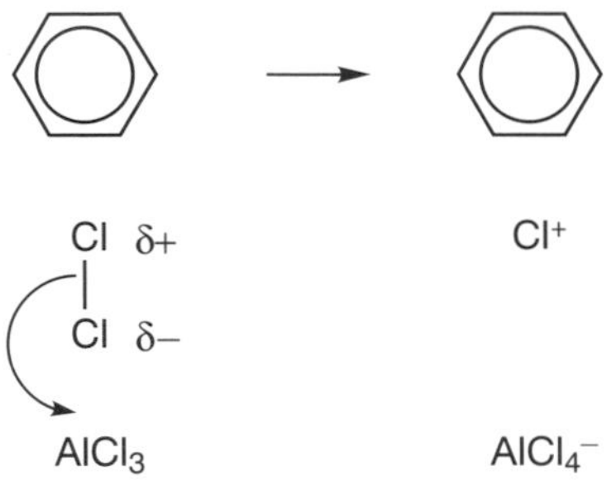

Then:

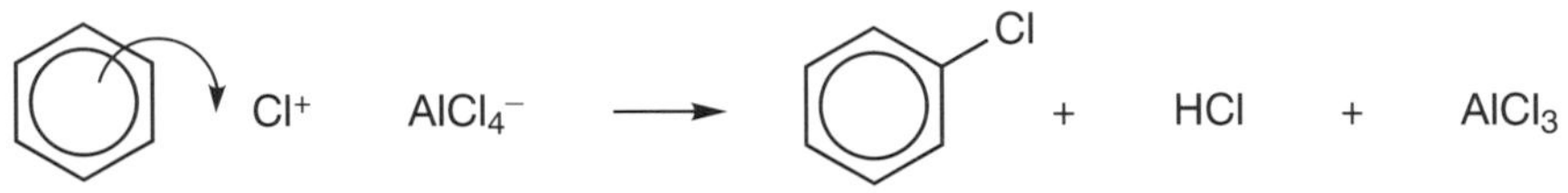

Aluminium chloride reacts violently with water, so the reaction must be carried out under anhydrous conditions.

Friedel-Crafts reactions

Aluminium chloride can also be used as a catalyst to help polarise *halogen-containing organic molecules* and cause them to substitute in a benzene ring. This type of reaction is called a **Friedel-Crafts reaction**, after its discoverers.

If benzene is warmed with chloromethane and anhydrous aluminium chloride, a substitution reaction occurs and methylbenzene is formed:

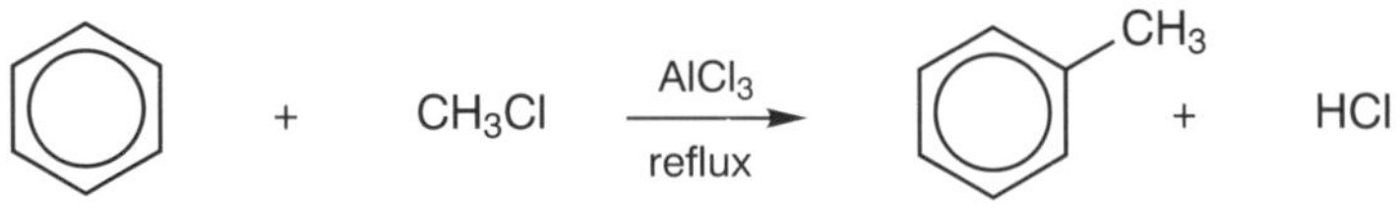

As in the reaction with halogens, the aluminium chloride helps to polarise the chloromethane molecule:

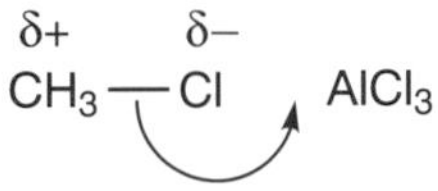

The positively charged carbon atom then reacts with the benzene ring to form methylbenzene. Because an alkyl group has been introduced into the ring, the reaction is called **alkylation**.

A similar reaction takes place when benzene is treated with an acyl chloride (or an acid anhydride) and aluminium chloride. Here the reaction is an **acylation**:

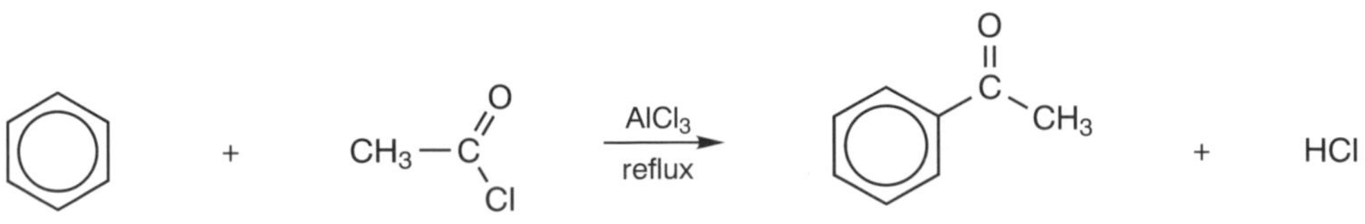

acyl chloride

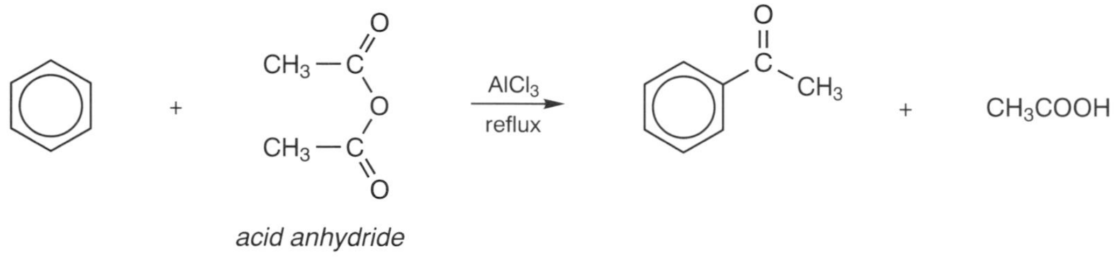

Friedel-Crafts reactions are particularly useful to synthetic chemists because they provide a way of adding carbon atoms to the benzene ring and building up side-chains.

Summary

Benzene is an important starting material for the synthesis of many useful compounds like dyes, pharmaceuticals and perfumes. Electrophilic substitution reactions provide ways of introducing different functional groups into the ring. These groups may then be modified further to build up more complicated molecules. The important electrophilic substitution reactions are summarised in Figure 16. They are *general* reactions and work on other arenes as well as benzene itself.

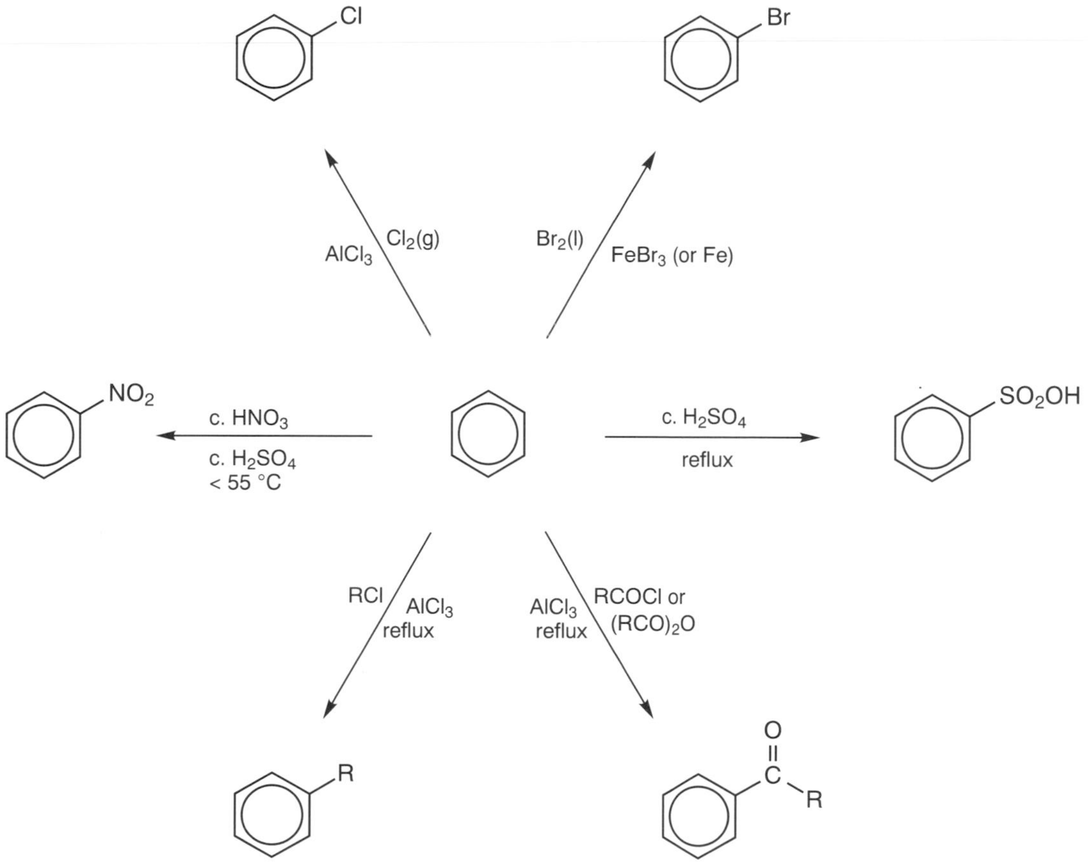

Figure 16 Some important electrophilic substitution reactions of benzene

PROBLEMS FOR 12.4

1 Write down the structure of the main product of each of the following reactions.

a

Cl₂ / AlCl₃

b

c. HNO₃
c. H₂SO₄
< 55 °C

c

CH₃CHClCH₃
AlCl₃
reflux

d

CH₃CH₂COCl
AlCl₃
reflux

2 Predict the major products of each of the following reactions. In each case, 1 mole of the compound reacts with 1 mole of Br_2 molecules.

a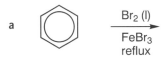

Br₂ (l)
FeBr₃
reflux

b

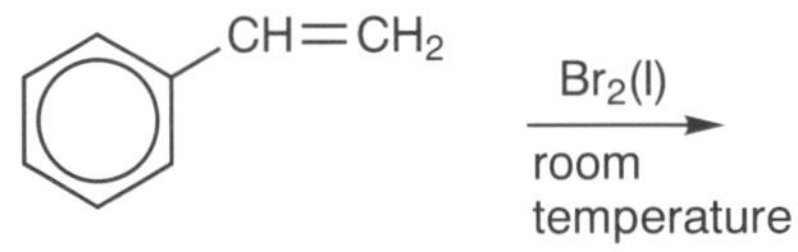

CH=CH₂

Br₂(l)
room temperature

3 Napthalene undergoes electrophilic substitution reactions in the same way as benzene. Two monosubstituted products can be obtained. In one, substitution is at the 1-position; in the other it is at the 2-position.

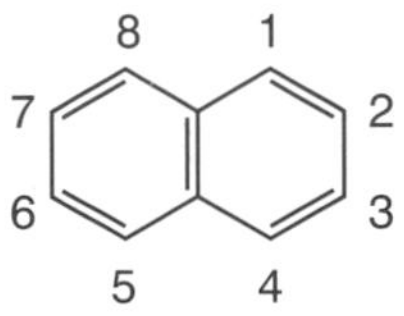

```
   8   1
 7       2
 6       3
   5   4
```

Write a *balanced equation* to show the reaction which takes place when naphthalene is heated with:

a   a nitrating mixture at 50 °C to give a 1-substituted product

b   concentrated sulphuric acid at 160 °C to give a 2-substituted product.

4 Iodine is too unreactive to substitute benzene even in the presence of a catalyst. However, iodobenzene can be made by treating benzene with iodine(I) chloride (ICl).

a   Write a balanced equation for this reaction.
b   Explain why a catalyst is not needed.
c   Why is chlorobenzene not formed in this reaction?

5 Benzene undergoes an addition reaction with hydrogen in the same way as alkenes do, but considerably higher temperatures are needed. A special nickel catalyst is used which is very active.

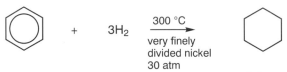

300 °C
very finely divided nickel
30 atm

cyclohexane

a   Explain why special conditions are needed for the hydrogenation of benzene.
b   The first steps in the manufacture of nylon-6,6 involves the catalytic hydrogenation of phenol to produce cyclohexanol.

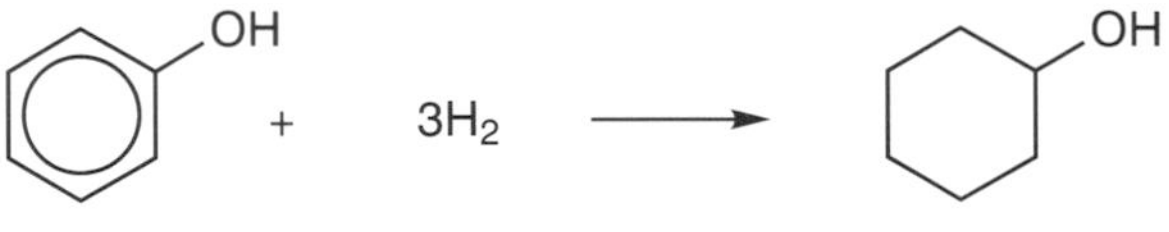

OH   OH

+   3H₂   ⟶

phenol   *cyclohexanol*

A nylon plant hydrogenates 800 tonnes of phenol a day. Calculate

i   the mass of hydrogen used by the plant for this purpose each day

ii the volume of this hydrogen measured at room temperature and pressure. (A mole of molecules of any gas has a volume of 24 dm³ at room temperature and pressure.)

ORGANIC CHEMISTRY: MODIFIERS

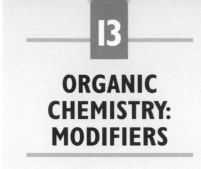

Chloroethane
Halogenoalkane

Figure 1 Chloroethane is a typical halogenoalkane

Organic halogen compounds

Organic halogen compounds have one or more halogen atoms (F, Cl, Br or I) attached to a hydrocarbon chain. They do not occur much in nature, but they are useful for all sorts of human purposes, so chemists make and use them a lot.

As with all functional groups, the halogen atom modifies the properties of the relatively unreactive hydrocarbon chain. The simplest examples are the **halogenoalkanes**, with the halogen atom attached to an alkane chain (Figure 13.1).

Naming halogenoalkanes

The halogenoalkanes are examples of homologous series: the chloroalkanes, the bromoalkanes and so on. They are named after the parent alkanes, using the same basic rules as for naming alcohols – except in this case the halogen atom is added as a *prefix* to the name of the parent alkane. Thus, the molecule $CH_3CH_2CH_2Cl$ is called *1-chloropropane* while $CH_3CHClCH_2Cl$ is called *1,2-dichloropropane*. The more complicated molecule, $CH_2ClCH_2CHBrCH_3$, is called *3-bromo-1-chlorobutane*.

Notice that the prefixes bromo- and chloro- are listed in alphabetical order. Notice too that the numbers used to show the positions of the bromine and chlorine atoms are the lowest ones possible: 3 and 1, rather than 2 and 4.

As you would expect, the properties of halogenoalkanes depend on which halogen atoms they contain.

Physical properties.

The carbon-halogen bond is polar, but not polar enough to make a big difference to the physical properties of the compounds. For example, all halogenoalkanes are immiscible with water. Their boiling points depend on the size and number of halogen atoms present: the bigger the halogen atom and the more halogen atoms there are, the higher the boiling point, as you can see from Table 1.

The influence of halogen atoms on the boiling point is important when it comes to designing halogen compounds for particular purposes. If you want a compound with a high boiling point, you have to include a larger halogen atom like Cl or Br rather than a smaller one like F: but it is these larger atoms that can cause the greatest environmental damage. These considerations are very important in the design of replacements for CFCs.

Compound	State at 298 K	Boiling point/K
CH_3F	g	195
CH_3Cl	g	249
CH_3Br	g	277
CH_3I	l	316
CH_2Cl_2	l	313
$CHCl_3$	l	335
CCl_4	l	350
C_6H_5Cl	l	405

Table 1 Boiling points of some organic halogen compounds

What kind of reactions do you get with halogenoalkanes?

Reactions of halogenoalkanes involve breaking the C—Hal bond (Hal stands for any halogen atom). The bond can break homolytically or heterolytically (see **Section 6.3**).

Homolytic fission

Homolytic fission forms radicals. One way this can occur is when radiation of the right frequency (visible or ultra-violet) is absorbed by the halogenoalkane. For example, with chloromethane:

$$CH_3\text{—}Cl + h\nu \rightarrow CH_3{}^\bullet + Cl^\bullet$$

This kind of reaction occurs when halogenoalkanes reach the stratosphere, where they are exposed to intense ultra-violet radiation. This is what forms the chlorine radicals which cause so much trouble for ozone in the stratosphere.

Heterolytic fission

Heterolytic fission is more common under normal laboratory conditions when reactions of halogenoalkanes tend to be carried out in a polar solvent such as ethanol, or ethanol and water. The C—Hal bond is already polar, and in the right situation it can break, forming a negative halide ion and a positive **carbocation**. For example, with 2-chloro-2-methylpropane:

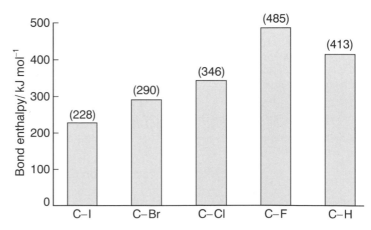

2-chloro-2-methyl propane   carbocation   chloride ion

Sometimes, ions are not formed by simple bond fission in this way. Instead, heterolytic bond fission is brought about by another, negatively charged substance reacting with the positively charged carbon atom, causing a substitution reaction – more about this later in this section.

Importance of reaction conditions

For many molecules, the conditions under which the reaction is carried out can determine how a bond breaks. For example, when bromomethane is dissolved in a polar solvent, such as a mixture of ethanol and water, the C—Br bond breaks heterolytically to form ions. However, when it reacts in a non-polar solvent, such as hexane, or in the gas phase at high temperature or in the presence of light, the C—Br bond breaks homolytically.

Different halogens, different reactivity

Whether homolytic or heterolytic fission occurs, all reactions of halogenoalkanes involve breaking the C—Hal bond. The stronger the bond is, the more difficult it is to break. Figure 2 gives the bond enthalpies of the four different types of C—Hal bond, and also the C—H bond for comparison. Remember, the higher the bond enthalpy, the stronger the bond.

Figure 2 Some bond enthalpies

The great strength of the C—F bond makes it very difficult to break, so fluoro compounds are very unreactive. As you go down the halogen group, the C—Hal bond gets weaker and weaker, so the compounds get more and more reactive. Chloro compounds are fairly unreactive, and once they have been released into the troposphere they stay there quite a long time. This means that compounds such as CFCs can stay around long enough to get into the stratosphere and wreak havoc on the ozone layer.

Bromo and iodo compounds are fairly reactive, which makes them useful as intermediates in synthesising other organic compounds.

Substitution reactions of halogenoalkanes

Substitution reactions are typical of halogenoalkanes. A substitution reaction takes place between halogenoalkanes and hydroxide ions, in which the halogenoalkane is hydrolysed to form an alcohol. With bromobutane, for example:

$$CH_3CH_2CH_2CH_2Br + OH \rightarrow CH_3CH_2CH_2CH_2OH + Br^-$$

The C—Br bond is polar, like all C—Hal bonds.

The oxygen atom of the hydroxide ion is negatively charged.

The positive charge on the carbon atom attracts the negatively-charged oxygen of the hydroxide ion. A lone pair of electrons on the O atom forms a bond with the C atom as the C—Br bond breaks.

Notice that this reaction involves heterolytic fission – ions, rather than radicals, are formed. A free carbocation is not formed in this case, because the HO⁻ ion attacks at the same time as the C—Br bond breaks – it's a smooth, continuous process.

Note too that we have used **curly arrows** again to show the movement of electrons during the reaction. In this case a full-headed arrow is used to show the movement of *a pair* of electrons. (Compare this with the half-arrow used to show the movement of single electrons in radical reactions.)

Halogenoalkanes give substitution reactions with many different reagents, as well as hydroxide ions. What is needed is a group carrying a pair of electrons to start forming a bond to the carbon atom.

Attacking groups like these, which can donate a pair of electrons to a positively charged carbon atom to form a new covalent bond, are called **nucleophiles**. They can be either anions like ⁻OH or molecules with a lone pair of electrons available for bonding, for example, H_2O. Table 2 shows some common nucleophiles.

The carbon atom attacked by the nucleophile may be part of a carbocation and carry a *full* positive charge, or it may be part of a neutral molecule and carry a *partial* positive charge as a result of bond polarisation.

If we write X⁻ as a general symbol for any anionic nucleophile, the **nucleophilic substitution** process can be described by:

Name and formula	Structure, showing lone pairs
hydroxide ion, HO⁻	$H-\ddot{O}:^-$
cyanide ion, CN⁻	$^-\ddot{C}\equiv N$
ethanoate ion, CH_3COO^-	$CH_3-C-\ddot{O}:^-$ with =O
ethoxide ion, $C_2H_5O^-$	$CH_3CH_2-\ddot{O}:^-$
water molecule, H_2O	$H-\ddot{O}-H$
ammonia molecule, NH_3	$H-\ddot{N}(H)-H$

Table 2 Some common nucleophiles

In general, when any nucleophile X⁻ reacts with a general halogenoalkane RHal, the reaction that occurs is

$$RHal + X^- \rightarrow RX + Hal^-$$

Water as a nucleophile

Nucleophiles don't need to have a full negative charge: it is possible for a neutral molecule to act as a nucleophile, provided it has a lone pair of electrons. For example, the water molecule has a lone pair on the oxygen atom, so water can act as a nucleophile and attack a halogenoalkane molecule such as 1-bromobutane. The substitution reaction goes in two stages. First H_2O attacks the halogenoalkane:

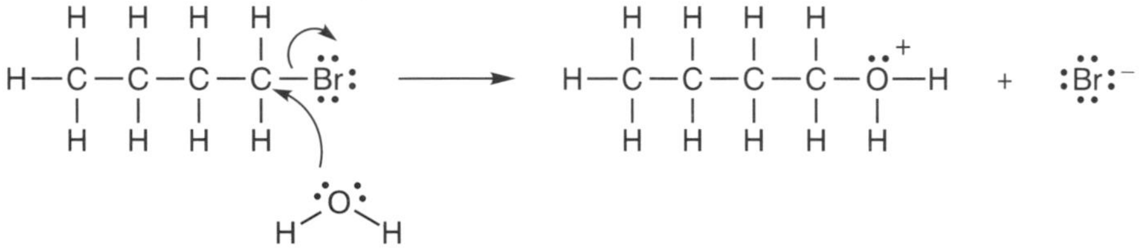

Then the resulting ion loses H^+ to form an alcohol:

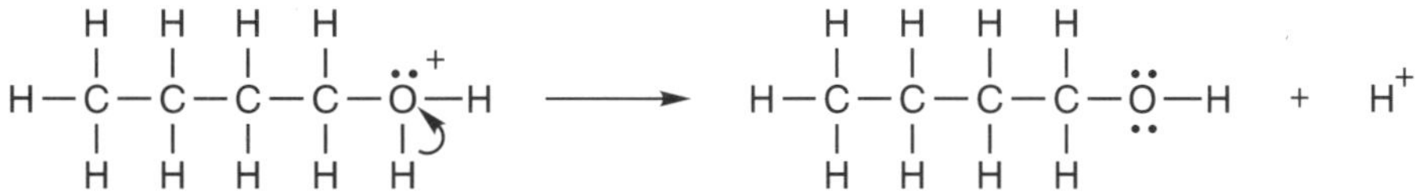

The overall equation for the reaction of water with a general halogenoalkane RHal is:

$$RHal + H_2O \rightarrow ROH + H^+ + Hal^-$$

Ammonia as a nucleophile

Ammonia, NH_3, can act as a nucleophile in a similar way, with the lone pair of electrons on the N atom attacking the halogenoalkane. The product is an **amine** with an NH_2 group. The overall equation is

$$RHal + NH_3 \rightarrow RNH_2 + HHal$$

Using nucleophilic substitution to make halogenoalkanes

When a halogenoalkane reacts with ⁻OH ions, a nucleophilic reaction occurs and an alcohol is formed (see above). We can use the *reverse* of this reaction to produce a halogenoalkane from an alcohol, if we choose the reaction conditions carefully. It is another nucleophilic substitution reaction, but this time the nucleophile is Hal⁻.

For example, we can make 1-bromobutane using a nucleophilic substitution reaction between butan-1-ol and Br⁻ ions. The reaction is done in the presence of a strong acid, and the first step involves bonding between H^+ ions and the O atom on the alcohol:

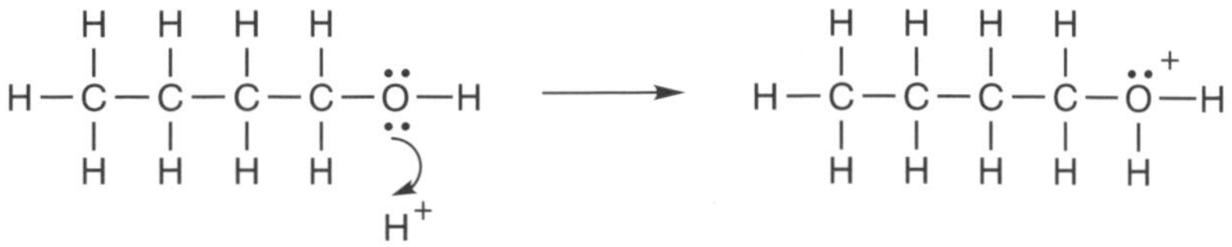

This gives the C atom to which the O is attached a partial positive charge. It is now more readily attacked by Br^- ions, forming bromobutane.

The overall equation for the reaction is:

$$CH_3CH_2CH_2CH_2OH + H^+ + Br^- \rightarrow CH_3CH_2CH_2CH_2Br + H_2O$$

You can use this reaction to prepare bromobutane in **Activity A4.2**.

PROBLEMS FOR 13.1

1 When 1-chloropropane is refluxed with aqueous sodium hydroxide solution a nucleophilic substitution reaction occurs, forming propan-1-ol.
  a Write a balanced equation to show the overall reaction.
  b Explain why this is classed as a substitution reaction.
  c Write down the structure of the attacking nucleophile, showing the charge and any lone pairs of electrons.
  d Draw a labelled diagram of the apparatus you would use to carry out this reaction.
  e How would you obtain a pure sample of propan-1-ol from the reaction mixture?

2 a Table 2 shows the structure of common nucleophiles. Write a balanced equation to show the nucleophilic substitution reaction which takes place between each of the following pairs of compounds. Your equations should show the structures of the reactants and products clearly.
  Example: bromomethane and CN^- ions
  CH_3—$Br + CN^- \rightarrow CH_3$—$CN + Br^-$
    i iodoethane and ^-OH ions
    ii bromoethane and $C_2H_5O^-$ ions
    iii 2-chloropropane and CH_3COO^- ions
  b What type of organic compound is formed in **i**?

3 1-Bromoethane reacts with concentrated ammonia solution when heated in a sealed tube.
  a Write an equation for the nucleophilic substitution reaction which takes place.
  b Using the reaction of halogenoalkanes with water as a guide (page 246), show the mechanism for this reaction.

4 Chloromethane has a lifetime in the troposphere of about 1 year, which allows enough time for some of it to be transported into the stratosphere where it helps to destroy ozone.
  In contrast, iodomethane has a tropospheric lifetime of about 8 days and very little reaches the stratosphere. This is because iodomethane is rapidly broken down by light in the troposphere. This is called photolysis.
  a Write an equation for the photolysis of iodomethane.
  b Explain why iodomethane is photolysed in the troposphere whereas chloromethane is only photolysed when it reaches the stratosphere.

5 Suggest the reagents and the conditions you would use to prepare a sample of each of the following compounds:

  a
  $CH_3CH_2CH_2CH_2CH_2OH$ (from *1-bromopentane*)

  b
  CH_3CHCH_3 (from *2-chloropropane*)
    |
    NH_2

  c
  $CH_3CH_2CH_2OH$ (from *1-bromopropane*)

6   Long strips of a type of seaweed called *laminaria* are used by some people to forecast the weather. The same seaweed was once used as a source of iodine.

     The seaweed concentrates iodine from seawater, by turning the iodine to iodomethane. Iodomethane is volatile, and you would expect some of it to escape into the atmosphere.

     However, iodomethane can only be detected in the atmosphere in tiny amounts. One possible theory is that some of it reacts with the chloride ions in sea water to generate chloromethane – the principal natural chlorine-containing substance of the atmosphere. Photodissociation of chloromethane is the major *natural* source of chlorine atoms in the stratosphere.

a   **i**   Write an equation for the reaction of iodomethane with chloride ions.
     ii   What is the role of the chloride ion in this process?

b   Suggest two reasons why bromomethane would be unlikely to react to the same extent with seawater under similar circumstances.

c   **i**   Write an equation for the photodissociation of chloromethane.
     ii   Use a value for the C—Cl bond enthalpy to calculate the minimum frequency of the photon of radiation which would cause the photodissociation of chloromethane.
     iii   To which part of the spectrum does this frequency correspond?

13.2 *Alcohols and ethers*

There are two isomers with the molecular formula C_2H_6O:

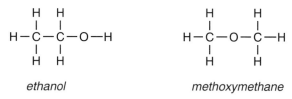

ethanol   *methoxymethane*

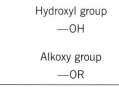

Hydroxyl group
—OH

Alkoxy group
—OR

Ethanol belongs to a homologous series called **alcohols.** All members of the series contain the hydroxyl functional group —OH and show similar chemical properties. They differ only in the length and structure of the hydrocarbon chain.

     Methoxymethane is a member of a different homologous series, called **ethers**. All ethers contain the alkoxy functional group, —OR, and show similar chemical properties.

Alcohols

General formula of
alcohols is R—OH

The names of alcohols
end in -ol

Alcohols are derived from alkanes by substituting an —OH group for an —H atom. They are named from the parent alkane by omitting the final *-e* and adding the ending *-ol*. For example,

CH_3OH	methanol
CH_3CH_2OH	ethanol
isomers $\{$ $CH_3CH_2CH_2OH$	propan-1-ol
$CH_3CH(OH)CH_3$	propan-2-ol
$CH_3CH_2CH_2CH_2OH$	butan-1-ol

For alcohols containing more than two carbon atoms, isomeric compounds are possible. To distinguish between these, it is necessary to label the position of the OH group. The hydrocarbon-chain is always numbered from the end which gives the lowest number for the position of the functional group. Thus, $CH_3CH_2CH_2CH_2OH$ is named butan-1-ol, not butan-4-ol.

Some alcohols, particularly biologically occurring ones, contain more than one —OH group in their molecules. They are known as **polyhydric** alcohols. Look how they are named.

$CH_2(OH)CH_2OH$ ethane-1,2-diol
$CH_2(OH)CH(OH)CH_2OH$ propane-1,2,3-triol

Compounds in which the —OH group is directly attached to a benzene ring are called **phenols**. The presence of the aromatic ring modifies the properties of the —OH group, so phenols are best studied separately from the alcohols.

phenol

Physical properties of alcohols

You can think of alcohols as being derived from water by replacing one of the H atoms by an alkyl group.

       water   an *alcohol*

Like water molecules, alcohol molecules are polar because of the polar O—H bond. In both water and alcohols, there is a special sort of strong attractive force *between the molecules* due to **hydrogen bonds** (Figure 3).

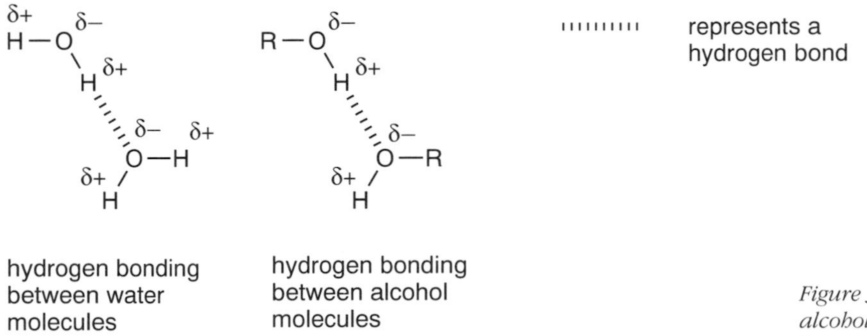

|||||||||| represents a hydrogen bond

hydrogen bonding between water molecules

hydrogen bonding between alcohol molecules

*Figure 3 Hydrogen bonding in water and alcohols. Hydrogen bonds and other intermolecular forces are discussed in more detail in **Section 5.3**.*

Hydrogen bonds are not as strong as covalent bonds, but are stronger than other attractive forces *between* covalent molecules. When a liquid boils, these forces must be broken so the molecules escape from the liquid to form a gas. This explains why the boiling points of alcohols are higher than those of corresponding alkanes with similar relative molecular mass (M_r). For example, ethanol ($M_r = 46$) is a liquid, while propane ($M_r = 44$) is a gas at room temperature.

Hydrogen bonding *between alcohol and water molecules* (see Figure 4) explains why the two liquids mix together.

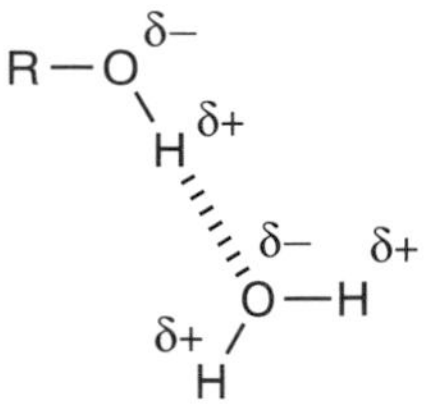

Figure 4 Hydrogen bonding between water molecules and alcohol molecules

Table 3 shows the solubility of some alcohols in water. As the hydrocarbon chain becomes larger, the influence of the —OH group on the properties of the molecule becomes less important. So the properties of the higher alcohols get more and more like those of the corresponding alkane.

Name	Formula	Solubility/g per 100 g water
methanol	CH_3OH	miscible in all proportions
ethanol	CH_3CH_2OH	
propan-1-ol	$CH_3CH_2CH_2OH$	
butan-1-ol	$CH_3CH_2CH_2CH_2OH$	8.0
pentan-1-ol	$CH_3CH_2CH_2CH_2CH_2OH$	2.7
hexan-1-ol	$CH_3CH_2CH_2CH_2CH_2CH_2OH$	0.6

Table 3 Solubility of alcohols in water

Ethers

Ethers are derived from alkanes by substituting an alkoxy group (—OR) for an H atom. For example,

CH_3CH_2—O—CH_2CH_3 ethoxyethane
$CH_3CH_2CH_2$—O—CH_3 methoxypropane

> General formula of ethers R—O—R′

Note that the longer hydrocarbon chain is chosen as the parent alkane.

Physical properties of ethers

You can think of ethers as being derived from water by replacing *both* the H atoms by alkyl groups.

H—O
    \
     H

R—O
    \
     R′

water   *an ether*

Ether molecules are only slightly polar and the attractive forces *between* molecules are relatively weak. There are no H atoms attached to the oxygen to form hydrogen bonds between ether molecules.

The boiling point of an ether is similar to that of the alkane with corresponding relative molecular mass. Like alkanes, the lower ethers are very volatile and dangerously flammable.

Ethers are only slightly soluble in water, but mix well with other non-polar molecules such as alkanes.

PROBLEMS FOR 13.2

1   Draw out the structural formulae of alcohols which are isomers of butan-1-ol.

2   How many other isomers of butan-1-ol are there which are not alcohols?

3   Name the following alcohols:
    a $CH_3CH_2CH(OH)CH_2CH_2CH_2CH_3$
    b $CH_3CH(OH)CH(OH)CH_3$

    c   ⬡—OH

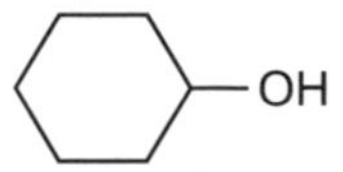

4 Look at the following compounds:

A $CH_3OCH_2CH_2CH_3$

B $CH_3CH(OH)CH_2CH_3$

C

D

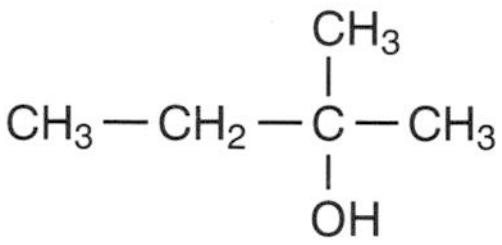

E $CH_3CH(OH)CH_2CH(OH)CH_3$

F

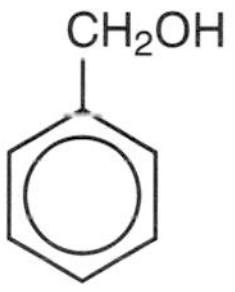

a Which compound(s) is (are) alcohols?
b Which compound(s) is(are) ethers?
c Which compound(s) is(are) phenols?
d Which compound(s) is(are) diols?
e Which compounds are isomers?
f Which compound do you think will be the most volatile?
g Which compound would you expect to be the most soluble in water?

5 Here are the boiling points and relative molecular masses (M_r) of a number of substances:

Substance	Boiling point/°C	M_r
water, H_2O	100	18
ethane, CH_3CH_3	–88.5	30
ethanol, CH_3CH_2OH	78	46
butan-1-ol, $CH_3CH_2CH_2CH_2OH$	117	74
ethoxyethane, $CH_3CH_2OCH_2CH_3$	35	74

Use ideas about bonding and forces between molecules to explain why
a ethanol has a higher boiling point than ethane
b water has a higher boiling point than ethanol
c butan-1-ol has a higher boiling point than ethanol
d butan-1-ol has a higher boiling point than ethoxyethane.

13.3 *Carboxylic acids and their derivatives*

This section provides an introduction to carboxylic acids and their derivatives. You will find more details in later sections. Carboxylic acids contain the **carboxyl** group

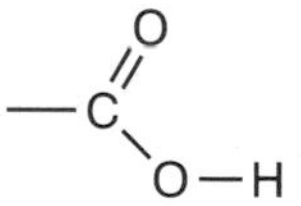

carboxyl group

They have the general formula **RCOOH**. They are named from the parent alkane by omitting the final *-e* and adding the ending *–oic acid*. When you count the carbon atoms to find the parent alkane, remember to include the one in the carboxyl group. for example,

HCOOH	methanoic acid	(formic acid)
CH_3COOH	ethanoic acid	(acetic acid)
CH_3CH_2COOH	propanoic acid	

The first two members of the series, methanoic acid and ethanoic acid are often still called by their older names, formic acid and acetic acid, so it is as well to know these too.

When two carboxyl groups are present, the ending **-dioic acid** is used. For example,

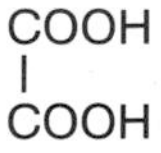

ethanedioic acid

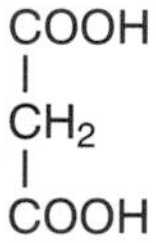

propanedioic acid

A carboxyl group can be attached to a benzene ring. For example,

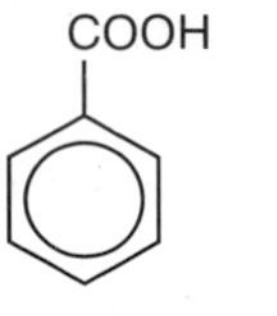

benzenecarboxylic acid
also known as *benzoic acid*

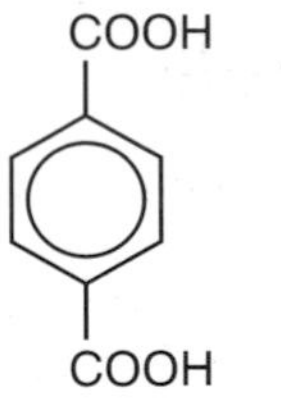

benzene-1,4-dicarboxylic acid

You will learn more about the chemical reactions of carboxylic acids in **Section 13.4**.

The —OH group in the carboxyl group can be replaced by other groups to give a whole range of **carboxylic acid derivatives**. Some examples of acid derivatives are shown in Table 4. You will meet these again in several parts of the course, so there will be plenty of opportunities to get used to recognising them. The table shows the sections which deal with each derivative in more detail.

Acid derivative	Dealt with in Section(s)	Example
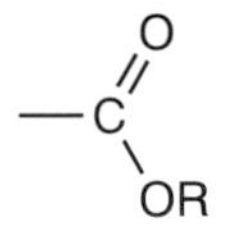 ester	**13.5** and **14.2**	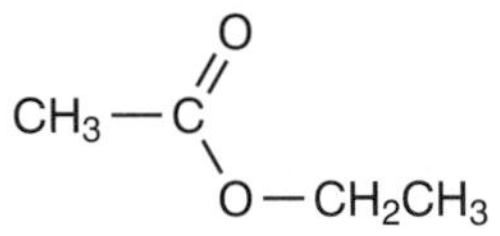 ethyl ethanoate
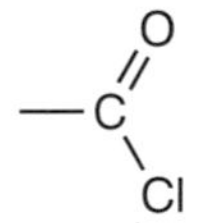 acyl chloride	**13.5** and **14.2**	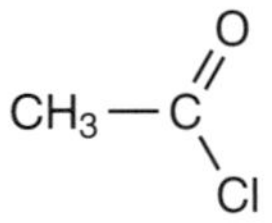 ethanoyl chloride
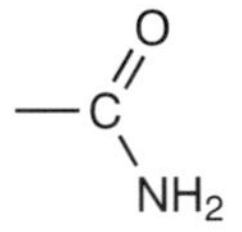 amide	**13.7**	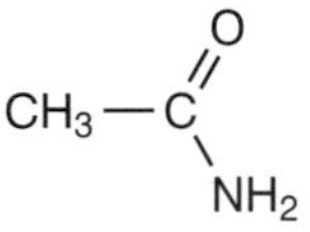 ethanamide
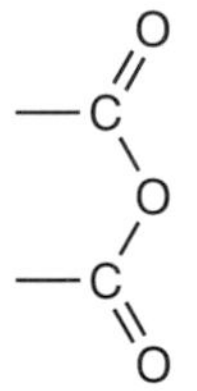 acid anhydride	**13.5** and **14.2**	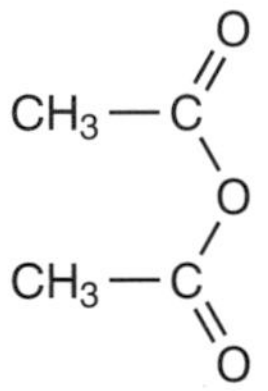 ethanoic anhydride

Table 4 Some examples of acid derivatives

PROBLEMS FOR 13.3

1 Name the following carboxylic acids:
   a $CH_3(CH_2)_6COOH$
   b $HOOC(CH_2)_3COOH$
   c

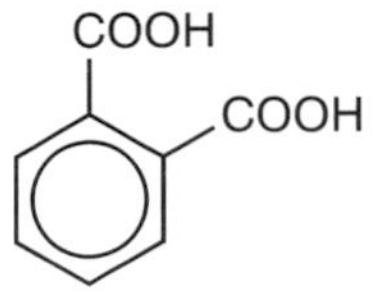

2 Write out the structural formula of
   a   the methyl ester of propanoic acid
   b   the acyl chloride derivative of benzoic acid
   c   the amide of butanoic acid
   d   the acid anhydride formed from propanoic acid
   e   the acyl chloride of hexanedioic acid

3 Draw out the structural formulae of any other esters which are isomeric with your answer to question **2a**. Try to name any isomer you draw.

13.4 The —OH group in alcohols, phenols and acids

The hydroxy group, —OH can occur in three different environments in organic molecules:

- attached to an alkane chain in **alcohols**. Alcohols are of three types, primary, secondary and tertiary, according to the position of the OH group (see Table 5).

Type of alcohol	Position of OH group	Example
primary	at end of chain: 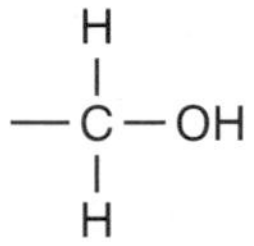 H | —C—OH | H	propan-1-ol 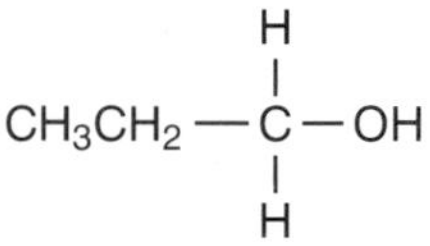 H | CH_3CH_2—C—OH | H
secondary	in middle of chain: 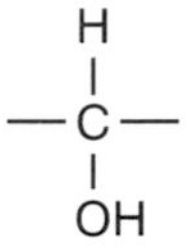 H | —C— | OH	propan-2-ol 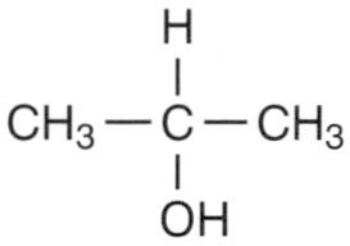 H | CH_3—C—CH_3 | OH
tertiary	attached to a carbon atom which carries no H atoms: 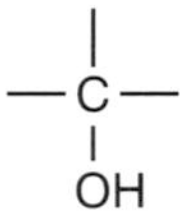 | —C— | OH	2-methylpropan-2-ol 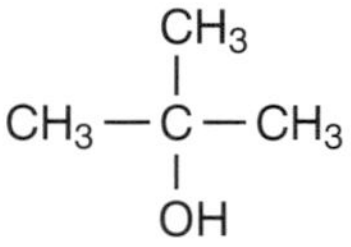 CH_3 | CH_3—C—CH_3 | OH

Table 5 Primary, secondary and tertiary alcohols

- attached to a benzene ring in **phenols**, for example

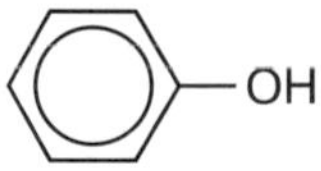

phenol

(Although phenols look similar to alcohols they behave very differently. It is generally true that functional groups behave differently when attached to an aromatic ring from when they are attached to an alkane group.)

- as part of a carboxyl group in **carboxylic acids**, for example

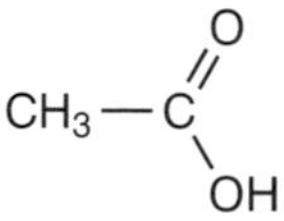

ethanoic acid

In this section we will compare the way the —OH group behaves in these three different kinds of compounds.

Acidic properties

The —OH group can react with water like this:

$$R—OH + H_2O \rightleftharpoons R—O^- + H_3O^+$$

where R stands for the group of atoms which makes up the rest of the molecule.

Water itself does this to a very small extent in the reaction

$$H—OH + H_2O \rightleftharpoons H—O^- + H_3O^+$$

So, at any one time a small number of water molecules donate H^+ ions to other water molecules: water behaves as a weak acid. A similar reaction occurs with ethanol, but to a lesser extent. The equilibrium lies further to the left, and ethanol is a weaker acid than water.

With phenol, the equilibrium lies further to the right than in water: phenol is slightly more acidic than water. Carboxylic acids are even more acidic, though still weak. The order of acid strength is

ethanol < water < phenol < carboxylic acids

It is the stability of the anion formed from the acid which decides how strong the acid is. If the negative charge on the oxygen can be shared with other atoms, the anion will be more stable and more of it will be made. In the anion derived from alcohols, no such sharing is possible. However, in phenols and in carboxylic acids the electric charge gets spread out by a process called *delocalisation*. This involves a smearing out of the electrons over the anion (Figure 5). There is more about delocalisation in **Section 12.3**.

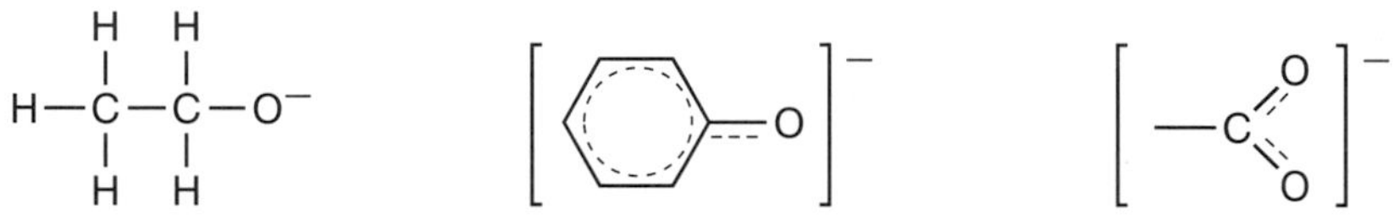

Very unstable; charge located totally on O, therefore ethanol is neutral (pH of solution =7)

More stable; charge spread onto benzene ring, therefore phenol is a very weak acid (pH of solution = 5–6)

More stable; charge spread across —— COO group, therefore carboxylic acids are weak acids (pH of solution = 3–4)

Figure 5 The strength of an acid depends on the stability of its anion

Phenols and carboxylic acids are strong enough acids to react with strong bases, such as sodium hydroxide, to form salts. For example,

ethanoic acid   *sodium ethanoate*

Only with solutions of carboxylic acids is the concentration of H_3O^+ ions great enough to give carbon dioxide when reacted with carbonates

$$CO_3^{2-} + 2H_3O^+ \rightarrow CO_2 + 3H_2O$$

So carboxylic acids make carbonates fizz, but alcohols and phenols do not.

Oxidation

The —OH group can be oxidised by strong oxidising agents such as acidified potassium dichromate(VI). The basic reaction is

The orange dichromate(VI) ion, $Cr_2O_7^{2-}$, is reduced to green Cr^{3+}.

Notice that in this reaction, two atoms of hydrogen are being removed: one from the oxygen atom, and one from the carbon atom. Oxidation of the —OH group will not take place unless there is an hydrogen atom on the carbon atom to which the —OH is attached.

The product is a **carbonyl** compound: an aldehyde or a ketone (see box). The type of product you get depends on the type of compound you start with.

Alcohols

Primary alcohols such as ethanol, are oxidised to aldehydes, but the aldehyde is then itself oxidised to a carboxylic acid. The aldehyde can be extracted as an intermediate product if the reaction is done carefully. With ethanol:

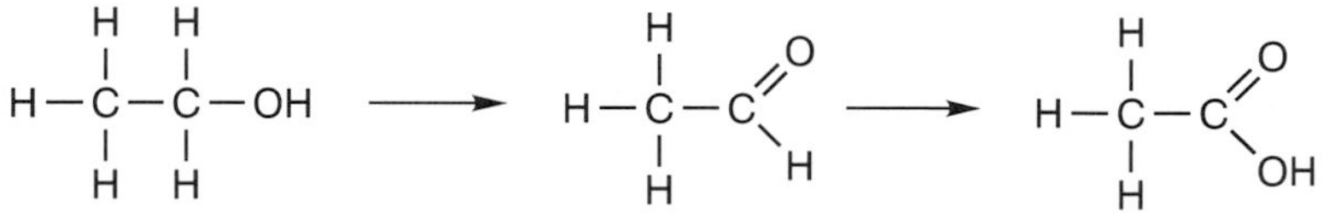

ethanol   ethanal   ethanoic acid

Secondary alcohols, such as propan-2-ol , are oxidised to ketones. For example:

propan-2-ol   propanone

Aldehydes and ketones

Adehydes and ketones contain the **carbonyl group**, $-C-$.
$$\underset{O}{\overset{\parallel}{}}$$

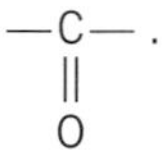

In an **aldehyde**, the carbonyl group is at the *end* of an alkane chain, so the functional group is

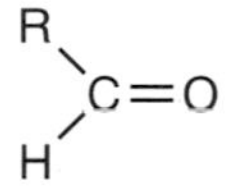

Aldehydes are named using the suffix -*al*.

For example:

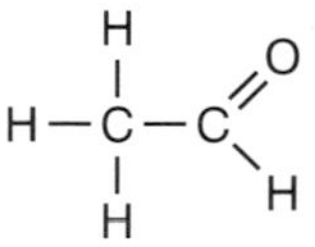

   CH_3CHO *ethanal*

In a **ketone**, the carbonyl group is *inside* an alkane chain, so the functional group is

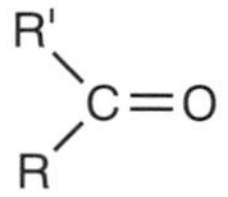

Ketones are named using the suffix -*one*.

For example:

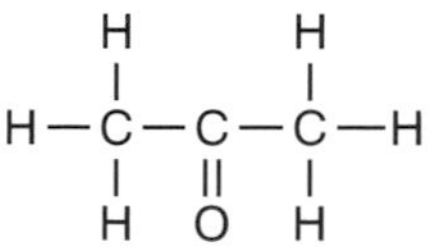

   CH_3COCH_3 *propanone*

It is difficult to oxidise the ketone further than this, because to do so would involve breaking a C—C bond.

Tertiary alcohols such as 2-methylpropan-2-ol, are difficult to oxidise because they do not have a hydrogen atom on the carbon atom to which the —OH group is attached.

Phenols and carboxylic acids

Although these compounds have an —OH group, they are difficult to oxidise because they do not have a hydrogen atom on the carbon atom to which the —OH group is attached.

The iron(III) chloride test

Some groupings of atoms can become closely associated with metal ions and form complexes (an example you have met is the dark blue complex which ammonia forms with Cu^{2+} ions).

The C=C—OH group (it is called the 'enol' group. Can you see why?) can form a purple complex with neutral Fe^{3+} ions. Only phenol and its derivatives have such an arrangement and are the only ones to give a colour with iron(III) chloride. Similar complexes are used to make the colours of some inks.

Ester formation

See **Section 13.5**.

PROBLEMS FOR 13.4

1 a i Name the following compounds:

A

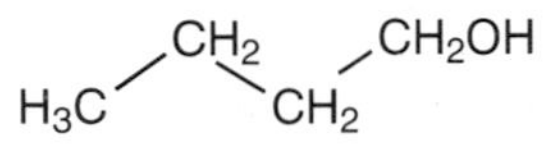

C

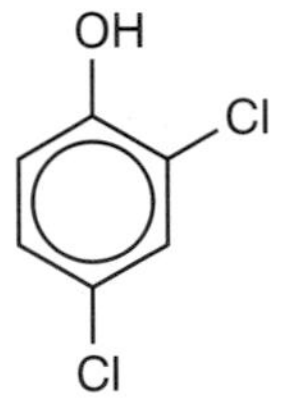

B

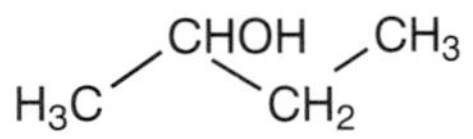

ii Draw structures for the following compounds:
D butanal
E butanoic acid
F butanone
G 3-methylpentan-3-ol

b Which one of the above compounds A to G
i is a secondary alcohol
ii is an aldehyde
iii is a phenol
iv is a ketone
v is an aliphatic alcohol which is not easily oxidised on heating with acidified potassium dichromate(VI)
vi produces a purple colour with neutral aqueous iron(III) chloride
vii gives carbon dioxide with sodium carbonate
viii produces a carboxylic acid on refluxing with excess acidified potassium dichromate(VI)?

2 Write equations for the reactions of aqueous sodium hydroxide with the following compounds. In each case write out a full structure for the organic product.

a 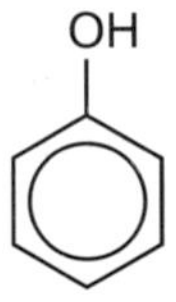 **b**

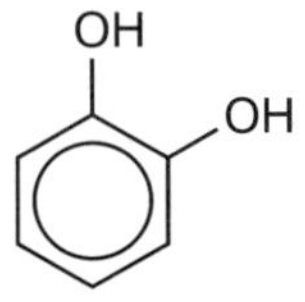

c

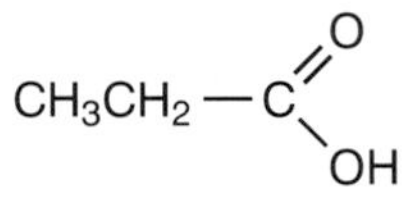

d

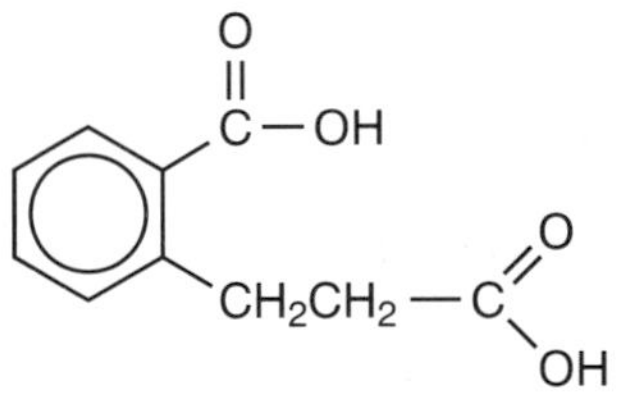

e

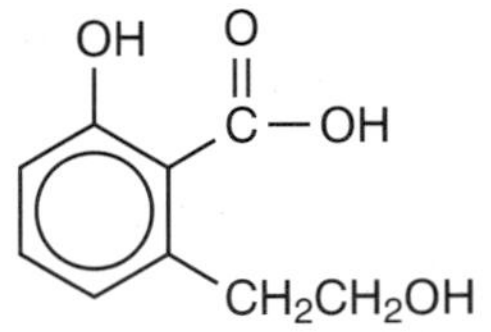

3 Draw structures for the compounds produced by reacting the following substances with acidified potassium dichromate(VI) solution.

a 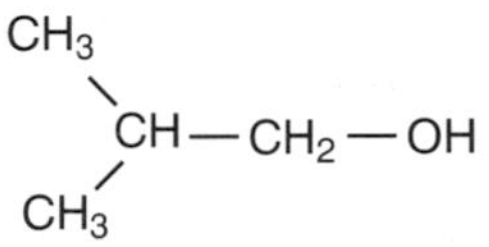 **b**

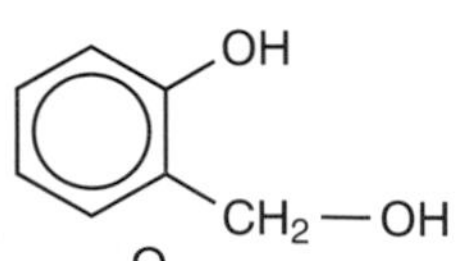

c

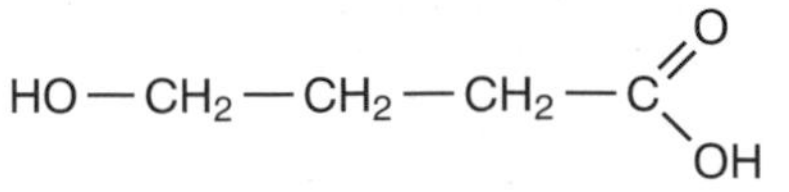

13.5 *Esters*

What are esters?

Esters are formed when an alcohol reacts with a carboxylic acid.

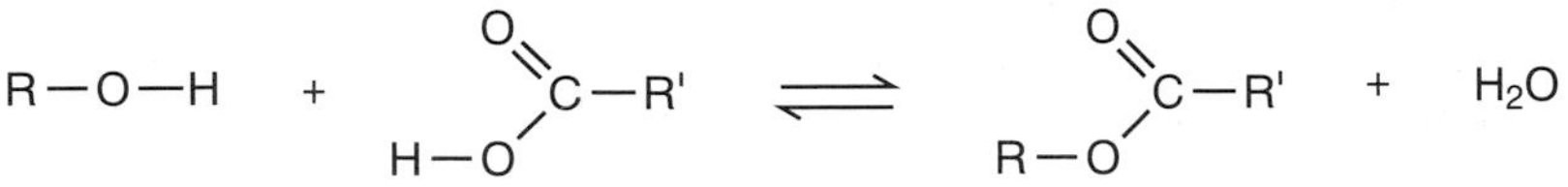

(R is an *alkyl group* such as CH_3 or C_2H_5)

Because water is produced, just like when we breathe on a cold surface, the process was christened 'condensation' and the name has stuck. A **condensation reaction** can be thought of as two molecules reacting together to form a larger molecule with the elimination of a small molecule such as water.

Notice that this is an *equilibrium* reaction, as shown by the ⇌ symbol in the equation. The reaction is easily reversed: the reverse reaction is called **ester hydrolysis** (see below).

The esterification reaction occurs extremely slowly unless an acid catalyst is present. Concentrated sulphuric acid is often used. Removing the water from the reaction as it is formed helps to drive the equilibrium to the right.

Esters have strong sweet smells which are often floral or fruity. They are widely used as solvents, for example in some glues, and many naturally occurring esters are responsible for well known fragrances. Some examples are given in Table 6.

Ester	Fragrance
ethyl methanoate	raspberries
3-methylbutyl ethanoate	pears
ethyl 2-methylbutanoate	apples
phenylmethyl ethanoate	jasmine

Table 6 Some ester fragrances

Naming esters

Esters are named after the alcohol and acid from which they are derived (Figure 6):

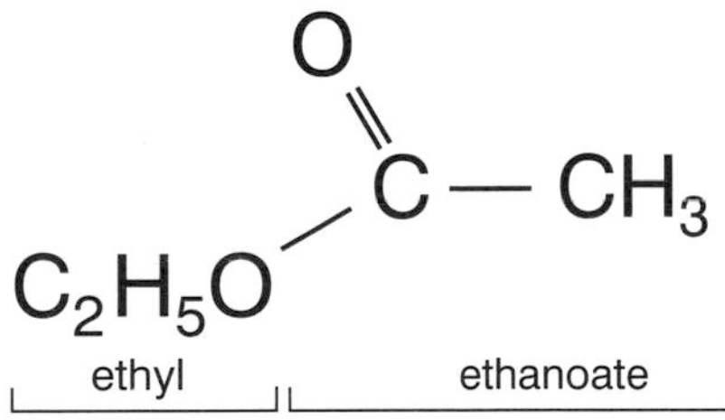

Figure 6 How to name an ester

ethanol and ethanoic acid give ethyl ethanoate
phenol and benzoic acid give phenyl benzoate.

You have to be careful to get the 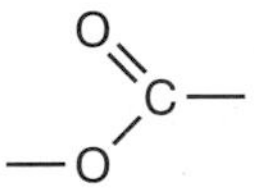 group the right way round.

For example, ethyl ethanoate

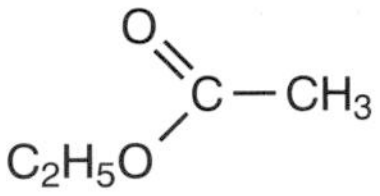

is very different from its isomer, methyl propanoate

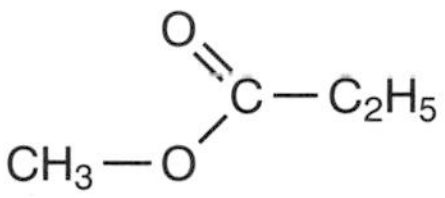

Making models of these two esters will help you understand how they are named.

Polyesters

Polyesters are condensation polymers (**Section 13.7**). They are made by reacting a *diol* (containing two —OH groups) with a *dicarboxylic acid* (containing two —COOH groups). For example, a common polyester is made from ethane-1,2-diol and benzene-1,4-dicarboxylic acid.

$$HOCH_2CH_2OH$$

ethane-1,2-diol

benzene-1,4-dicarboxylic acid

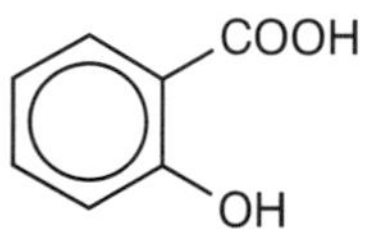

Esters from 2-hydroxybenzoic acid

2-hydroxybenzoic acid

A look at the structure of 2-hydroxybenzoic acid shows that there are two ways of esterifying it. Either the —OH group could be reacted with a carboxylic acid, or the —COOH group could be reacted with an alcohol. *Aspirin* is the product of esterifying the —OH group to form 2-ethanoylhydroxybenzoic acid. It is quite soluble in water, so it can be absorbed into the bloodstream through the stomach wall.

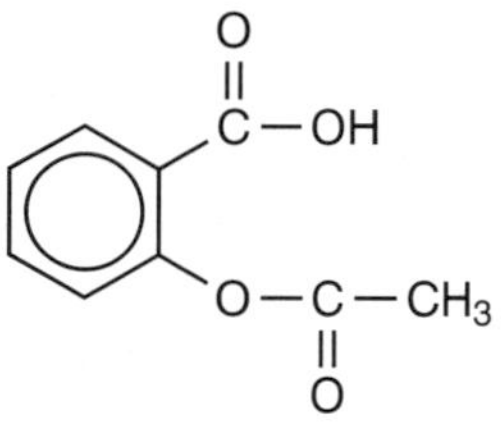

aspirin

The product of reacting the —COOH group with methanol is called methyl 2-hydroxybenzoate. This is better known as *oil of wintergreen* which is used as a liniment. It is soluble in fat rather than water so it is absorbed through the skin. Like aspirin, it reduces pain and swelling.

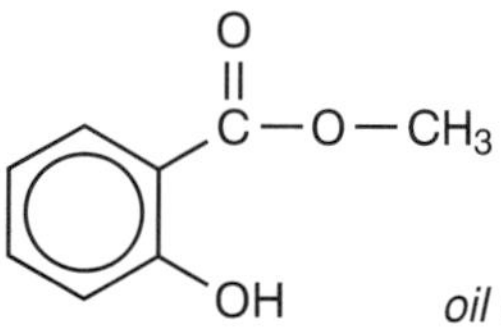

oil of wintergreen

Acylating agents

The —OH group in phenol is less reactive to esterification than the —OH of ethanol so it needs a more vigorous reagent to esterify it. When ethanoic acid is involved in esterification the process is sometimes known as **ethanoylation**. A more vigorous ethanoylating agent than ethanoic acid is **ethanoic anhydride**, made by eliminating a molecule of water between two ethanoic acid molecules:

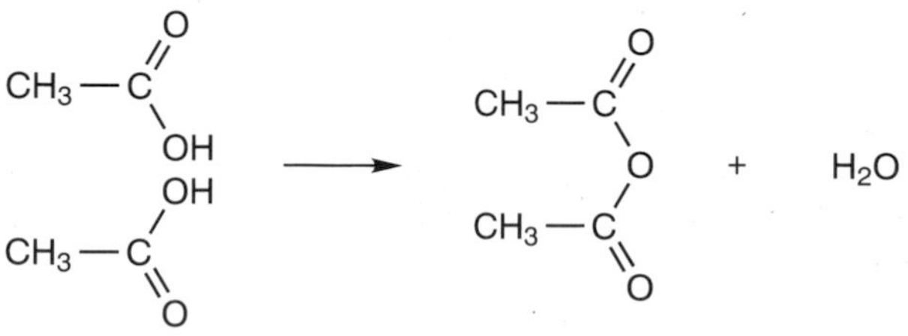

ethanoic anhydride

The equation for the reaction of ethanoic anhydride with 2-hydroxybenzoic acid is

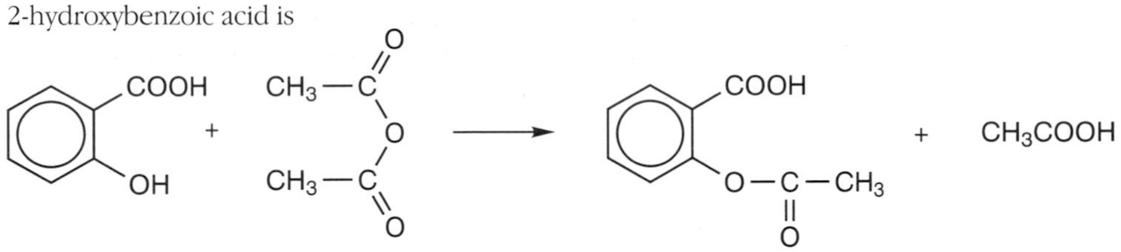

Ethanoic anhydride is often used as an ethanoylating agent because it is reactive but not too unpleasant or dangerous. A much more reactive ethanoylating agent is ethanoyl chloride:

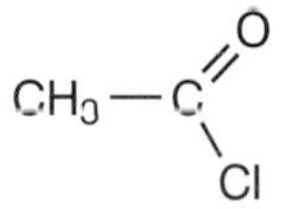

ethanoyl chloride, an acyl chloride

Ethanoyl chloride is toxic and hazardous to use because it is so reactive. In this equation it is reacting with phenol to produce phenyl ethanoate.

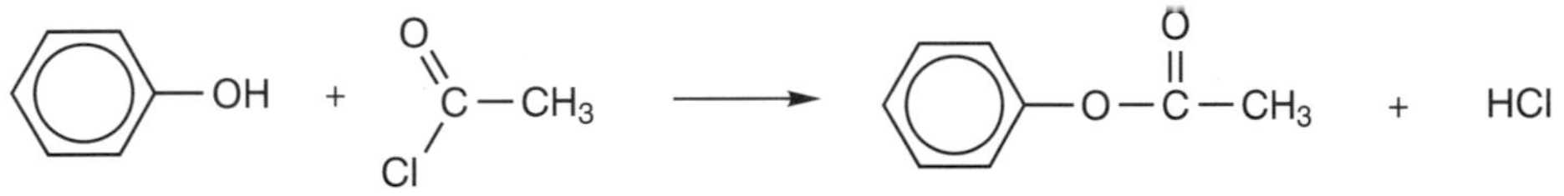

Ethanoic anhydride and ethanoyl chloride are members of a general group of reagents called **acylating agents.** Acylating agents substitute an acyl group, R—CO, for the H on an —OH group. The *only* way to esterify a phenol is by using an acylating agent. Alcohols can be esterified either by using an acylating agent or by reacting with a carboxylic acid in the presence of an acid catalyst.

Ester hydrolysis

The reverse of esterification corresponds to the breakdown of an ester by water. In other words, it is a **hydrolysis**. On their own, water and an ester react very slowly, but the process can be speeded up by catalysis.

A catalyst is effective for both directions of a reversible reaction, so sulphuric acid (or any other acid) will do. Alkalis, like sodium hydroxide solution, also work in the case of ester hydrolysis.

When alkali is used, the hydrolysis does not produce a carboxylic acid, but a *carboxylate salt*. For ethyl ethanoate we can write the reaction as:

$$C_2H_5—O—CO—CH_3 + HO^- \rightarrow C_2H_5OH + CH_3COO^-$$

ethyl ethanoate   *ethanol*   *ethanoate ion*

PROBLEMS FOR 13.5

1   Look at the esters in Table 6, page 257.
   a   Write down the names of the alcohols and acids from which they are derived.
   b   Write down the structures of these alcohols, acids and esters.
   c   How are 3-methylbutyl ethanoate and ethyl 2-methylbutanoate related?

2   a   Name and draw the structures for the esters formed when the following are refluxed with a concentrated sulphuric acid catalyst:
      i   propan-2-ol and propanoic acid
      ii ethanoic acid and ethane-1,2-diol.

   b   Write balanced equations for the reactions between:
      i   ethanoyl chloride and 2-methylpropan-2-ol
      ii equimolar quantities of ethanol and phthalic anhydride,

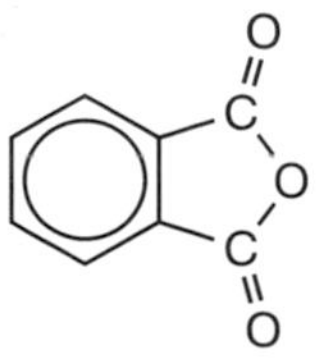

phthalic anhydride

3 Look at the structure of the molecule of 2-hydroxybenzoic acid on page 258.
  a Could the —OH and —COOH groups on one 2-hydroxybenzoic acid molecule react together? Explain your answer.
  b Could two molecules of 2-hydroxybenzoic acid react together to form a 'di-ester'?
    Building models might help you answer these questions.

4 4-Hydroxybutanoic acid has the structure $HOCH_2CH_2CH_2COOH$.
  a Individual molecules of 4-hydroxybutanoic acid can undergo 'internal esterification'. Draw a structure for this ester. It will help if you build a model.
  b The oxidation product of 4-hydroxybutanoic acid can undergo 'internal anhydride formation'. Draw a structure for the anhydride. Again, a model will help.

5 Look at the section on 'Polyesters' on page 258. Draw the structure of a polymer chain formed by joining two molecules of ethane-1,2-diol and two molecules of benzene-1,4-dicarboxylic acid.

6 a When ethyl ethanoate is hydrolysed with water enriched with oxygen-18, $H_2{}^{18}O$, the oxygen-18 appears in the ethanoic acid and not in the ethanol.

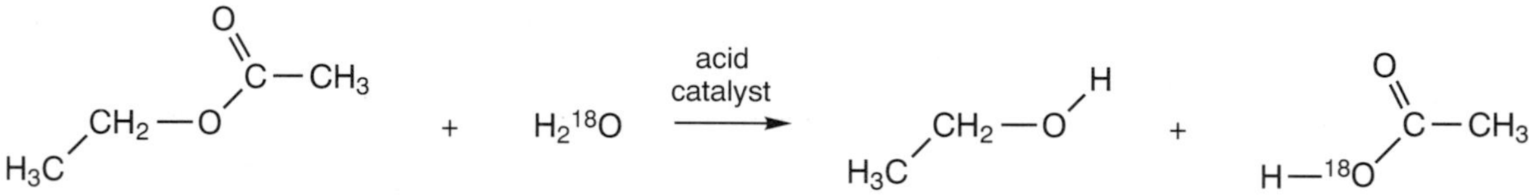

  i Use this information to identify which ester bond is broken during hydrolysis of the ester.
  ii Suggest why it is this bond that is broken in preference to any other bond.
  b The following compound is an example of a lactide

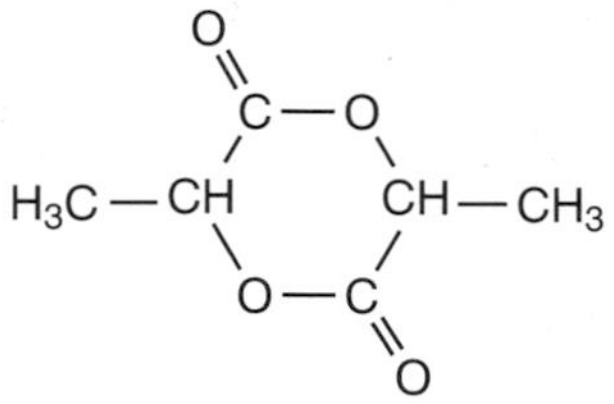

  i On refluxing this compound with water and an acid catalyst, only one product results. Draw the structure of this product and name it.
  ii Write an equation for the hydrolysis of the lactide, marking the bonds which are broken.

13.6 Oils and fats

The chemical structure of oils and fats

Oils and fats provide an important way of storing chemical energy in living systems. Most of them have the same basic structure. The only difference is that oils are liquid at room temperature whereas fats are solid.

Most oils and fats are **esters** of propane-1,2,3-triol (commonly called *glycerol*) with long chain carboxylic acids, RCOOH. In *palmitic acid*, for example, R is a $CH_3(CH_2)_{14}$— group.

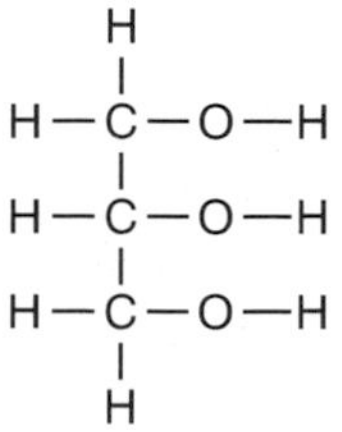

propane-1,2,3-triol (glycerol) 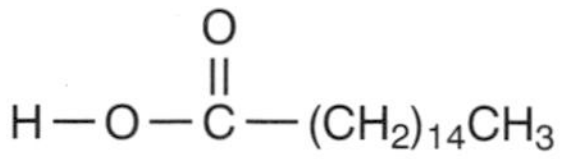   palmitic acid

Because glycerol has three alcohol groups in each molecule, *three* carboxylic acid molecules can form ester linkages with each glycerol molecule to form a **triester** (sometimes called a **triglyceride**). The triester formed from glycerol and palmitic acid has the following structure:

You can remind yourself of the structure and chemistry of esters by reading **Section 13.5**.

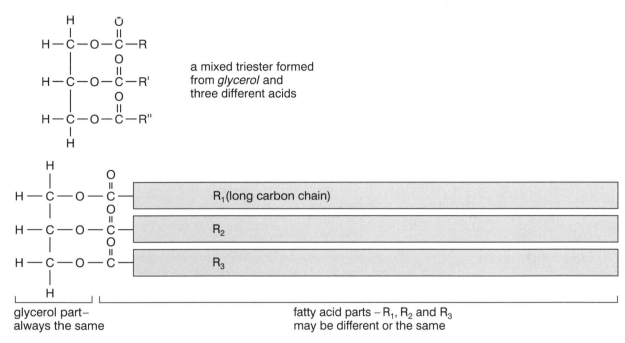

triester formed from *glycerol* and *palmitic acid*

The triesters found in natural oils and fats are often **mixed triesters** in which the three acid groups are not all the same. Figure 7 summarises the structure of triesters.

a mixed triester formed from *glycerol* and three different acids

R$_1$ (long carbon chain)

R$_2$

R$_3$

glycerol part – always the same

fatty acid parts – R$_1$, R$_2$ and R$_3$ may be different or the same

Figure 7 The general structure of the triesters found in fats and oils

Fats and fatty acids

The carboxylic acids usually have unbranched hydrocarbon chains and often contain 16 or 18 carbon atoms. They are sometimes called **fatty acids** because of their origin. The alkyl groups, R, R' and R" are either fully saturated or contain one or more double bonds. Table 7 shows some common fatty acids. There are many others, some of which contain 3 or 4 double bonds.

Structure	Traditional name	Origin of name
$CH_3(CH_2)_{14}COOH$	palmitic acid	palm oil
$CH_3(CH_2)_{16}COOH$	stearic acid	suet (Greek: *stear*)
$CH_3(CH_2)_7CH{=}CH(CH_2)_7COOH$	oleic acid	olive oil
$CH_3(CH_2)_4CH{=}CHCH_2CH{=}CH(CH_2)_7COOH$	linoleic acid	oil of flax (Latin: *linum*)

Table 7 Common fatty acids

Many doctors believe that the more saturated fats – the ones with few double bonds – tend to cause blockage of blood vessels and so may lead to heart disease. They believe that a diet containing *polyunsaturated* fats with many double bonds may be healthier. Figure 8 shows the skeletal formulas of typical saturated and unsaturated fatty acids.

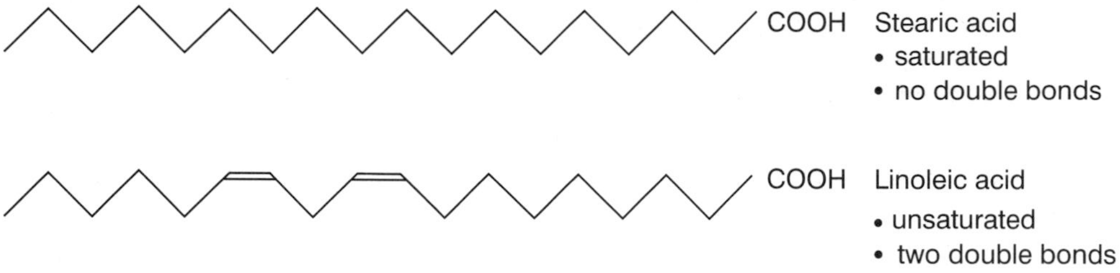

COOH   Stearic acid
• saturated
• no double bonds

COOH   Linoleic acid
• unsaturated
• two double bonds

Figure 8 A typical saturated and unsaturated fatty acid

The modern systematic names for fatty acids can get quite long (for example, oleic acid is octadec-9-enoic acid), so they are still known by their old names. These names are often derived from one of the oils or fats from which they are obtained.

Any natural oil or fat contains a mixture of triesters. The nature of the acids present affects the properties and determines whether the substance is a liquid oil or a solid fat. You can investigate the effect of the structure of fatty acids on the properties of oils in **Activity CD4.2**.

The *proportions* in which the acid groups occur are more or less constant for a particular oil or fat. Human body fat, for example, contains mainly four acids:

oleic acid	47%
palmitic acid	24%
linoleic acid	10%
stearic acid	8%

An oil or fat can be identified by breaking it down into glycerol and its fatty acids and then measuring the amount of each acid present.

Like all esters, oils and fats can be split up by **hydrolysis**. This is usually done by heating the oil or fat with concentrated sodium hydroxide solution to give glycerol and the sodium salts of the acids. For example

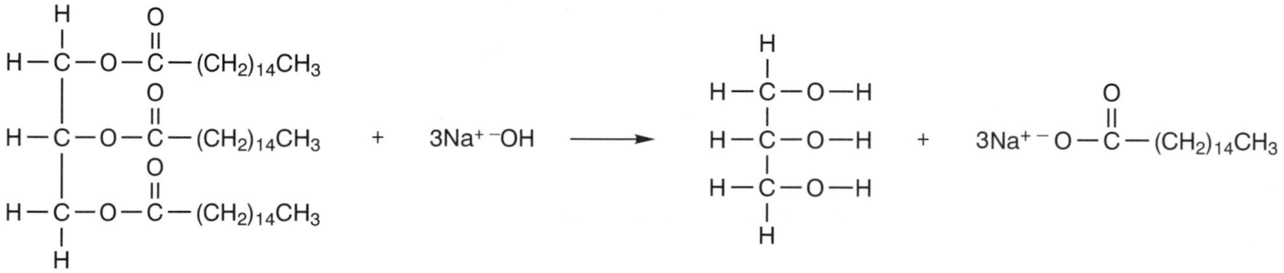

triester formed from *glycerol* and *palmitic acid*   *glycerol*   *sodium palmitate*

This is how soap is made. Soaps are sodium or potassium salts of fatty acids and they are made by heating oils and fats with sodium or potassium hydroxide. What oils do you think *Palmolive* soap is made from?

The free fatty acids can be released from the sodium salts by adding a dilute mineral acid such as hydrochloric acid.

PROBLEMS FOR 13.6

1 a What structural feature of a molecule identifies a compound as an ester?

b Write down the full structural formula of the triester made from glycerol and ethanoic acid. Make a model of this triester.

c Give an equation to show what happens when this ester is heated with hot concentrated sodium hydroxide solution.

2 Below is the structure of a triester present in a naturally occurring oil.

$$
\begin{array}{l}
H-\overset{\displaystyle H}{\underset{\displaystyle |}{C}}-O-\overset{\displaystyle O}{\underset{\displaystyle \|}{C}}-(CH_2)_7CH{=}CH(CH_2)_7CH_3 \\[6pt]
H-\overset{\displaystyle |}{\underset{\displaystyle |}{C}}-O-\overset{\displaystyle O}{\underset{\displaystyle \|}{C}}-(CH_2)_7CH{=}CHCH_2CH{=}CH(CH_2)_4CH_3 \\[6pt]
H-\overset{\displaystyle |}{\underset{\displaystyle |}{C}}-O-\overset{\displaystyle O}{\underset{\displaystyle \|}{C}}-(CH_2)_7CH{=}CH(CH_2)_7CH_3 \\[6pt]
\overset{\displaystyle |}{\underset{\displaystyle }{H}}
\end{array}
$$

The triester was hydrolysed by heating with hot concentrated sodium hydroxide solution, and the resulting solution was neutralised by addition of hydrochloric acid.

Name the hydrolysis products obtained. How many moles of each product would be obtained from 1 mole of the triester?

3 Explain the difference between the following terms which often appear on food labels:

a saturated fat

b monounsaturated fat

c polyunsaturated fat.

4 Animal fat is made up almost entirely of saturated fats. One compound present in sheep fat produces only glycerol and stearic acid on hydrolysis. Draw out the structural formula of this compound (using the shortened notation for the acid chain as in Problem **2**)

5 A food company buys a vegetable oil and converts it to margarine by treatment with hydrogen in the presence of a metal catalyst. During the process some of the double bonds in the oil are hydrogenated.

a Name a suitable catalyst and give conditions under which the hydrogenation could be carried out.

b Assume that the oil contains only the triester in problem **2**.

i What is the M_r of the triester in the oil?

ii 1 tonne (1000 kg) of the oil requires 4.90 kg of hydrogen to make margarine. What percentage of the double bonds in the oil have been hydrogenated?

13.7 *Amines, amides and amino acids*

What are amines?

Amines are the organic chemistry relatives of ammonia. Their structures resemble ammonia molecules in which alkyl groups take the place of one, two or all three hydrogen atoms.

$$
\begin{array}{cccc}
H-\overset{\displaystyle |}{\underset{\displaystyle |}{N}}-H & CH_3-\overset{\displaystyle |}{\underset{\displaystyle |}{N}}-H & CH_3CH_2CH_2-\overset{\displaystyle |}{\underset{\displaystyle |}{N}}-H & C_2H_5-\overset{\displaystyle |}{\underset{\displaystyle |}{N}}-C_2H_5 \\[4pt]
H & H & CH_3 & C_2H_5
\end{array}
$$

| ammonia | methylamine (a primary amine) | methylpropylamine (a secondary amine) | triethylamine (a tertiary amine) |

Notice three points about amines:

- amines with *one* alkyl group are called **primary amines**; **secondary amines** have *two* alkyl groups, and so on;
- the alkyl groups in a secondary or tertiary amine need not all be the same;
- the names of amines follow the usual rules: alkyl groups are arranged in alphabetical order. The prefix *amino* is sometimes also used to indicate the NH$_2$ group. For example:

$$
CH_3CH_2-\overset{\displaystyle }{\underset{\displaystyle |}{C}}H-CH_2CH_3 \\
NH_2
$$

3-aminopentane

Amines with low relative molecular masses are gases or volatile liquids. The volatile amines also resemble ammonia in having strong smells. The characteristic smell of decaying fish comes from amines such as ethylamine and trimethylamine. Rotting animal flesh gives off the diamines $H_2N(CH_2)_4NH_2$ and $H_2N(CH_2)_5NH_2$, which are sometimes called by the names putrescine and cadaverine respectively.

Properties of amines

The properties of amines are similar to the properties of ammonia, but modified by the presence of alkyl groups. Most of the properties are due to the lone pair of electrons on the nitrogen atom.

The bonding around the nitrogen atom of an amine is similar to that in ammonia: three pairs of electrons form localised covalent bonds while the other two electrons form a lone pair (Figure 9).

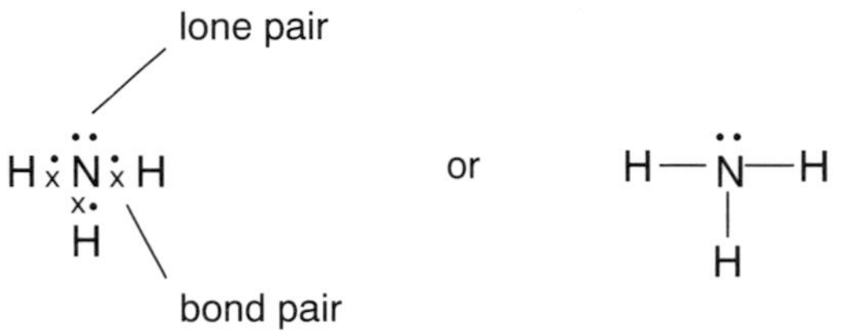

Figure 9 The arrangement of electrons in the ammonia molecule

The lone pair electrons are responsible for ammonia being
- very soluble in water
- a base
- a ligand
- a nucleophile.

We find these properties in amines too.

Solubility of amines

Like ammonia, amines can form hydrogen bonds to water:

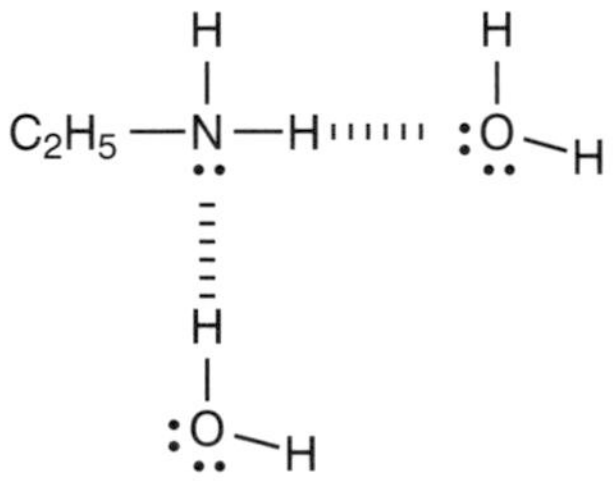

Figure 10 Hydrogen bonding between ethylamine and water molecules

Because of this strong attraction between amine molecules and water molecules, amines with small alkyl groups are soluble. Amines with larger alkyl groups are less soluble because the alkyl groups disrupt the hydrogen bonding in water.

Amines as bases

The lone pair on the nitrogen atom can take part in dative bonding. When the electron pair is donated to an H^+, ammonia acts as an H^+ acceptor and is a *base*.

$$NH_3(aq) + H_2O(l) \rightleftharpoons NH_4^+(aq) + {}^-OH(aq)$$
ammonium ion

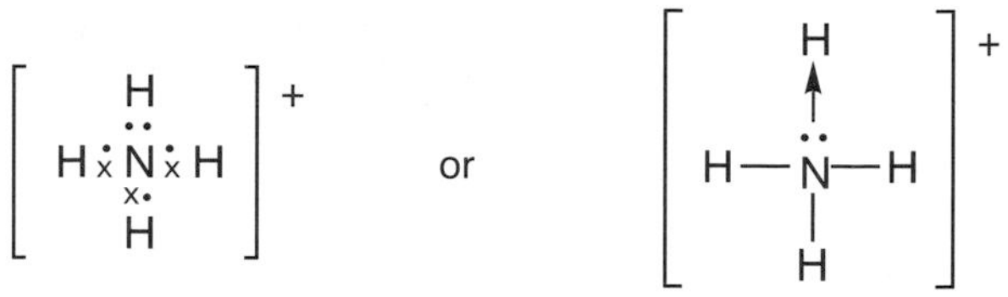

$$\left[\begin{array}{c} H \\ \overset{\cdot\cdot}{H \overset{\times}{\underset{\times\bullet}{N}} H} \\ H \end{array} \right]^{+} \quad \text{or} \quad \left[\begin{array}{c} H \\ \uparrow \\ H - \overset{\cdot\cdot}{N} - H \\ | \\ H \end{array} \right]^{+}$$

Figure 11 Dative bond formation between N and H^+ in the ammonium ion

A similar thing happens with amines. Like ammonia, they can accept H^+ from water. Using propylamine as an example, the reaction with water is

$$C_3H_7NH_2(aq) + H_2O(l) \rightleftharpoons C_3H_7NH_3^+(aq) + {}^-OH(aq)$$
$$\textit{propylammonium ion}$$

The presence of hydroxide ions makes the solution alkaline. So solutions of amines, like ammonia solution, are alkaline.

Like ammonia, amines also react with acids. The H_3O^+ ions in acidic solutions are more powerful H^+ donors than water. Their reaction with amines goes to completion and the solution therefore loses its strong amine smell. For example,

$$C_2H_5NH_2(aq) + H_3O^+(aq) \rightarrow C_2H_5NH_3^+(aq) + H_2O(l)$$
ethylamine   *ethylammonium ion*

Amines as ligands

(Ligands and complex ions are explained in **Section 11.7**).

Ammonia is an effective ligand because the lone pair of electrons on the N atom can bond to metal ions. For example, with copper, ammonia forms a deep blue complex ion,

$$[Cu(NH_3)_4 (H_2O)_2]^{2+}.$$

Complex ions containing ammonia ligands have their counterparts in amine chemistry. For example, adding butylamine to aqueous copper(II) sulphate produces a dark blue complex ion

$$[Cu(C_4H_9NH_2)_4 (H_2O)]^{2+}$$

whose structure is similar to

$$[Cu(NH_3)_4(H_2O)_2]^{2+}.$$

Amines as nucleophiles

You saw in **Section 13.1** that ammonia can act as a nucleophile, with the lone pair of electrons on the N atom attacking electrophiles such as halogenoalkanes.

$$RCl + NH_3 \rightarrow RNH_2 + HCl$$

The product of this reaction is an amine; but amines can themselves behave as nucleophiles, because they have a lone pair of electrons on the N atom, just the same as ammonia. So amines do substitution reactions with halogenoalkanes, to form secondary and tertiary amines.

$$R'Cl + RNH_2 \rightarrow RR'NH + HCl$$

Notice that in these reactions, a molecule of HCl is also formed.

A similar reaction occurs when we use an acyl chloride instead of a halogenoalkane. (You met acyl chlorides in **Section 13.3**, when you looked at their reaction with alcohols.) This time the product is an **amide**. Figure 12 shows the reactions of ammonia and a primary amine with ethanoyl chloride, a simple acyl chloride. The reactions are very vigorous even at room temperature, because acyl chlorides are very reactive.

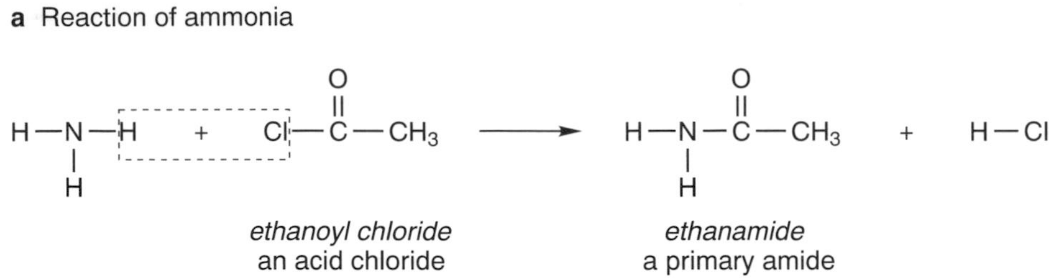

a Reaction of ammonia

$$H-\underset{\underset{H}{|}}{N}-H \quad + \quad Cl-\overset{\overset{O}{\|}}{C}-CH_3 \quad \longrightarrow \quad H-\underset{\underset{H}{|}}{N}-\overset{\overset{O}{\|}}{C}-CH_3 \quad + \quad H-Cl$$

ethanoyl chloride
an acid chloride

ethanamide
a primary amide

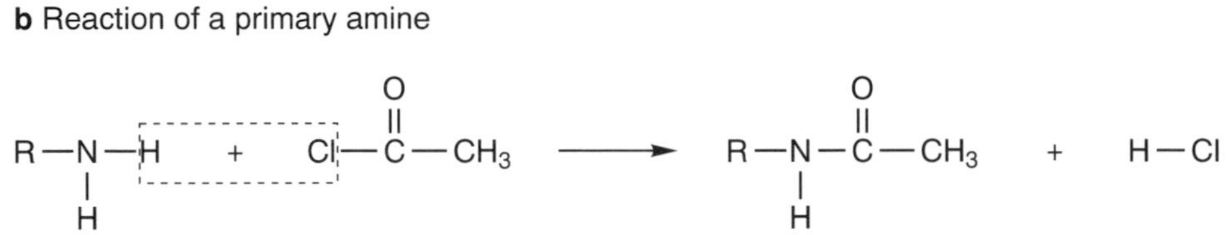

b Reaction of a primary amine

$$R-\underset{\underset{H}{|}}{N}-H \quad + \quad Cl-\overset{\overset{O}{\|}}{C}-CH_3 \quad \longrightarrow \quad R-\underset{\underset{H}{|}}{N}-\overset{\overset{O}{\|}}{C}-CH_3 \quad + \quad H-Cl$$

Figure 12 Formation of amides by the reaction of an acyl chloride with **a** *ammonia and* **b** *an amine*

a secondary amide

Amides

Amides contain the group

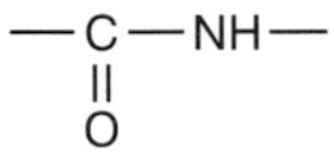

$$-\underset{\underset{O}{\|}}{C}-NH-$$

Primary amides have the formula

$$R-\underset{\underset{O}{\|}}{C}-NH_2$$

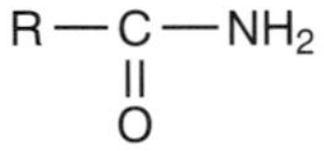

They are formed by the reaction of ammonia with an acyl chloride, as in Figure 12a above.
Secondary amides have the formula

$$R-\underset{\underset{O}{\|}}{C}-NHR'$$

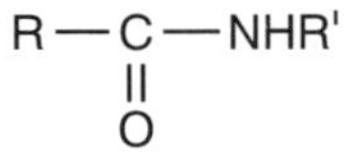

They can be formed by the reaction of an amine with an acyl chloride, as in Figure 12b above.

When the secondary amide group occurs in proteins, it is called a **peptide** group. The secondary amide group forms the link in the condensation polymers called *polyamides* or *nylons* (see page 268).

Hydrolysis of amides

The C—N bond in amides can be broken by hydrolysis – reaction with water. The reaction is catalysed by either acid or alkali.

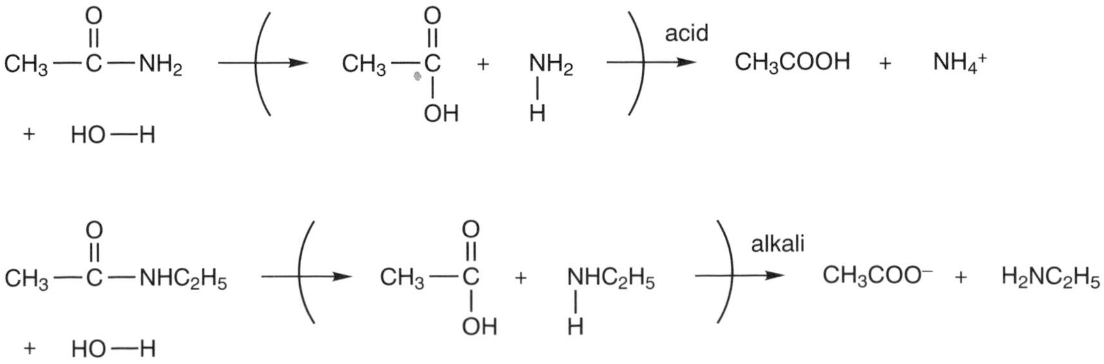

$$CH_3-\overset{\overset{O}{\|}}{C}-NH_2 \quad \longrightarrow \quad \left(CH_3-\underset{\underset{OH}{|}}{\overset{\overset{O}{\|}}{C}} \quad + \quad \underset{\underset{H}{|}}{N}H_2 \right) \overset{acid}{\longrightarrow} \quad CH_3COOH \quad + \quad NH_4^+$$

$$+ \quad HO-H$$

$$CH_3-\overset{\overset{O}{\|}}{C}-NHC_2H_5 \quad \longrightarrow \quad \left(CH_3-\underset{\underset{OH}{|}}{\overset{\overset{O}{\|}}{C}} \quad + \quad \underset{\underset{H}{|}}{N}HC_2H_5 \right) \overset{alkali}{\longrightarrow} \quad CH_3COO^- \quad + \quad H_2NC_2H_5$$

$$+ \quad HO-H$$

Both processes require reflux of the amide with the aqueous acid or alkali.

The products depend upon whether an acid or alkali catalyst is used. Ammonia reacts with acid, so if an acid catalyst is used, the product contains ammonium ions. If an alkali catalyst is used, the carboxylic acid loses an H^+ and the product contains carboxylate anions.

An important example is the hydrolysis of the peptide group in proteins. The breakdown of proteins in the laboratory is routinely carried out by boiling with moderately concentrated hydrochloric acid to hydrolyse the amide C—N bond. In living organisms, the hydrolysis of proteins is catalysed by enzymes rather than acid or alkali.

Amino acids

Amino acids contain the amino group together with the carboxylic acid group. The α-**amino acids** are particularly important in living systems. Figure 13 shows the general structure of an α-amino acid.

α-carbon: the first carbon atom
attached to the –COOH group

amino group H_2N — C — COOH   acid group

R
|
H_2N — C — COOH
|
H

Naming amino acids

The systematic name for α-amino acids like the ones shown in Figure 13 is *2-aminocarboxylic acids*. However, the name 'α-amino acid' is more commonly used.

Each amino acid has a systematic name, but normally its shorter common name is used instead. Thus the amino acid whose formula is

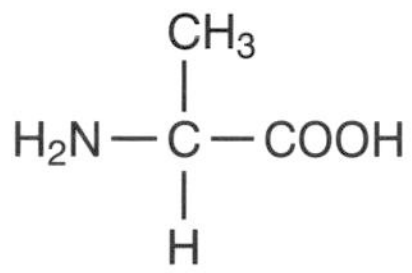

$$CH_3$$
|
$$H_2N — C — COOH$$
|
H

has the systematic name *2-aminopropanoic acid*, but its common name *alanine* is normally used.

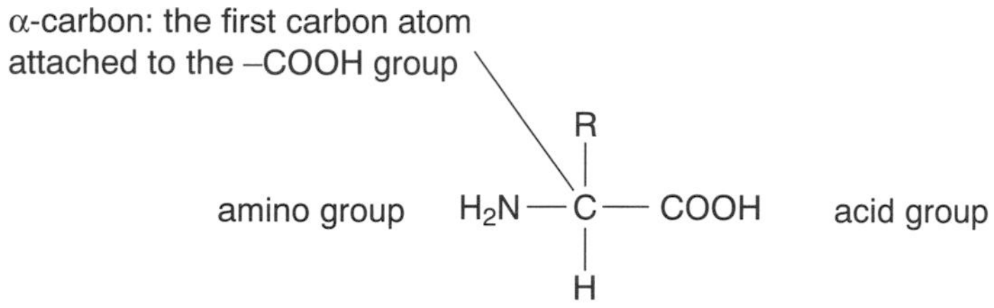

Figure 13 The generalised structure of an α-amino acid

Amino acids are examples of **bifunctional compounds** – compounds with *two* functional groups. The properties of bifunctional compounds are sometimes simply the same as the properties of the two separate functional groups. This is not the case with amino acids because the functional groups interact. The proton donating —COOH and proton accepting —NH_2 groups can react with one another, forming **zwitterions** – particles containing both anionic and cationic groups (Figure 14).

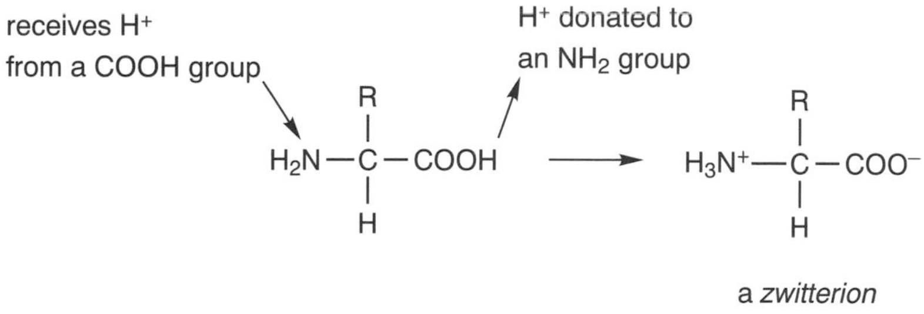

receives H^+
from a COOH group

H^+ donated to
an NH_2 group

R
|
H_2N—C—COOH
|
H

⟶

R
|
H_3N^+—C—COO$^-$
|
H

a *zwitterion*

Figure 14 How an amino acid forms a zwitterion

An aqueous solution of an amino acid consists mainly of zwitterions, with very few free molecules. Amino acids are very soluble in water because they are ionic.

Unless there is an extra —COOH or —NH_2 group in the molecule (as there is in some naturally occurring amino acids), they are neutral in aqueous solution.

Adding small quantities of acid or alkali to an amino acid solution causes little change to the pH because the zwitterions neutralise the effect of the addition.

HO^- + H_3N^+—CHR—COO$^-$ ⟶ H_2O + H_2N—CHR—COO$^-$

H_3O^+ + H_3N^+—CHR—COO$^-$ ⟶ H_2O + H_3N^+—CHR—COOH

So, amino acids exist in three different ionic forms, depending on the pH of the solution they are in:

$$H_3N^+\!-\!CHR\!-\!COOH \qquad\qquad H_3N^+\!-\!CHR\!-\!COO^- \qquad\qquad H_2N\!-\!CHR\!-\!COO^-$$

in acid solution   in neutral solution   in alkaline solution

Solutions which can withstand the addition of acid or alkali are called **buffer solutions** (see **Section 8.3**).

Condensation polymers involving the NH₂ group

When an —NH₂ group reacts with the —COOH group in a carboxylic acid, a **secondary amide group** is formed, with the structure —CONH—. In the process, a molecule of water is eliminated, so it is a condensation reaction (**Section 13.5**).

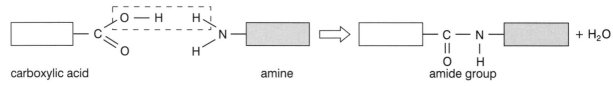

Figure 15

If we use *di*amines and *di*carboxylic acids which have reactive groups in *two* places in their molecules, we can make a polymer chain, in which monomer units are linked together by amide groups (Figure 16). The process is called **condensation polymerisation** because the individual steps are condensation reactions.

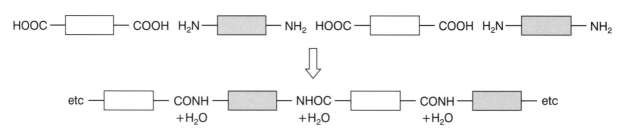

Figure 16

Examples of a diamine and a dicarboxylic acid that can be made to polymerise in this way are

$H_2NCH_2CH_2CH_2CH_2CH_2CH_2NH_2$ 1,2-diaminohexane
$HOOCCH_2CH_2CH_2CH_2COOH$ hexanedioic acid

Because the group linking the monomer groups together is an amide group, these polymers are called **polyamides**. More usually, though, they are known as **nylons**.

The industrial preparation of nylon from a diamine and a dicarboxylic acid is quite slow. It is easier to demonstrate the process in the laboratory if an *acyl chloride* derivative of the acid is used. Thus 1,6-diaminohexane and decanedioyl dichloride react readily. The equation is

$$nH_2N(CH_2)_6NH_2 \qquad + \qquad nClCO(CH_2)_8COCl$$
1,6-diminohexane   *decanedioyl chloride*

$$\downarrow$$

$$-[NH(CH_2)_6\!-\!NH\!-\!CO\!-\!(CH_2)_8CO]_n\!- \; + \; 2nHCl$$

Notice that in this condensation reaction a molecule of HCl is eliminated instead of a molecule of H_2O.

Proteins are also condensation polymers containing the —CONH— group. In proteins this group is called the **peptide** group, and the monomers are amino acids.

1   Name the amines with the following structures
  a $C_2H_5NH_2$
  b $(CH_3)_2NH$
  c $CH_3-\!\!\overset{\displaystyle |}{\underset{\displaystyle CH_3}{CH}}\!\!-NH_2$
  d $(CH_3)_2NC_2H_5$

2   Draw structures for the following amines
  a   propylamine
  b   phenylamine
  c   diethylmethylamine
  d   butylethylmethylamine

3   Write equations for the reactions of the following substances:

  a

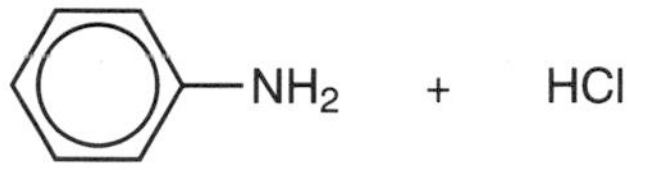

—NH₂   +   HCl

  b

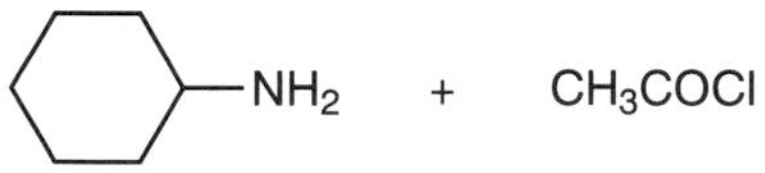

—NH₂   +   CH₃COCl

  c

N—H   +   C₂H₅COCl

4   Draw structures for the products of the following hydrolysis reactions.

  a $CH_3CH_2CONH_2 \xrightarrow[\text{reflux}]{\text{HCl(aq)}}$

  b

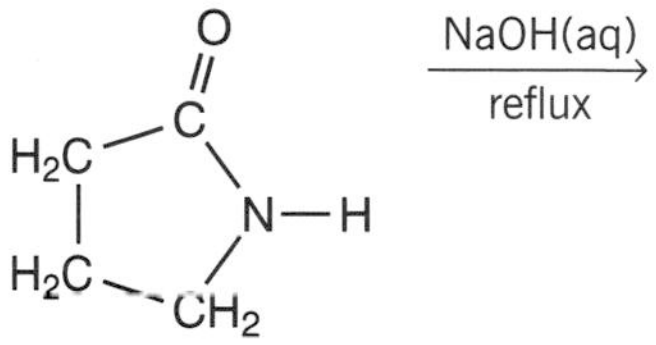

$\xrightarrow[\text{reflux}]{\text{NaOH(aq)}}$

5   Which of the reactions of amines described in this section could not be undergone by a tertiary amine like triethylamine? Briefly explain your answer.

6   The general structure of an α-amino acid is

$$H_2N-\!\!\overset{\displaystyle R}{\overset{\displaystyle |}{CH}}\!\!-COOH$$

Write equations for the following reactions

  a   alanine ($R = CH_3$) with hydrochloric acid
  b   serine ($R = CH_2OH$) with sodium hydroxide solution
  c   lysine ($R = (CH_2)_4NH_2$) with hydrochloric acid
  d   aspartic acid ($R = CH_2COOH$) with sodium hydroxide solution.

13.8 *Azo compounds*

Azo compounds

Azo compounds contain the $-N\!=\!N-$ group.

   azo group

Compounds in which the groups R and R' are arene groups are more stable than those in which R and R' are alkyl groups. This is because the $-N\!=\!N-$ group is stabilised by becoming part of an extended delocalised system involving the arene groups. These aromatic azo compounds are highly coloured and many are used as dyes.

They are formed as a result of a **coupling reaction** between a **diazonium salt** and a **coupling agent**.

Diazonium salts

The only stable diazonium salts are aromatic ones, and even these are not particularly stable. Benzenediazonium chloride has the following structure:

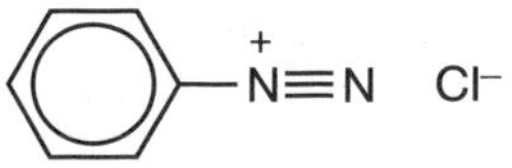

   benzenediazonium chloride

Diazonium salts are unstable because they tend to lose the $-\overset{+}{N}\equiv N$ group as $N_2(g)$. However, the presence of the electron-rich benzene ring stabilises the $-\overset{+}{N}\equiv N$ group. Even so, in aqueous solution, benzenediazonium chloride decomposes above about 5 °C. The solid compound is explosive. For this reason, diazonium salts are prepared in ice-cold solution and used immediately.

Diazonium salts are prepared by adding a cold solution of sodium nitrite, sodium nitrate(III), ($NaNO_2$) to a solution of an arylamine in concentrated acid below 5 °C. This type of reaction is known as **diazotisation**.

The acid (usually hydrochloric acid or sulphuric acid) reacts with sodium nitrite to form unstable nitrous acid (nitric(III) acid):

$$NaNO_2(aq) + HCl(aq) \rightarrow HNO_2(aq) + NaCl(aq)$$

sodium   *nitrous*
nitrite   *acid*

The nitrous acid then reacts with the arylamine. For example,

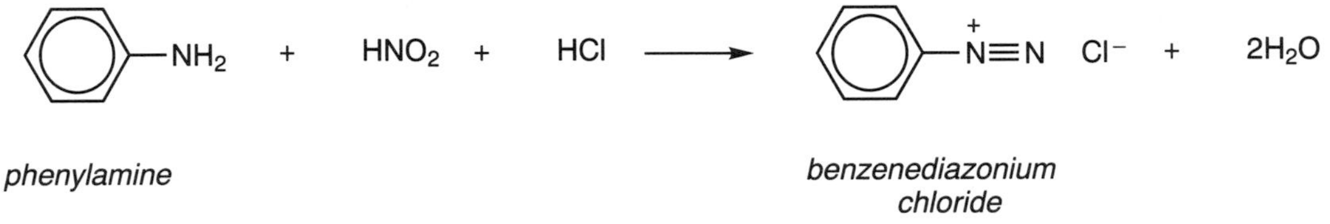

Diazo coupling reactions

In a diazo coupling reaction, a diazonium salt reacts with another compound containing a benzene ring, called a **coupling agent**. The diazonium salt acts as an electrophile and reacts with the benzene ring of the coupling agent. Figure 17 summarises a general coupling reaction.

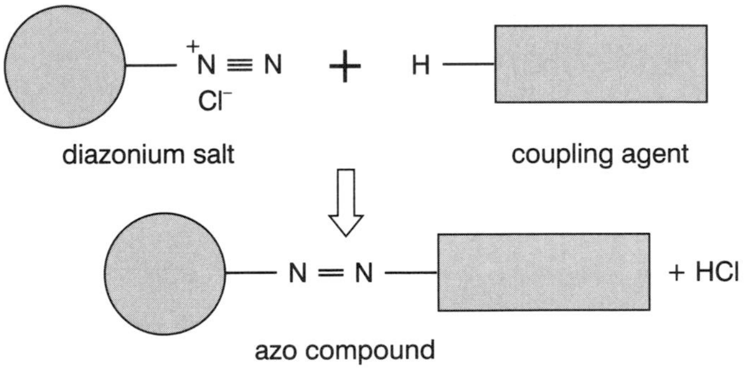

Figure 17 A generalised coupling reaction

When the ice-cold solution of the diazonium salt is added to a solution of a coupling agent, a coloured precipitate of an azo compound immediately forms. Many of these coloured compounds are important dyes.

Coupling with phenols

When a solution of the benzenediazonium salt is added to an alkaline solution of phenol, a yellow-orange azo compound is formed:

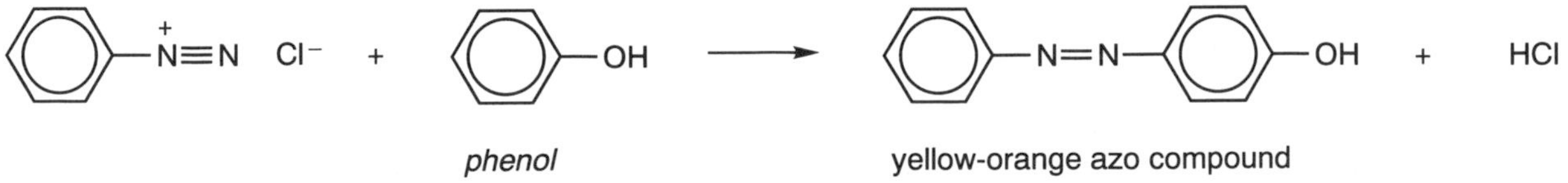

With an alkaline solution of naphthalen-2-ol, a red azo compound is precipitated:

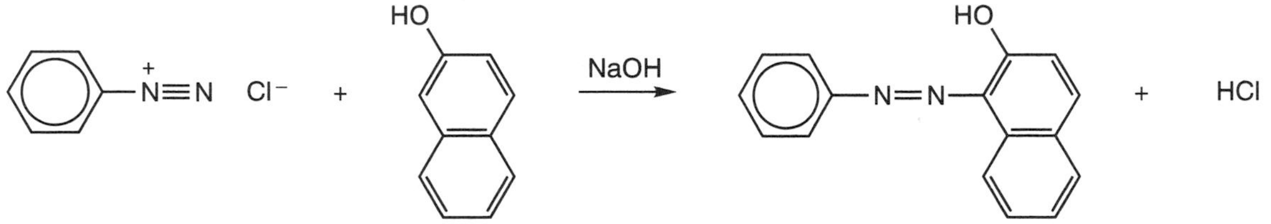

naphthalen-2-ol   red azo compound

Coupling with amines

Diazonium salts also couple with arylamines like phenylamine.

phenylamine   yellow azo compound

Many different azo compounds can be formed by coupling different diazonium salts with a whole range of coupling agents. Unlike diazonium salts, the azo compounds are stable, so their colours do not fade.

PROBLEMS FOR 13.8

1 Write the structures of the products you would expect to be formed at each stage when:
   a phenylamine is dissolved in sulphuric acid
   b sodium nitrite solution is added to the cooled solution in **a**
   c the product from **b** is added to an alkaline solution of phenol.

2 a Draw the structure of the diazonium salt that would form when 4-aminophenol is treated with sodium nitrite in the presence of cold hydrochloric acid.
   b Write equations for the coupling reaction of this diazonium salt with
      i phenol
      ii naphthalen-2-ol
      iii phenylamine.

3 Give the structures of the diazonium compounds and coupling agents you would need to make each of the azo compounds shown on the right. (The coupling agent usually contains a phenol group or an amine group attached to an arene ring system.)

4 Suggest why diazo coupling reactions only occur with coupling agents which have especially electron-rich benzene rings, such as phenols and arylamines.

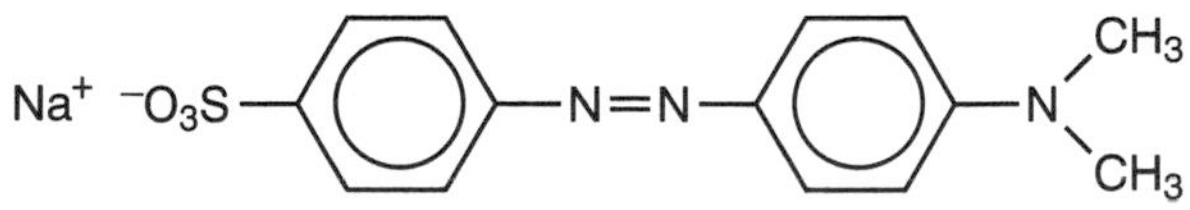

Methyl Orange

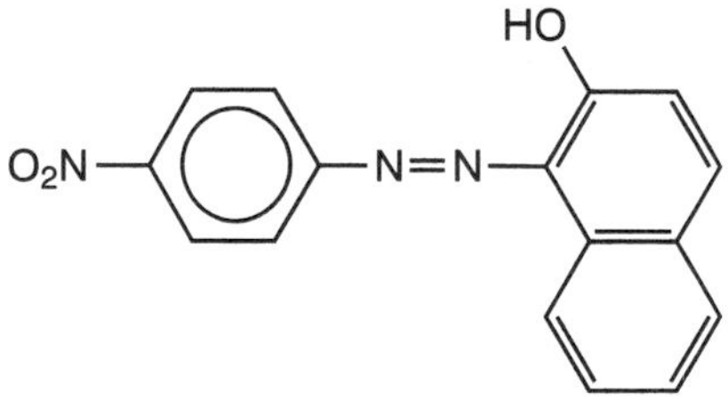

Para Red

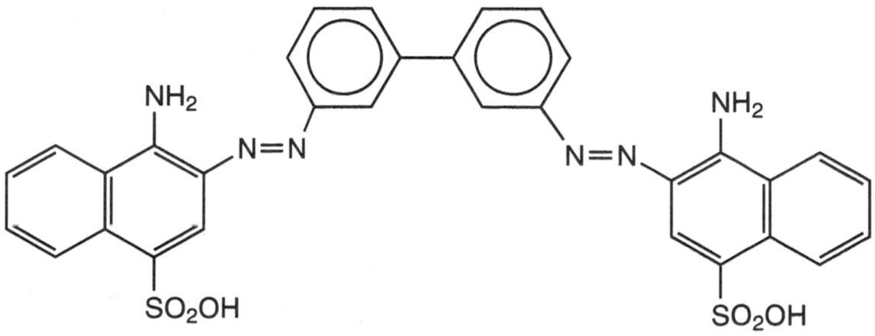

Congo Red

14 ORGANIC SYNTHESIS

14.1 Planning a synthesis

Why make organic compounds?

About seven million organic compounds are known to chemists, and more are constantly being made or discovered. Fortunately, their behaviour can be understood in terms of the **functional groups** they contain, and the way these functional groups are arranged in space.

The carbon framework gives a molecule its shape, but is usually fairly unreactive. It is the functional groups which govern the way a molecule reacts chemically. Each group usually undergoes the same type of reactions, whatever carbon framework it is attached to.

For example, the hydroxyl group, —OH, and the alkene group, C=C, react in a characteristic way, whether they are found in simple molecules like ethanol and ethene, or in the more complex steroid, *cholesterol* (Figure 1).

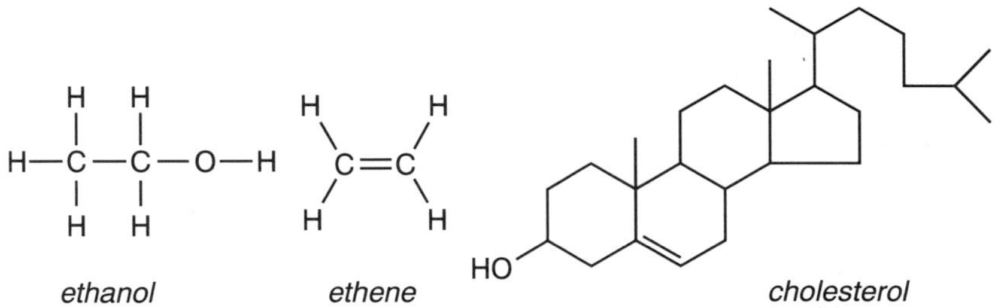

ethanol   ethene   cholesterol

Figure 1 Ethanol, ethene and cholesterol. Cholesterol is a natural product, but can be synthesised in the laboratory. It was first synthesised by the American Nobel Prizewinner R B Woodward in 1952. The synthetic route involved more than 40 steps!

New compounds are made by a process of **synthesis**, in which more complex structures are built up from simpler starting materials. The synthetic chemist is a sort of molecular architect, who plans and carries out strategies for making new and useful substances. There may be many steps in the synthesis, each step involving the preparation of a new compound from the previous compound. However, the basic principles are the same in each preparation (see box).

Chemists make new compounds for many different reasons. At one time, it was the only sure way to confirm the structure of a compound. Many complex syntheses of natural products were undertaken for this reason. Once they thought they knew the structure of a natural compound, chemists would synthesise it, then check that the synthetic compound had the same properties as the natural one.

Nowadays, most new substances are made in the hope that they will be useful in everyday life. In the pharmaceutical industry, for every medicine that becomes commercially available, many thousands of new compounds are prepared and tested. Minor variations in the chemical structure of molecules, such as in the penicillin range of antibiotics, can result in significant changes in their biological activity.

Planning a synthesis

The starting point in planning a synthesis is to examine the required compound itself. This compound is called the **target molecule**. By looking at the functional groups it contains, chemists can work backwards, through a logical sequence of reactions, until suitable **starting materials** can be found.

Carrying out an organic preparation

There are normally three stages involved in preparing an organic compound.

Organic reactions rarely give one pure product: there are usually two or more products from competing side-reactions, and unreacted starting material. The reaction mixture must be 'worked up' to extract the desired product in impure form; this is then purified.

The three states are

1 **Reaction**

2 **Extraction of product**

3 **Purification of product.**

After purification, the product is usually tested to confirm its purity.

The preparation of bromobutane (**Activity A4.2**), includes all these stages.

The experimental techniques that are commonly used in organic preparations – and which you should know – are

heating under reflux

distillation

vacuum filtration

fractional crystallisation

measurement of melting point and boiling point

paper and thin-layer chromatography.

Several steps may well be necessary, and a number of **intermediates** may have to be prepared and purified on the way to the target molecule. For example, the synthesis of ethylamine from ethene involves two steps:

| ethene | intermediate | ethylamine |
| starting material | | target molecule |

Acceptable starting materials must be cheap and readily available. Most hydrocarbons containing six or fewer carbon atoms are obtained from petroleum refining. These, and simple compounds made from them, are good starting points for synthesis. Sometimes a readily available natural product from a plant or animal source may be used.

In a more complex synthesis involving several steps, there will often be more than one possible route to the target molecule. Then, choices will have to be made. The preferred route is usually the one with the fewest steps – but this may not always be the case. Sometimes, other factors are more important, such as the cost of the starting materials and the reagents, the time involved, disposal of waste materials and possible safety and health hazards.

Another important factor to consider is the **overall yield** of the synthesis.

Working out the yield

Organic reactions rarely give a 100% yield of the required product; in other words, you rarely obtain the amount of product you would calculate from the equation. This is because most reactions of organic compounds are accompanied by side reactions which lead to the formation of by-products.

For example, the hydrolysis of 1-bromobutane with aqueous alkali gives butan-1-ol as the major product.

$$C_4H_9Br \quad + \quad NaOH \longrightarrow C_4H_9OH \quad + \quad NaBr$$

1-bromobutane   *butan-1-ol*
M_r 137 M_r 74

Suppose you plan to carry out the hydrolysis, starting with 10.0 g 1-bromobutane. If all the 1-bromobutane was converted into butan-1-ol as shown in the equation, you would expect to get

$(\frac{74}{137} \times 10.0\,g) = 5.4\,g$ of the alcohol.

In fact, when you carry out the reaction, you will find that you obtain less. In one experiment, 4.2 g of butan-1-ol were obtained. In this case, the yield of butan-1-ol is

$\frac{4.2\,g}{5.4\,g} \times 10.0 = 78\%$ of the maximum yield which could, in theory, be
obtained. (Yields are usually given to the nearest whole number.)

It is often not necessary to write out a completely balanced equation to calculate the maximum yield for a reaction. You need only the M_r of the original material, the M_r of the product, and the mole ratio of the reactant and product.

A reaction yield of more than 90% of the maximum possible would usually be considered excellent, but even a yield of 50% may be adequate.

In a synthesis with several steps, a few steps with low yields can have a disastrous effect on the overall yield, and the material may effectively disappear before the synthesis is complete.

Suppose a synthesis involves three steps, each of which gives a 90% yield of product:

$$A \xrightarrow{\;90\%\;} B \xrightarrow{\;90\%\;} C \xrightarrow{\;90\%\;} D$$

The starting compound for each step will be the product of the previous step. So, the **overall yield** for the series of reactions from A to D is

$$\frac{90}{100} \times \frac{90}{100} \times \frac{90}{100} = \frac{73}{100}, \quad \text{or} \quad 73\%$$

However, if each step resulted in a 50% conversion to the required product, the overall yield after 3 stages would be only

$$\left(\frac{50}{100} \times \frac{50}{100} \times \frac{50}{100} \right) = \frac{125}{100}, \quad \text{or} \quad 12.5\%$$

Getting the right isomer

The yield can be even lower when there are several isomers of a product molecule formed, and only one is wanted.

This is always a problem in substitution reactions involving benzene rings in which there is already a group present. In the nitration of methylbenzene, for example, three isomeric products are formed:

methylbenzene   nitration   1-methyl-2-nitrobenzene   1-methyl-4-nitrobenzene   1-methyl-3-nitrobenzene

main products   minor product

In this case, the products are all solids, and the required isomer can be separated by fractional crystallisation or by chromatography.

Separating isomers can sometimes be difficult and time-consuming. This is particularly true in the case of **optical isomers**, so a pure D or L form of a molecule is usually more difficult to prepare.

Choosing the reagents

An organic chemist chooses the reactions needed for a synthesis from a vast 'tool-kit' of reaction of functional groups.

If there is more than one functional group in the molecule, it is important to check, in each step, that the reagents do not react with the other groups present. This may influence the order in which the steps are carried out.

There is often no single correct solution to a synthetic problem, and several alternatives may well be equally viable. The preferred route will take into account all the relevant factors, and may not be one which seemed most likely at first.

Before you try your hand at designing a synthesis, make sure that you are familiar with the main reactions of the functional groups you have met throughout the course. These are summarised in **Section 14.2**.

PROBLEMS FOR 14.1

1 Compound **Z** can be prepared from the starting material, compound **W**, by two alternative routes. Route I (**W → X → Y → Z**) involves 3 steps. Route II (**W → V → Z**) involves 2 steps. The yields for each of the steps is shown below. Which route has the highest overall yield?

Route I

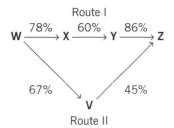

Route II

2 Phenylamine can be made in two steps from benzene as shown below:

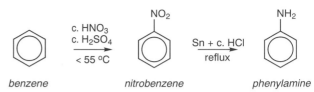

benzene   nitrobenzene   phenylamine

Nitration of benzene gives an 85% yield of nitrobenzene. The second step gives an 80% yield of phenylamine. If you carry out the synthesis, starting with 20.0 g benzene, what mass of phenylamine would you obtain?

3 Even when there are no side-reactions in an organic reaction, a 100% yield of product is rarely obtained. Suggest reasons why this might be.

14.2 *A summary of organic reactions*

The following sections summarise the important reactions of the various functional groups you have met throughout the course. *These are the reactions you should remember.* You should be able to use them correctly to convert one functional group into another or to design the synthesis of an organic molecule.

Alkenes

Simple alkenes, such as ethene and propene, are obtained by cracking petroleum fractions. They are much more reactive than the corresponding alkanes, and make good starting materials for a synthesis. You can see the main reactions of alkenes in Figure 2.

Figure 2 Reactions of alkenes

$$-CH_2-CH_2- \quad \xrightarrow[\text{sunlight}]{Cl_2(g)} \quad -CH_2-CHCl- \quad + \quad HCl(g)$$

$H_2(g)$
finely divided Ni
at 150 °C and 5 atm.
(or Pt at room temp
and 1 atm.)

$$\underset{\substack{Br \quad Br}}{-HC-CH-} \quad \xleftarrow[\substack{\text{organic solvent} \\ \text{room temp}}]{Br_2} \quad \boxed{-HC=CH- \quad \textit{alkene}} \quad \xrightarrow[\text{room temp}]{\text{conc HBr(aq)}} \quad \underset{\substack{H \quad Br}}{-HC-CH-}$$

$H_2O(g)$ / phosphoric
acid catalyst
300 °C and
60 atm.

trace $O_2(g)$
200 °C and
1500 atm.

$$\underset{\substack{H \quad OH}}{-HC-CH-}$$

$$\left(\!\!\begin{array}{cc} | & | \\ HC & -CH \\ | & | \end{array}\!\!\right)_{\!\!n}$$

addition polymer

All the reactions of alkenes in Figure 2 are **addition** reactions. The reaction may proceed by an *ionic mechanism* or it may involve *radicals* (**Section 12.2**).

For example, addition polymerisation reactions involve radicals, whereas the addition of bromine is an **electrophilic addition** reaction in which the alkene reacts with a polarised bromine molecule.

$$\text{>C=C<}$$
$$\text{Br } \delta+$$
$$|$$
$$\text{Br } \delta-$$

Addition of an unsymmetrical molecule, such as HBr, to an alkene may lead to two possible isomeric products.

For example, when propene reacts with a concentrated aqueous solution of HBr, two addition products are formed. The isomer with a central Br atom is the main product, but it must be separated from small amounts of the other isomer.

$$CH_3 - CH = CH_2 \xrightarrow{\text{HBr(aq)}} CH_3 - CH_2 - CH_2 - Br \quad + \quad CH_3 - CH - CH_3$$

propene   Br

                                                          main product

Halogenoalkanes

Halogenoalkanes are useful synthetic intermediates because they readily undergo a large variety of **nucleophilic substitution** reaction (**Section 13.1**). Three important examples are shown in Figure 3.

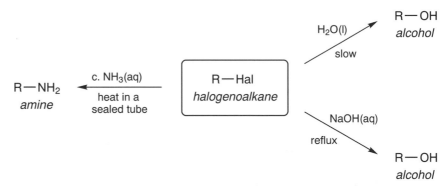

Figure 3 Reactions of halogenoalkanes

These reactions are useful because they allow one functional group to be converted into another.

Alcohols

Alcohols undergo a variety of reactions which can be useful in synthesis. (**Section 13.2** deals with the structure and physical properties of alcohols; **Section 13.4** looks at their chemical reactions.)
The important reactions of alcohols are summarised below: you will have met some of these reactions before, but some will be new to you.

Oxidation of alcohols

Alcohols are oxidised by warming with an oxidising agent such as acidified potassium dichromate(VI) solution. The product depends on the type of alcohol and on the way the oxidation is carried out.

Primary alcohols give first aldehydes, and then carboxylic acids.

$$RCH_2OH \xrightarrow[\text{reflux}]{Cr_2O_7{}^{2-}/H^+(aq)} RCHO \xrightarrow[\text{reflux}]{Cr_2O_7{}^{2-}/H^+(aq)} RCOOH$$

primary alcohol   *aldehyde*   *carboxylic acid*

The half-equations for the oxidation reactions are:

$$R-CH_2OH \longrightarrow R-CHO + 2H^+ + 2e^-$$

$$R-CHO + H_2O \longrightarrow R-COOH + 2H^+ + 2e^-$$

The reduction half-equation is the same in each case:

$$Cr_2O_7{}^{2-} + 14H^+ + 6e^- \longrightarrow 2Cr^{3+} + 7H_2O$$

orange solution   green solution

To make the aldehyde, the oxidising agent is dripped slowly into the hot alcohol and the aldehyde distilled off as soon as it is formed, before it has time to be oxidised further to the acid.

To make the carboxylic acid the alcohol is heated under reflux with an excess of the oxidising agent to make sure the reaction goes to completion.

With *secondary alcohols*, the oxidation stops at the ketone. The ketone has no hydrogen atoms on the carbon atom of the ketone group, so it cannot easily undergo further oxidation.

Tertiary alcohols have no hydrogen atoms on the carbon atom attached to the —OH group and are much more difficult to oxidise.

Under more vigorous oxidising conditions, both ketones and tertiary alcohols will react, but the oxidation then involves the breaking of carbon–carbon bonds and usually leads to a mixture of products.

Dehydration of alcohols

Many alcohols readily eliminate a molecule of water to give an alkene. This is called a **dehydration** reaction, and is an example of an **elimination**. You can think of an elimination reaction as being the reverse of an addition reaction.

For example, ethene is formed when ethanol vapour is passed over alumina (Al_2O_3) at 300 °C:

$$CH_3CH_2OH \xrightarrow[\text{300 °C}]{Al_2O_3(s)} CH_2{=}CH_2 + H_2O$$

ethanol   *ethene*

Alcohols can also be dehydrated by heating with concentrated sulphuric acid.

Reaction of alcohols with HBr

Alcohols can be converted to bromoalkanes by heating with hydrobromic acid, HBr(aq). This is prepared *in situ* (in the reaction mixture) by the reaction of NaBr with concentrated sulphuric acid. For details of the preparation of 1-bromobutane from butan-1-ol, see **The Atmosphere**, **Activity A4.2**.

$$CH_3CH_2CH_2CH_2OH + HBr \longrightarrow CH_3CH_2CH_2CH_2Br + H_2O$$

butan-1-ol   *1-bromobutane*

This is a **nucleophilic substitution** reaction. The alcohol group must first be protonated in the strongly acid solution before it can be displaced by the Br⁻ nucleophile:

$$C_4H_9-O-H \xrightarrow{H^+(aq)} C_4H_9-\overset{H}{\underset{+}{O}}-H \xrightarrow{Br^-(aq)} C_4H_9-Br + H_2O$$

Esterification

An alcohol reacts with a carboxylic acid in a **condensation** reaction to give an ester. The alcohol and the carboxylic acid are heated under reflux in the presence of a few drops of concentrated sulphuric acid, acting as a catalyst. To get the best yield of ester, the water is distilled off as it forms, driving the equilibrium to the right.

$$R—OH \quad + \quad R'—COOH \quad \underset{}{\overset{\text{c. } H_2SO_4 \text{ catalyst}}{\rightleftharpoons}} \quad R—O—\overset{\displaystyle O}{\overset{\displaystyle \|}{C}}—R' \quad + \quad H_2O$$

Esters can also be made by the reaction of alcohols with acylating agents such as acyl chlorides R'COCl and acid anhydrides (R'CO)$_2$O (see **Section 13.5**).

The main reactions of primary alcohols are shown in Figure 4.

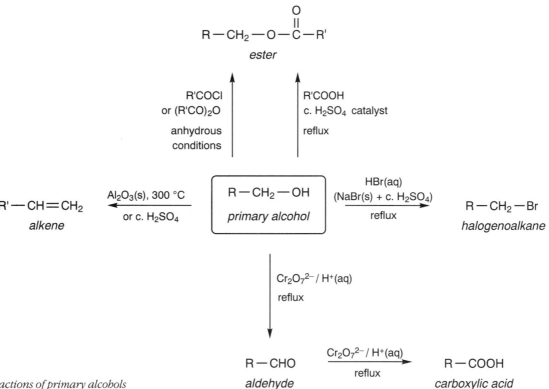

Figure 4 Reactions of primary alcohols

Carboxylic acids and related compounds

Carboxylic acids themselves are not particularly reactive, but they can be converted into the more reactive *acyl chlorides* and *acid anhydrides*. These compounds are very useful as synthetic intermediates or as reagents in synthesis.

You will find more information about acids and their derivatives in **Sections 13.3** and **13.5**.

Acyl (acid) chlorides

Acyl chlorides, such as ethanoyl chloride, are readily prepared from carboxylic acids. They are more reactive than chloroalkanes, such as chloroethane; the chlorine is much more readily displaced as chloride ion by nucleophiles.

$$CH_3—\overset{\displaystyle O}{\overset{\displaystyle \|}{C}}\diagdown_{Cl} \qquad\qquad CH_3CH_2—Cl$$

ethanoyl chloride   *chloroethane*

Acyl chlorides undergo a variety of **nucleophilic substitution** reactions. Most are readily hydrolysed by cold water. Ethanoyl chloride, for example, fumes in moist air as droplets of ethanoic acid and hydrochloric acid are formed.

$$CH_3COCl + H_2O \rightarrow CH_3COOH + HCl$$

Acyl chlorides react rapidly with alcohols and phenols to form esters:

$$CH_3COCl + C_2H_5OH \rightarrow CH_3COOC_2H_5 + HCl$$
$$CH_3COCl + C_6H_5OH \rightarrow CH_3COOC_6H_5 + HCl$$

and with ammonia and amines to form amides:

$$CH_3COCl + NH_3 \rightarrow CH_3CONH_2 + HCl$$
$$CH_3COCl + RNH_2 \rightarrow CH_3CONHR + HCl$$

These reactions take place readily at room temperature.

Most acyl halides react violently with water, so acylation reactions are usually carried out under strictly anhydrous conditions. Benzoyl chloride is an exception. It is much less reactive than ethanoyl chloride and can be used in aqueous solutions: see **Activity MD5.1**.

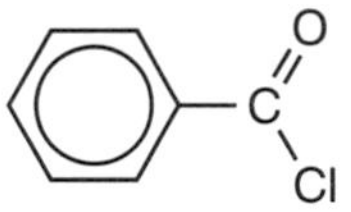

benzoyl chloride

Acid anhydrides

Acid anhydrides are also acylating agents and react in the same way as acyl halides with water, alcohols and amines to form amides, although not quite so vigorously. Like acyl chlorides, they must be used under anhydrous conditions. Acylation reactions using acid anhydrides often require heating under reflux.

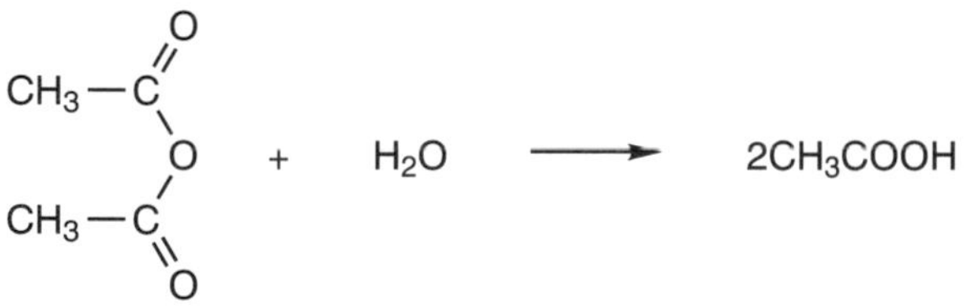

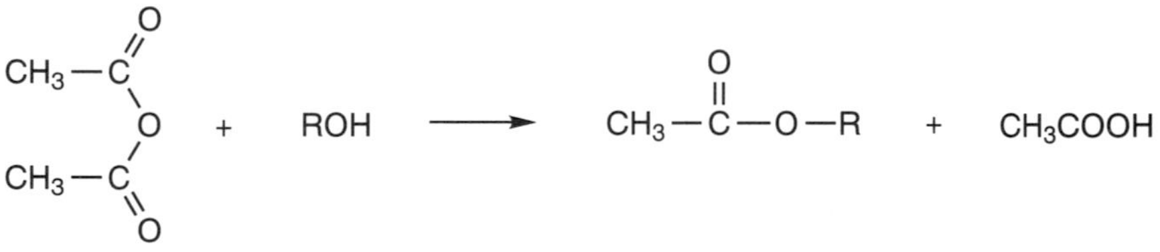

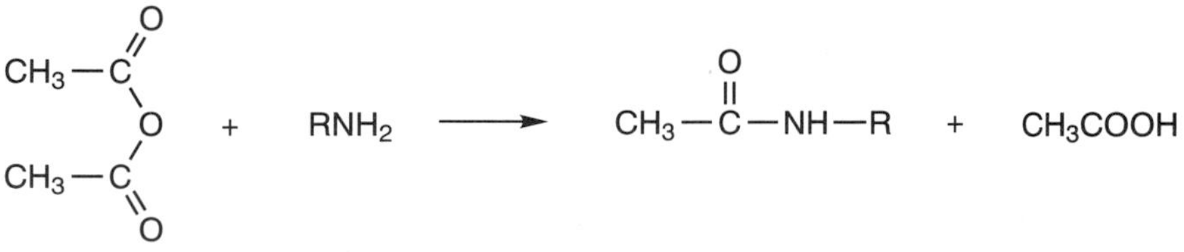

You can see the main reactions of carboxylic acids and their derivatives in Figure 5.

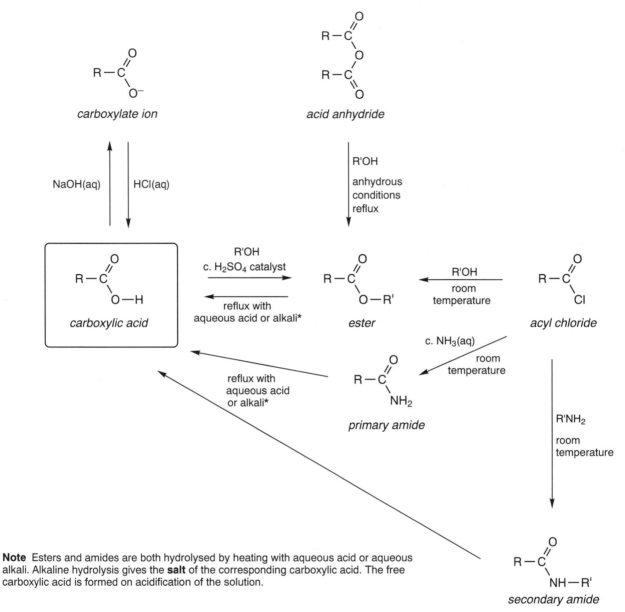

Note Esters and amides are both hydrolysed by heating with aqueous acid or aqueous alkali. Alkaline hydrolysis gives the **salt** of the corresponding carboxylic acid. The free carboxylic acid is formed on acidification of the solution.

Figure 5 The reactions of carboxylic acids and some related compounds

Arenes

Electrophilic substitution reactions provide ways of introducing different functional groups into the benzene ring (**Section 12.4**). The groups may then be modified further to build up more complex molecules. The main reactions of arenes are shown in Figure 6.

If you use excess reagent, you may get di- and tri- substituted products. Often a mixture of products is formed and you will have to separate the compound you want.

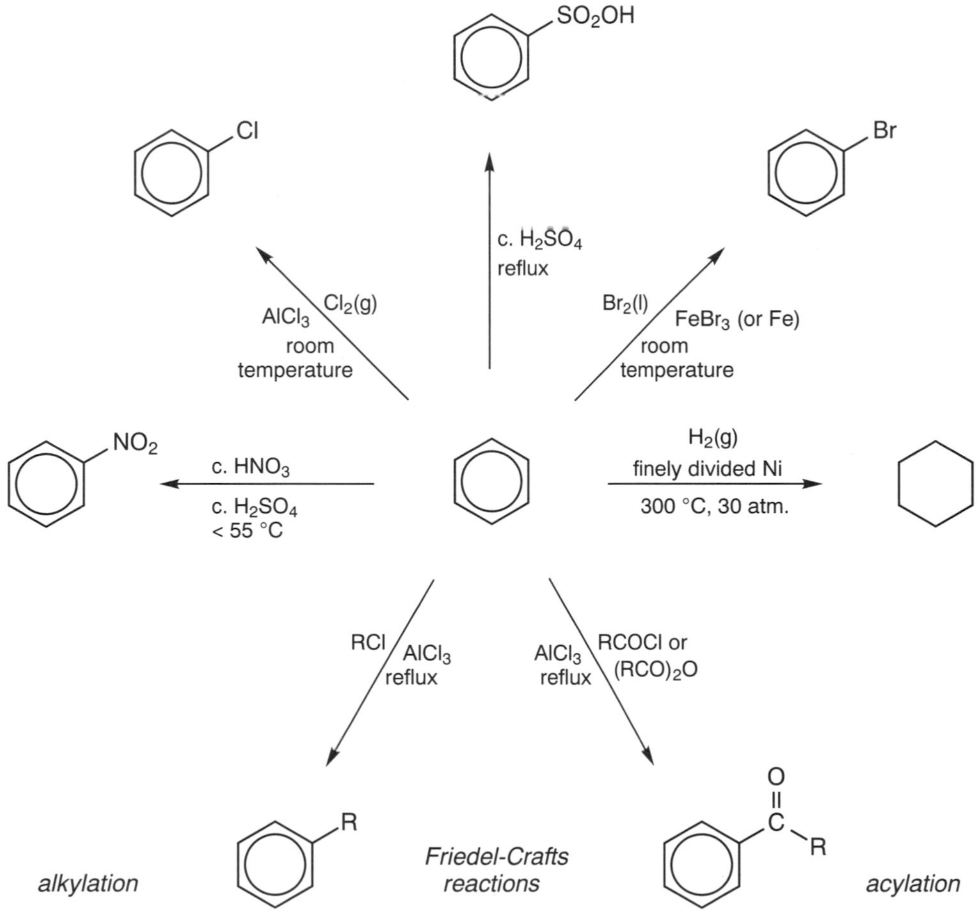

PROBLEMS FOR 14.2

1 Name the reagent and give the conditions you would use to bring about the following conversions.

a

$$CH_3CH=CHCH_3 \longrightarrow CH_3CH_2CHBrCH_3$$

b

$$CH_3CH_2CHBrCH_3 \longrightarrow CH_3CH_2CH(OH)CH_3$$

c

$$CH_3CH_2CH(OH)CH_3 \longrightarrow CH_3CH_2COCH_3$$

d

$$CH_3CH_2CH(OH)CH_3 \longrightarrow CH_3CH_2CH(OCOCH_3)CH_3$$

2 Treatment of methylbenzene with chlorine gas *in the presence of sunlight* at room temperature gives chloromethylbenzene.

chloromethylbenzene

a What *type* of reaction mechanism do you think is involved?
b Suggest a possible mechanism for the reaction.
c Describe how you would convert chloromethylbenzene into

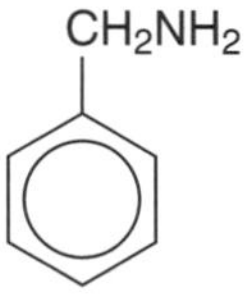

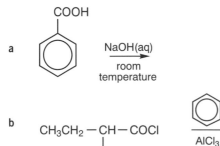

CH₂NH₂

3 Write down the structural formula of the main product(s) of each of the following reactions. (It is not necessary to write a *full* structural formula showing all the bonds. Your formula should be enough to indicate the structure clearly.)

a

COOH

$$\xrightarrow[\text{temperature}]{\text{NaOH(aq)} \atop \text{room}}$$

b

$$CH_3CH_2-\underset{\underset{CH_3}{|}}{CH}-COCl \quad \xrightarrow[\text{reflux}]{\text{AlCl}_3}$$

c

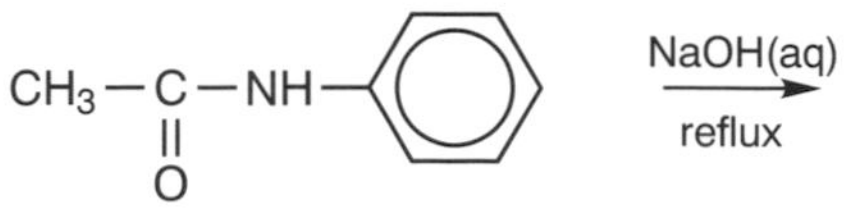

$$CH_3-\underset{\underset{O}{\|}}{C}-NH-\!\!\!\bigcirc \quad \xrightarrow[\text{reflux}]{\text{NaOH(aq)}}$$

d

OH

$$\xrightarrow[\text{300 °C}]{\text{Al}_2O_3\text{(s)}}$$

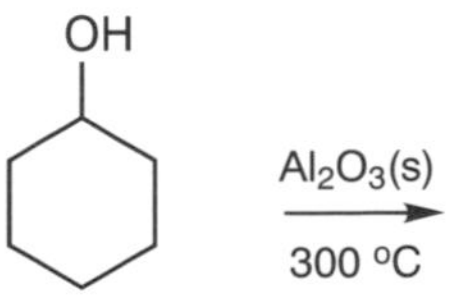

4 Choose two words, one from list A and one from list B, to describe the mechanism of each of the reactions given below.

A	B
electrophilic	substitution
nucleophilic	addition
radical	elimination

Write down the structural formula of the main product in each reaction.

a

$$\xrightarrow[\text{room temperature}]{\text{Cl}_2\text{(g)} \atop \text{AlCl}_3}$$

b

$$\xrightarrow[\text{room temperature}]{\text{Br}_2 \atop \text{organic solvent}}$$

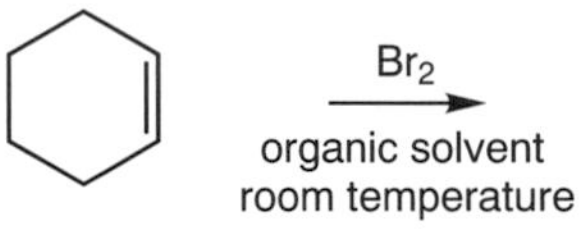

c $CH_4 \quad \xrightarrow[\text{sunlight}]{\text{Cl}_2\text{(g)}}$

d $CH_3CH_2CH_2Br \quad \xrightarrow[\text{reflux}]{\text{NaOH(aq)}}$

e

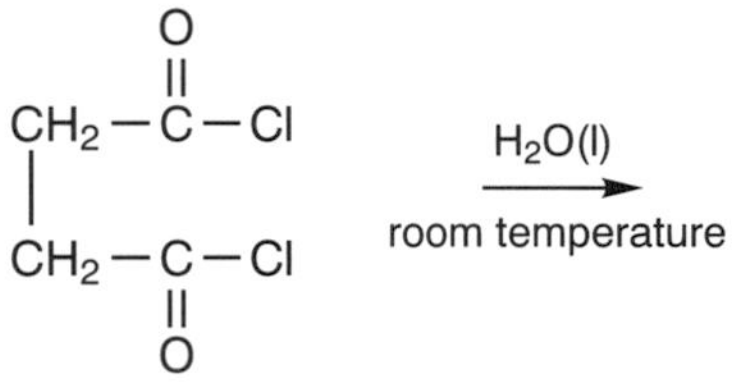

$$\underset{\underset{O}{\|}}{\underset{CH_2-C-Cl}{\overset{O}{\overset{\|}{CH_2-C-Cl}}}} \quad \xrightarrow[\text{room temperature}]{\text{H}_2\text{O(l)}}$$

5 Cyclohexanone is an intermediate in the production of nylon. Write down a synthetic route for the preparation of cyclohexanone from phenol, giving the reagents and essential conditions for each stage, and the name and the structural formula of the intermediate.

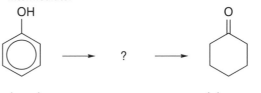

phenol   ?   *cyclohexanone*

6 Cocaine was the first effective local anaesthetic to be used in minor surgery. Unfortunately, it can over-stimulate the patient, and it is dangerously addictive. Modern local anaesthetics have to avoid these problems. Some of them are synthetic derivatives of 4-aminobenzoic acid. The simplest of these is benzocaine. This is often the active component of ointments to relieve the pain caused by sunburn.

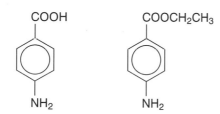

4-aminobenzoic acid   *benzocaine*

Design a synthesis of benzocaine, starting from 4-nitrobenzoic acid.

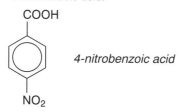

4-nitrobenzoic acid

(You may find the 'tool-kit' in **Activity MD3.1** helpful.)

7 As a pharmaceutical chemist, you are interested in making minor modifications to the structure of penicillin F, in the hope of extending its range of effectiveness.

Penicillins can be made in the laboratory by reacting 6-aminopenicillanic acid (6-APA) with a suitable acyl chloride. For penicillin F, the acyl chloride is $CH_3—CH_2—CH=CH—CH_2—COCl$, which can be made from the corresponding carboxylic acid.

6-APA + $CH_3—CH_2—CH=CH—CH_2—COCl$
$\rightarrow$ penicillin F + HCl

Your aim is to synthesise a range of acyl chlorides to react with 6-APA to make some modified penicillins. You decide to work with the acid, $CH_3—CH_2—CH=CH—CH_2—COOH$, rather than the more reactive acyl chloride. Use your knowledge of organic reactions to show how you would convert the acid into the following compounds. In each case, give the essential conditions and write a balanced equation for the reaction.

a the saturated acid
$CH_3—CH_2—CH_2—CH_2—CH_2—COOH$

b the halogenoacid
$CH_3—CH_2—CHBr—CH_2—CH_2—COOH$

c the hydroxyacid
$CH_3—CH_2—CH(OH)—CH_2—CH_2—COOH$

d the unsaturated acid
$CH_3—CH=CH—CH_2—CH_2—COOH$
(This may not be the only product of your reaction.)

INDEX

Note: **CS** = Chemical Storylines; **CI** = Chemical Ideas;
    AA = Activities and Assessment

absolute temperature **CI** 47
absorbance **AA** SS1
absorption spectrum
  CS 198; stars 11–12; water 122
  CI 108–11; transition metals 218
  AA US3.1b
abstract writing **AA** A9
ACE inhibition **CS** 264–6
ACE modelling **AA** MD4
acid anhydrides **CI** 279–80
acid dissociation constant **CI** 151
acid rain
  CS 48, 236
  AA analysis M2.4
acid-base pairs **CI** 148
acid-base properties (period 3) **CI** 92
acidity constant **CI** 151
acids
  CS weak 249–50
  CI concentration/strength distinguished 148, 152;
    conjugate 148; defined 147; strong and weak 150–2
  AA strong/weak A8.2, O4.2
activation enthalpy
  CS and catalysis 155; and reaction rate 62
  CI 52, 173–4, 182–3, 186; and catalysis 189
acyl chlorides
  CS 141
  CI 278-9
  AA MD3.1
acylation
  CS for penicillin manufacture 269
  CI 240; agents 258–9, 278–9
  AA reactions MD3.1
addition polymerisation
  CI 230–1
  AA PR2.1
addition reaction **CI** alkenes 276; of ethene 228–30
adsorption **CS** 198
agonistic molecules **CS** 263
alcohol
  CS blood concentration 254–7; as drug 254, 258–61;
    effect on GABA 260–1; as food 253–4
  CI 246
alcohols
  CS energy in 19; as oxygenates 29
  CI 248–50, 276–8; condensation reaction 278;
    elimination reaction 277; –OH group 253;
    oxidation 254-5; polyhydric 249;
    reaction with HBr 277; substitution reaction 277
  AA DF4.5; ethanol DF1.1; A4.1; –OH group WM3;
    oxidation MD1.1; viscosity PR5.4
Alcolmeter **CS** 255–6
aldehydes
  CI 254–5, 277
  AA MD3.1
alizarin **CS** 216–18
alkali, defined **CI** 147
alkali metal halides **AA** M1.3
alkane series, physical property changes **AA** DF3.1

alkanes
  CS bromination 116; from cracking 27; isomerisation 26
    in petrol 21, 22; for petrol blending 25-6
  CI 221–6; and alcohol synthesis 248;
    boiling points 74; branched 23; cracking 225-6;
    cycloalkanes 224; properties 225; reactions 225-6;
    shape 224–5; structure 223
  AA auto-ignition DF3.3; branched DF4.1;
    cracking DF4.4; naming DF4.1;
    octane numbers DF4.2; straight chain DF4.1;
    vapour, heat action DF4.4
alkenes
  CI 226, 227–31, 275–6; addition reactions 276;
    cycloalkenes 227; shape 228; *see also* ethene
  AA DF4.4; as hydrocarbon CD5.1;
    polymerisation PR2.1; test for DF4.4
alkoxy group **CI** 250
alkyl groups
  CI 223
  AA DF4.1
alkylation
  CI 240
  AA reactions MD3.1, MD3.2
alkynes **CS** 88, 90
alloys
  CS 159
  CI 208
alpha particles **CI** 15, 16
aluminium, in steelmaking
  CS 164
  AA SS2.4
aluminium atoms, zeolites **AA** DF4.3
amides
  CS 141
  CI 265, 266–7, 279
amines
  CS in nylon creation 92
  CI 263–5; coupling with diazo salts 271; reaction with
    acyl chlorides 279; substitution reactions 265
  AA EP2.1
amino acids
  CS 138–40; essential 140; formation 14–15
  CI 267–8
  AA EP2.1; α-amino acids EP2.4
amino groups
  CS 138
  AA MD3.1
aminopenicillanic acid (6-APA) **CS** 268–9
ammonia
  CS as refrigerant 66
  CI 200; as nucleophile 246; reaction with acyl
    chlorides 279; *see also* amines
ammonia leach, with roasted ores **AA** M2.3
ammonia manufacture kinetics **CI** 185–6
ammonia synthesis **CS** 191
ammonium ions
  CS and nitrification 188
  AA AA3.1
amphoteric
  CS tin 52, 174
  CI 92, 148
analgesic, *see* aspirin